T0179600

DEPLOYING IPv6 IN 3GPP NETWORKS

DEPLOYING IPv6 IN 3GPP NETWORKS

EVOLVING MOBILE BROADBAND FROM 2G TO LTE AND BEYOND

Jouni Korhonen

formerly Nokia Siemens Networks, now Renesas Mobile, Finland

Teemu Savolainen

Nokia Research Center, Finland

Jonne Soininen

Renesas Mobile, Finland

A John Wiley & Sons, Ltd., Publication

This edition first published 2013
© 2013 John Wiley and Sons Ltd

Registered office
John Wiley & Sons Ltd, The Atrium, Southern Gate, Chichester, West Sussex, PO19 8SQ, United Kingdom

For details of our global editorial offices, for customer services and for information about how to apply for permission to reuse the copyright material in this book please see our website at www.wiley.com.

Library of Congress Cataloging-in-Publication Data

Savolainen, Teemu.
  Deploying IPv6 in 3GPP networks : evolving mobile broadband from 2G to LTE and beyond / Teemu Savolainen, Jouni Korhonen, Jonne Soininen.
    pages cm
  Includes bibliographical references and index.
  ISBN 978-1-118-39829-6 (cloth)
  1. Long-Term Evolution (Telecommunications) 2. Cell phone systems. 3. Mobile computing. 4. TCP/IP (Computer network protocol) I. Korhonen, Jouni. II. Soininen, Jonne. III. Title.
  TK5103.48325.S28 2013
  621.3845'6–dc23

                                                                              2012050393

A catalogue record for this book is available from the British Library.
ISBN: 9781118398296

Set in 10/12pt Times by Laserwords Private Limited, Chennai, India
Printed and bound in Singapore by Markono Print Media Pte Ltd

Disclaimer

This book is based on the authors' personal experiences in the technical field and public standards documents created by 3GPP, IETF, and other standards defining organizations. The opinions and views of the authors are solely those of the authors and do not necessarily represent the views of organizations where the authors work. Throughout this book the authors have attempted to make it clear when something is an opinion or a view of the authors. Some of the examples, feature lists, and identified ambiguities may not apply universally to all deployments and products.

The publisher and the authors make no representations or warranties with respect to the accuracy or completeness of the contents of this work and specifically disclaim all warranties, including without limitation warranties of fitness for a particular purpose. No warranty may be created or extended by sales or promotional materials. The advice and strategies contained herein may not be suitable for every situation. This work is sold with the understanding that the publisher is not engaged in rendering legal, accounting, or other professional services. If professional assistance is required, the services of a competent professional person should be sought. Neither the publisher nor the authors shall be liable for damages arising herefrom. The fact that an organization or Website is referred to in this work as a citation and/or a potential source of further information does not mean that the author or the publisher endorses the information the organization or web site may provide or recommendations it may make. Further, readers should be aware that Internet web sites listed in this work may have changed or disappeared between when this work was written and when it is read.

This book is dedicated to the next generation
Jouni, Teemu, Jonne

Contents

Foreword

I have been fortunate to have first been involved in the Internet when it was still a relatively small research project at ARPA. I worked at Bolt, Beranek, and Newman in Cambridge, Massachusetts, on the Arpanet and the early Internet starting in 1978. It was definitely being at the right place at the right time. I was able to work with the people who really did invent the Internet. I was part of the research groups that evolved into what is now the organization that standardizes the core Internet protocols: the Internet Engineering Task Force (IETF).

In the early 1990s we realized that there was going to be a problem with the size of the address space used in the current version of the Internet protocol, IPv4. We could see that the Class-B part of the IPv4 address space was rapidly being consumed. Note that this was well before the World Wide Web (WWW) was a factor in growth, the Internet largely consisted of connecting universities and research organizations. Applications were pretty basic. Even then we were seeing rapid growth, though in hindsight it was only a hint as to what was to happen later.

This resulted in starting a project in the IETF to create a new version of the Internet Protocol. This was called the IP next generation (IPng) program. A number of different approaches were considered. A lot of effort was invested in all of the proposals and as you would expect a very vigorous debate ensued. In the end, the protocol we now call IP version 6 (IPv6) was selected. Stephen Deering and I led this effort.

The development of a new version of the Internet Protocol solved two problems, one technical and one political. The political problem, that is easy to forget now, is that at the time the TCP/IP Internet was not a sure thing. It wasn't supported by the large telecoms of the time, by governments, or by official standards bodies like ANSI and the ITU. While there was a general agreement that a new data network was desirable, there wasn't any agreement about what it should be based on. The TCP/IP Internet was then probably the least likely to become what was then called the 'Information Superhighway'. It was the 'dark horse', so to speak. In addition, most other standards groups, governments, and large telcos didn't even recognize the IETF because it was purely voluntary and didn't have any *de jure* underpinning. We weren't considered to be the 'grown-ups'.

The result was that, as more people started to hear that the TCP/IP Internet didn't have a technical future because it might run out of addresses soon, it became a significant political problem. The development and standardization of IPv6 fixed this political problem. We might not now have the current TCP/IP Internet if IPv6 had not been developed.

IPv6 also solved the technical problem of running out of IP addresses. This is the major problem that it was intended to solve and it is the main reason that IPv6 is being deployed today. I have concluded that a good way of evaluating new technologies (networking and otherwise) is if the problem they purport to solve stays the same. That is, is it a solution looking for a problem, or is it focused on a real problem? IPv6 is clearly an example of the latter. It was designed to solve the IPv4 address exhaustion problem and didn't try to evolve into solving some other problem. This is the reason that it is being deployed today.

As much as the Internet has grown from the early days, I think that the growth has only really begun. We have gone from the days where networked computers were very large and filled rooms to where a networked computer fits in a shirt pocket, from where many people shared a single large computer to where every person has many computers. This phase of the Internet is not complete as not everywhere in the world do people have access to computers and the Internet – but we are getting there. IPv6 is a necessary element in this continuing phase of Internet growth.

The next phase of Internet growth will be different and much larger. Instead of connecting people, it will to be connecting 'things'. The current phase of Internet growth is making the Internet broader and taller – the next phase will make it denser. We are moving toward a world where more and more 'things' are connected, devices that are not directly associated with people, for example, sensors, appliances, entertainment equipment, lightning controls, power distribution, and cars. Just about everything will have a computer in it and will be connected to the Internet. For this phase of Internet growth, IPv6 is essential.

Overall, IPv6 solves the problem of addressing in a much, much larger Internet. The Internet has changed the world in many ways; IPv6 will allow the Internet to continue growing and with that growth continue the benefits that it brings to the world. IPv6 running on cellular networks will have an important role in the continuing growth of the Internet.

I worked at Nokia from 1998 through 2009 in various groups and roles, and ended with the title of Nokia Fellow. I got to know Jouni, Teemu, and Jonne very well. I was honored to be asked to write the foreword to their book. They were very involved in the 3rd Generation Partnership Program (3GPP) standardization effort and were responsible for bringing IPv6 to 3GPP and making it part of the mobile protocol standards.

I believe that this book will make an important contribution to IPv6 deployment in mobile Internet devices.

In the IETF there is a toast we like to give. To paraphrase: 'Kudos to Jouni, Teemu, and Jonne for writing this useful book and to the universal deployment of IPv6.'

Robert Hinden, Palo Alto, California

Preface

The story of this book began in March 2010, when John Wiley and Sons Ltd approached Teemu with a request to review a book proposal. While reviewing the proposal, Teemu got an idea for a book, and thought it would be great fun to write such a book. And after all, it could not be that hard. The original idea was to write a book about 'IPv6 Multihoming', a topic we touch on in Chapter 6 of this book. In August 2010 Teemu and Jouni were working in a joint Finnish TEKES-funded research project called 'Wireless Broadband Access (WiBrA)', and Teemu came to ask if Jouni would be interested in co-authoring a book about 'Advanced IPv6 Multihoming'. Jouni was interested, but was also proposing a slightly different focus for the book. During 2010 and 2011 Wiley approached us periodically, and patiently reminded us to send a detailed proposal for a book. It took quite some time to get into grips with and actually write the book proposal, as we were both busy working with practical IPv6 matters of our employers, doing research under WiBrA, and spending time with IETF and 3GPP standardization activities. In retrospect pieces of the puzzle seem to have fallen into place rather nicely, as the time we did not manage to write a book proposal, was mostly spent on gaining actual experience and knowledge which have significantly affected the details of this book.

In fall of 2011 Teemu and Jouni got more into actually progressing the book, and approached Jonne with a request to be the third author. We all knew each other from the past, as we had all been working for Nokia or Nokia Siemens Networks at the same time. With the three of us we thought we would have wide enough skill-set and experience to write the book: Teemu had the background on the handset implementations, Jouni had the background on cellular network operations and network equipment implementation, and Jonne had the background from being a long-timer in network equipment implementation and IPv6. All of us had been active in 3GPP and IETF standardization, with Jonne being involved in 3GPP already when IPv6 got hammered into 3GPP standards. Together we had more than three decades of experience of IPv6.

With three of us together, we started planning to write a 150–200 page book, but it soon increased to close on 400 pages. At the same time the scope changed from multihoming into describing the basics of IPv6 in 3GPP networks, as we thought that there is more need to describe how IPv6 is implemented in 3GPP cellular systems, than focusing on advanced uses of the IPv6. The 400 pages was roughly the size of the book proposal that Wiley approved in February 2012. At the completion, the book had 398 pages.

We agreed on a schedule we thought would be doable and would bring the book to readers around mid 2013. Based on the experience elsewhere, nine months should be enough to complete a project anyway. The bold intent was to work on the book in our spare time, almost as a hobby, alongside our daily works duties. Perhaps influenced by the design of TCP, we got into slow start mode. We lived at different cities, and hence our working mode was effectively based on telephone, email, and an IPv6-only SVN-based version control tool that Jouni set up for us. Our face-to-face meetings were limited to IETF meetings, which we all participated in. At the midpoint of the project we had less than a third of the book written. This resulted in quite a busy August–October of 2012, when the bulk of the book was written and reviewed. We did not want to slip on the mid 2013 target, so we had to steal time from elsewhere – typically from sleeping hours.

Despite the long hours towards the completion of the book, we have found this topic fun and educative to write about. The IPv6 is a fascinating technology with many details and aspects. It provides topics of pure academic interest, engineering beauty, fixes and patches, politics and economics, research opportunities for the future, and overall it reflects how human beings work and build the world. From simple details we build very complex systems, which borderline being fully understandable for a single mind.

We hope that this book helps you to gain foot hold on IPv6 itself, and in particular in 3GPP systems. Reading this book should provide you with an knowledge framework of this technology, and thus help in applying the knowledge in the field, and also enable learning in more areas through references, literature, and elsewhere.

We wish you happy reading and a deeper love for acronyms.

Jouni, Teemu, Jonne

Acknowledgments

First of all, we would like to acknowledge our families for their support for, and bearing with, us during the writing of this book. Without our families' support it would not have been possible to generate the effort that this book required.

We would also like to thank our employers, Nokia, Nokia Siemens Networks, and Renesas Mobile, for not discouraging us to work from writing a book in our spare time, and also for providing us with years of work in the area of IPv6, standardization, and industry collaboration. Without being able to work on actual real-life issues, it would not have been possible to learn the things that form a major part of this book.

Major acknowledgments also go to the Nokia devices unit, which has provided us with IPv6 capable devices. Latest of the devices was the 21M-02 USB-modem that was capable of opening IPv4v6 type of PDP Context, and hence made interesting new testing scenarios possible. Before that, a long line of handsets, since around 2004, made it possible to gather experience on cellular IPv6 usage. We would also like to acknowledge Illka Keisala from TeliaSonera Finland for arranging us, via Finnish TEKES WiBrA-project, IPv6-enabled SIM-cards that made it possible to try out IPv6 also in roaming scenarios – free of charge.

Teemu sacrificed time from his three small children, Emil, Nea, and Elias, and spouse, Hanna, for creation of this book – mea culpa. Acknowledgments for helping Teemu to get to this point go to all managers, colleagues, and subordinates in Nokia who have provided the possibility to work with IPv6 implementations and standardization, made Teemu aware of interesting problems, and supported him in everyday work. Special acknowledgments go to Petri Vaipuro, who in 2001 hired Teemu to work on TCP/IPv6 implementation tasks and by so doing provided the opportunity to jump into this technology, and Juha Wiljakka, who taught Teemu the secrets of IPv6 standardization and IETF.

Jouni apologies, again, to his wife Hanna for being mentally absent during the furious writing sessions. It was all done at the expense of the family quality time. Jouni also acknowledges the highly skilled folks in Nokia Siemens Networks NeVe labs, Kari Tiirikainen and Mark Stoker among others, for providing him full access to fool around with the latest software releases and bearing with his newbie questions on the setup details. Gyorgy Wolfner and Giorgi Gulbani provided invaluable insight into 3GPP specification details over the years. The same gratitude also goes to Nokia Siemens Networks Smart-Labs for providing Jouni with the latest IPv6 enabled handsets and native IPv6 Internet access used in the TEKES WiBrA-project. Jouni also thanks Paulig for Presidentti coffee and all the caffeine he got out of it into his veins.

Jonne thanks his wife Anoush, and children Sofia and Matias for their excellent support during this book project, which limited the possibilities to really have quality family time, and doing anything else together other than stay at home. Jonne would also like to thank his manager, Erkki Yli-Juuti at Renesas Mobile, for the support to get through this project. Jonne would also like to thank Bob Hinden, Steve Crocker, Pertti Lukander, David Kessens, Mikko Puuskari, and Jaakko Rajaniemi for the guidance, understanding, collegiality, and support over the years spent learning the 3GPP technology, secrets of standardization, and especially during the transition from telecom to the Internet mindset. In addition, Jonne would like to thank Juha Wiljakka for the excellent cooperation and being a partner in crime in Nokia while working on IPv6 at a time when universal deployment of IPv6 was not quite as obvious as it is today.

Finally, we would like to thank friendly people from John Wiley and Sons Ltd, Laserwords Private Limited, and Archive Publications for their assistance in getting this project into covers and shelves. Special thanks go to Alexandra, Catherine, Claire, Krupa, Mark, Paul, Sandra, Sophia, Susan, and Teresa.

Acronyms

2G	2nd Generation
3G	3rd Generation
3GPP	3rd Generation Partnership Project
3GPP2	3rd Generation Partnership Project 2
4G	4th Generation
6LoWPAN	IPv6 over Low power Wireless Personal Area Networks
6RD	IPv6 Rapid Deployment on IPv4 infrastructures
6bone	6bone
6over4	IPv6 over IPv4 without explicit tunnels
6 to 4	Connection of IPv6 domains via IPv4 clouds
A	IPv4 address record
AAA	Authentication, Authorization and Accounting
AAAA	IPv6 address record
ACL	Access Control List
AD	Area Director
AfriNIC	African Network Information Center
AFTR	Address Family Transition Router
AH	Authentication Header
ALG	Application-Level Gateway
ANDSF	Access Network Discovery and Selection Function
API	Application Programming Interface
APN	Access Point Name
APNIC	Asia-Pacific Network Information Center
ARIB	Association of Radio Industries and Businesses
ARIN	American Registry for Internet Numbers
AS	Autonomous System
AT	ATtention
ATIS	Alliance for Telecommunications Industry Solutions
ATM	Asynchronous Transfer Mode
AuC	Authentication Center
AVP	Attribute Value Pair

B4	Basic Bridging BroadBand
BCP	Best Current Practice
BG	Border Gateway
BGP	Border Gateway Protocol
BIH	Bump-In-the-Host
BM-SC	Broadcast Multicast Service Centre
BMR	Basic Mapping Rule
BR	Border Relay
BSC	Base Station Controller
BSS	Base Station System
BSSGP	Base Station System GPRS Protocol
BTS	Base Transceiver Station
CALIPSO	Common Architecture Label IPv6 Security Option
CAMEL	Customized Applications for Mobile Network Enhanced Logic
CCSA	China Communications Standards Association
ccTLD	country code Top Level Domain
CDF	Charging Data Function
CDR	Charging Data Record
CER	Customer Edge Router
CGA	Cryptographically Generated Address
CGF	Charging Gateway Function
CGN	Carrier Graned NAT
CHAP	Challenge-Handshake Authentication Protocol
CIDR	Classless Inter-Domain Routing
CLAT	Client Side Translator
CN	Core Network
CoA	Care-of Address
CoAP	Constrained Application Protocol
CP	Control Plane
CPA	Certification Path Advertisement
CPE	Consumer Premises Equipment
CPNS	Converged Personal Network Service
CPS	Certification Path Solicitation
CPU	Central Processing Unit
CS	Circuit Switched
DAD	Duplicate Address Detection
DAF	Dual Address Bearer Flag
DCCP	Datagram Congestion Control Protocol
DHCP	Dynamic Host Configuration Protocol
DHCPv4	Dynamic Host Configuration Protocol version 4
DHCPv6	Dynamic Host Configuration Protocol version 6
DHCPv6PD	DHCPv6 Prefix Delegation
DMR	Default Mapping Rule
DNA	Detecting Network Attachment

DNS	Domain Name System
DNS64	DNS Extensions for Network Address Translation
DNSSEC	Domain Name System Security Extensions
DoS	Denial of Service
DPI	Deep Packet Inspection
DR	Delegating Router
DS-Lite	Dual Stack Lite
DSCP	Differentiated Services Code Point
DSL	Digital Subscriber Line
DS-MIPv6	Dual Stack Mobile IPv6
DSTM	Dual Stack Transition Mechanism
DUID	DHCP Unique IDentifier
DUID-EN	DUID vendor-assigned unique identifier based on Enterprise Number
DUID-LL	DUID Link-Layer address
DUID-LLT	DUID Link-Layer address plus Time
DUID-UUID	DUID Universally Unique IDentifier
E-UTRA	Evolved UMTS Terrestrial Radio Access
E-UTRAN	Evolved UMTS Terrestrial Radio Access Network
EA	Embedded Address
EAP	Extensible Authentication Protocol
ECN	Explicit Congestion Notification
EIR	Equipment Identity Register
eNodeB	Evolved Node B
EPC	Evolved Packet Core
EPS	Evolved Packet System
ESP	Encapsulating Security Payload
ETSI	European Telecommunications Standards Institute
FDDI	Fiber Distributed Data Interface
FMR	Forwarding Mapping Rule
FQDN	Fully Qualified Domain Name
FTP	File Transfer Protocol
GERAN	GSM/Edge Radio Access Network
GGSN	Gateway GPRS Support Node
GMM/SM	GPRS Mobility Management and Session Management
GPRS	General Packet Radio Service
GRE	Generic Routing Encapsulation
GRX	GPRS Roaming eXchange
GSM	Global System for Mobile Communications
GSMA	GSM Association
gTLD	generic Top Level Domain
GTP	GPRS Tunneling Protocol
GTP-C	GTP Control Plane
GTP-U	GTP User Plane

GTPv1	GPRS Tunneling Protocol version 1
GTPv1-C	GTP Control Plane version 1
GTPv2	GPRS Tunneling Protocol version 2
GTPv2-C	GTP Control Plane version 2
GUA	Global Unicast Address
HA	Home Agent
HLR	Home Location Register
HNP	Home Network Prefix
HoA	Home Address
HPLMN	Home PLMN
HSDPA	High Speed Downlink Packet Access
HSPA	High Speed Packet Access
HSS	Home Subscriber Server
HSUPA	High Speed Uplink Packet Access
HTTP	HyperText Transfer Protocol
I-WLAN	Interworking-WLAN
IAB	Internet Architecture Board
IAID	Identity Association IDentifier
IANA	Internet Assigned Number
IAOC	IETF Administrative Oversight Committee
IAPD	Identity Association for Prefix Delegation
ICANN	Internet Corporation for Assigned Names and Numbers
ICMP	Internet Control Message Protocol
ICMPv4	Internet Control Message Protocol version 4
ICMPv6	Internet Control Message Protocol version 6
IDN	Internationalized Domain Name
IE	Information Element
IEEE	Institute of Electrical and Electronics Engineers
IESG	Internet Engineering Steering Group
IETF	Internet Engineering Task Force
IFOM	IP Flow Mobility and Seamless WLAN Offload
IGD	Internet Gateway Device
IGF	Internet Governance Forum
IGP	Interior Gateway Protocol
IID	Interface IDentifier
IKEv2	Internet Key Exchange version 2
IMEI	International Mobile Equipment Identity
IMS	IP Multimedia Subsystem
IMSI	International Mobile Subscriber Identity
IoT	Internet of Things
IP	Internet Protocol
IPCP	Internet Protocol Control Protocol
IPIP	IP in IP tunneling
IPsec	Internet Protocol security

IPTV	Internet Protocol Television
IPv4	Internet Protocol version 4
IPv6	Internet Protocol version 6
IPV6CP	IPv6 Control Protocol
IPX	IP Packet eXchange – evolved GRX
IS-IS	Intermediate System to Intermediate System
ISATAP	Intra-Site Automatic Tunnel Addressing Protocol
ISC	Internet Systems Consortium
ISP	Internet Service Provider
L2TP	Layer 2 Tunneling Protocol
L2TPv3	Layer 2 Tunneling Protocol version 3
LAC	L2TP Access Concentrator
LACNIC	Latin America and Caribbean Network Information Center
LAN	Local Area Network
LCP	Link Control Protocol
LI	Legal Interception
LIPA	Local IP Access
LIR	Local Internet Registry
LLC	Logical Link Control
LMA	Local Mobility Anchor
LNS	L2TP Network Server
LTE	Long Term Evolution
M2M	Machine-to-Machine
MAC	Media Access Control
MAG	Mobile Access Gateway
MANET	Mobile Ad hoc NETworking
MAP	Mapping of Address and Port with Encapsulation or Translation
MBMS	Multimedia Broadcast Multicast Service
ME	Mobile Equipment
MIB	Management Information Base
MIPv6	Mobile IPv6
MLD	Multicast Listener Discovery
MLDv2	Multicast Listener Discovery version 2
MME	Mobile Management Entity
MMS	Multimedia Messaging
MN	Mobile Node
MP-BGP	Multi-Protocol Border Gateway Protocol
MPLS	MultiProtocol Label Switching
MS	Mobile Station
MSC	Mobile Switching Centre
MSISDN	Mobile Station International Subscriber Directory Number
MSS	Maximum Segment Size
MT	Mobile Terminal
MTC	Machine-Type Communications
MTU	Maximum Transmission Unit

NAPDEF	Network Access Point Definition
NAS	Non-Access Stratum
NAT	Network Address Translation
NAT-PMP	NAT Port Mapping Protocol
NAT-PT	Network Address Translation – Protocol Translation
NAT44	Network Address Translation from IPv4 to IPv4
NAT46	Network Address Translation from IPv4 to IPv6
NAT64	IPv4/IPv6 Network Address Translation
NBMA	Non-Broadcast Multiple Access
NCP	Network Control Protocol
ND	Neighbor Discovery
NDP	Neighbor Discovery Protocol
NFC	Near Field Communications
NNI	Network-to-Network Interface
NodeB	UMTS base station
NSP	Network Specific Prefix
NUD	Neighbor Unreachability Detection
OCS	Online Charging System
OECD	Organisation for Economic Co-operation and Development
OEM	Original Equipment Manufacturer
OFCS	Offline Charging System
OFDMA	Orthogonal Frequency-Division Multiple Access
OMA	Open Mobile Alliance
OS	Operating System
OSI	Open System Interconnect
OSPF	Open Shortest Path First
OSPFv2	Open Shortest Path First version 2
OSPFv3	Open Shortest Path First version 3
OUI	Organizationally Unique Identifier
P-CSCF	Proxy Call Session Control Function
PAA	PDN Address Allocation
PCC	Policy and Charging Control
PCEF	Policy and Charging Enforcement Function
PCG	Project Coordination Group
PCO	Protocol Configuration Option
PCP	Port Control Protocol
PCRF	Policy and Charging Rules Function
PD	Prefix Delegation
PDCP	Packet Data Convergence Protocol
PDN	Packet Data Network
PDP	Packet Data Protocol
PDU	Protocol Data Unit
PGW	Packet Data Network Gateway
PHB	Per-Hop Behavior

PIO	Prefix Information Option
PKI	Public Key Infrastructure
PLAT	Provider Side Translator
PLMN	Public Land Mobile Network
PMIP	Proxy Mobile IP
PMIPv6	Proxy Mobile IPv6
PMTUD	Path MTU Discovery
PNAT	Prefix NAT
POSIX	Portable Operating System Interface for uniX
PPP	Point to Point Protocol
PS	Packet Switched
PSID	Port-set Identifier
PSTN	Public Switched Telephony Network
PTB	Packet Too Big
PTR	Pointer Record
QoS	Quality of Service
RAB	Radio Access Bearer
RADIUS	Remote Authentication Dial In User Service
RAN	Radio Access Network
RANAP	Radio Access Network Application Part
RAT	Radio Access Technology
RAU	Routing Area Update
RDNSS	Recursive DNS Server
REST	REpresentational State Transfer
RF	Radio Frequency
RFC	Request For Comments
RH0	Type 0 Routing Header
RIO	Route Information Option
RIP	Routing Information Protocol
RIPE-NCC	Réseaux IP Européens Network Coordination Centre
RIPng	Routing Information Protocol next generation
RIR	Regional Internet Registry
RLC	Radio Link Control
RNC	Radio Network Controller
RoHC	Robust Header Compression
RPKI	Resource Public Key Infrastructure
RPL	Routing Protocol for Low-Power and Lossy Networks
RR	Requesting Router
RRC	Radio Resource Control
RTT	Round Trip Time
S4-SGSN	Serving Gateway Support Node with S4 interface
SA	Security Association
SAD	Security Association Database

SAE	System Architecture Evolution
SAE-GW	System Architecture Evolution Gateway
SaMOG	S2a Mobility based on GTP and WLAN access to EPC
SAVI	Source Address Validation Improvements
SCCP	Signaling Connection Control Part
SCTP	Stream Control Transmission Protocol
SDO	Standards Developing Organization
SEND	Secure Neighbor Discovery
SGSN	Serving Gateway Support Node
SGW	Serving Gateway
SIG	Special Interest Group
SIIT	Stateless IP/ICMP Translator
SIP	Session Initiation Protocol
SIPTO	Selective IP Traffic Offload
SLAAC	Stateless Address Autoconfiguration
SMS	Short Message Service
SNDCP	Subnetwork Dependent Convergence Protocol
SNMP	Simple Network Management Protocol
SPI	Security Parameters Index
SS7	Signaling System No. 7
SSID	Service Set Identifier
SSM	Source-Specific Multicast
TCP	Transport Control Protocol
TDMA	Time Division Multiple Access
TE	Terminal Equipment
TEID	Tunnel Endpoint Identifier
Teredo	Tunneling IPv6 over UDP through NATs
TFT	Traffic Flow Template
TIA	Telecommunications Industry Association
TLD	Top Level Domain
TLS	Transport Layer Security
TOS	Type of Service
TP	Transport Plane
TSG	Technical Specification Group
TTA	Telecommunications Technology Association
TTC	Telecommunication Technology Committee
TTL	Time To Live
UDP	User Datagram Protocol
UE	User Equipment
UI	User Interface
UICC	Universal Integrated Circuit Card
ULA	Unique Local Address
UMTS	Universal Mobile Telecommunications System
UN	United Nations

UNI	User-to-Network Interface
UP	User Plane
UPnP	Universal Plug and Play
URI	Uniform Resource Identifier
URL	Uniform Resource Locator
USB	Universal Serial Bus
UTRAN	UMTS Terrestrial Radio Access Network
VLAN	Virtual Local Area Network
VoIP	Voice over IP
VPLMN	Visited PLMN
VPN	Virtual Private Network
WAN	Wide Area Network
WCDMA	Wideband Code Division Multiple Access
WG	Working Group
WKP	Well-Known Prefix
WLAN	Wireless Local Area Network
XML	eXtended Markup Language

Glossary

2G and GSM Global Systems for Mobile Communications, 2G (second generation), is a digital wireless telephony technology originally developed by ETSI and maintained by 3GPP since 1998. The technologies before 2G, the so-called '1G' technologies, were analog systems.

3G Direct descendant from 2G and the first release made by 3GPP. 3G effectively brought improvements to speed and also introduced IPv6 for both 2G and 3G.

4G and LTE The 3GPP release-8 introduced the Evolved Packet System (EPS), which consists of Long Term Evolution (LTE) radio and Evolved Packet Core (EPC) network core. The whole system is referred to as 4G or LTE, depending on the source.

6bone A global IPv6 technology testbed that ran from 1996 to 2006. The 6bone network was mostly a virtual network run over the global Internet infrastructure to connect different IPv6 test networks. The purpose of this experiment was to test the interoperability of the IPv6 technology and the transition mechanism, and gather operational experience. The experiment was designed to be temporary from the start, and it was to be replaced by the Internet itself becoming IPv6 capable. The experiment officially ended on 6 June 2006 (6.6.6) when the routing of the experimental addresses ceased.

A resource record A DNS record type that contains one 32-bit IPv4 address, which maps to a hostname. Used to find an IPv4 address for a hostname.

AAAA resource record A DNS record type that contains one 128-bit IPv6 address, which maps to a hostname. Used to find an IPv6 address for a hostname.

Access Point Name (APN) Name of an access point in the 3GPP system. The APN points to an external network or service. The APN has two parts: the network identifier and the operator identifier. The structure of APN follows those of fully qualified domain names, which eventually resolves to an IP address of a GGSN or a PGW. The operator identifier part of an APN is not usually visible to the end user, and if absent can be filled by the network (e.g., by SGSN) during connection establishment.

Address autoconfiguration Hosts can automatically configure IPv6 addresses in stateless or stateful manner. The Stateless Address AutoConfiguration (SLAAC) works by hosts receiving one or more IPv6 prefixes in Router Advertisements and combining them with Interface Identifiers to create 128-bit full IPv6 addresses. The Stateful Address

AutoConfiguration works by hosts using DHCPv6 to request and receive full 128-bit IPv6 addresses from a DHCPv6 server.

Address formats IPv6 addresses always have some format in them. The basic formats indicate whether an address is unspecified, loopback, unicast, subnet anycast, or multicast, link-local or global, or unique or local. The advanced address formats embed additional information inside the address, which can be, for example, IPv4 addresses, port numbers or port ranges, and hardware identifiers.

Address prefix The high-order bits of an IPv6 address that determine the subnetwork the address belongs to. An individual link typically has a 64-bit prefix denoted as /64, while network operators are allocated a shorter prefix, such as /32.

Address scopes IPv6 addresses can be of link-local or global scope. The link-local scoped addresses can only be used for on-link communication, while global addresses can be used for any communications. IPv6 had a site-local scope, but it has been deprecated in favor of Unique Local Addresses (ULA). ULAs are considered to be of global scope, even though they are not globally routable.

Anycast address An address that identifies a set of interfaces, which typically are on different nodes. An IPv6 packet set to anycast address is delivered by the Internet's routing system to the nearest instance of an anycast address.

Application layer Layer seven in the OSI model. This layer transports application-specific content in an application specific manner. Examples of such protocols include HTTP and SIP.

Base Station Controller (BSC) Controls a group of base stations and handles mobility management between the BaseTransceiver Stations (BTSs) under its control.

Base Station System GRPS Protocol (BSSGP) Provides routing-related and Quality of Service (QoS)-related information between the BSC and the SGSN. The network service that the Gb uses is frame relay.

Base Transceiver Station (BTS) 2G radio network element that directly communicates with the mobile station over the air interface.

Bellhead A hardcore telecommunications engineer who despises anything except circuit switched networking or more recently anything but managed vertical solutions.

Control plane Takes care of the signaling in the network used for controlling the network and the mobile station. The control plane transports the messages that are needed to attach a subscriber to the network, authenticate and authorize the user, open and close connections, and for signal handovers.

Core network The Internet service provider's network between the radio access network and the transit network. In the case of 4G, the core is referred as the Evolved Packet Core (EPC) and includes MME, SGW, PDN-GW, and HSS. In the case of 2G and 3G, the core is referred as the packet core and includes SGSN, GGSN, and HLR.

Domain Name Service (DNS) Global name to address, and address to name translation service. The DNS allows human-readable names to be used for protocols and user interfaces, while allowing the Internet still to function with numbers with better computational characteristics. DNS is a global distributed database covering the whole Internet.

Domain Name System Security Extensions (DNSSEC) Extensions that allow signing of the DNS resource records and validation by the DNS resolvers. Protects against spoofing and tampering of the records on the path from authoritative name server to the resolver.

Duplicate Address Detection (DAD) The process by which an IPv6 node ensures that an IPv6 address that it is configuring for a network interface is not already in use on the link to which the node is attached.

eNodeB LTE base station that directly communicates with the mobile station over the air interface. The eNodeB, however, also includes some of the functionalities of the radio resource controller (BSC in 2G, and RNC in 3G).

Fall back Reverting to using something as an alternative to a preferred choice. In this book fall back refers to using one address family instead of an other (e.g., use of IPv4 if IPv6 has failed), or using less preferred bearers if more preferred ones have failed to work (e.g., using parallel IPv4 and IPv6 PDP Contexts if IPv4v6 PDP Context did not open).

Fragmentation and reassembly Fragmentation is a process for slicing outgoing large protocol data units into chunks that can be sent in IPv6 packets that fit into the MTU of the path. Reassembly is the process of combining chunks of data received in multiple IPv6 packets into the original protocol data unit.

Gateway GPRS Support Node (GGSN) The topological anchor point of the mobility management in the GPRS network. It is the gateway between the GPRS network and external networks, such as the Internet. From UE's IP stack point of view GGSN is the first-hop router.

Gb-interface The interface between the BSS and the core network – hence, between the BSC and the SGSN. This interface comprises both of the control and user plane parts.

General Packet Radio Service (GPRS) Packet data service for GSM network. 3G uses the same architecture, and hence both 2G and 3G are often referred to as GPRS.

Gi-interface The interface that connects the GPRS network to the external IP networks.

GSM radio protocols GSM Radio Frequency (RF), Media Access Control (MAC), Radio Link Control (RLC), and Logical Link Control (LLC) provide the GSM radio functionality including ciphering, reliability, and other radio procedures.

Gn/Gp-interface The interface between the SGSN and the GGSN. The Gn-interface is used when the Mobile Station (MS) is in the home network, whereas the Gp-interface is used when the subscriber is roaming in another network. Thus, the Gp-interface is the interface between two network operators. The Gn-interface is also used between SGSNs during a handover to exchange MS related information, and for forwarding user packets.

GPRS mobility management and session management Protocol for performing mobility management functions, and connection activation, modification, and deactivation.

GPRS Tunneling Protocol (GTP) Responsible for transporting the user's IP packets over the 2G, 3G, and 4G core networks, and controlling the associated tunnels. In the User Plane, the basic functionality of GTP is to uniquely identify the GTP-tunnel to which the user packets belong. The GTP-tunnel maps directly to the PDP Context or EPS bearer. In the control plane GTP is used for session and mobility management within the Core Network.

Gr-interface The interface between the SGSN and the HLR. The interface is based on SS7.

GSM See *2G and GSM.*

Home Location Register (HLR) Database on 2G/3G networks containing the subscriber data, which includes the authentication data used to authenticate the user, user profiles including what services the user is subscribed to, and the location where the user currently is.

Home Subscriber Server (HSS) The Home Subscriber Server is a combination of the AuC, HLR, and functionality to support the operator service infrastructure, the IMS. HSS replaced the HLR from the 3GPP Release-5 onwards with the introduction of the IMS. Replaces HLR in EPS.

Interface The interface is an overloaded term with meaning determined by the context. In the 3GPP context, it means the whole vertical protocol stack between two network elements – sometimes also called a reference point. In the IP context it means the network connection of an IP stack. In the software architecture context it means the defined interactions between two software modules, and sometimes specifically it is a short name for an application programming interface.

Interface IDentifier (IID) The host part of an IPv6 address. Typically 64 bits in length. The IID can be based on a hardware address, such as the IEEE 802 MAC, be manually or algorithmically generated. A random IID generation is often used for privacy purposes to prevent tracking based on IID.

Internet-Draft An individual or working group contribution to the IETF that may become RFC, may change drastically between subsequent versions, and will expire in six months if not renewed. An Internet-Draft can be referenced only as work-in-progress, as they are not stable documents by any means.

Internet Control Message Protocol version 6 (ICMPv6) A control protocol used, for example, for neighbor discovery, router discovery, communication of error situations, 'pinging', multicast listener discovery, mobile IPv6, RPL, and multicast router discovery.

Internet Engineering Task Force (IETF) The standardization organization responsible of key Internet protocols, such as IP, TCP, UDP, DNS, and HTTP.

Internet Protocol version 4 (IPv4) The Internet protocol that made the Internet a huge success and that provides the network layer services for almost all of the Internet as of today. Suffers from only having a 32-bit address space, and is therefore being replaced by IPv6.

Internet Protocol version 6 (IPv6) The next generation Internet protocol that provides a 128-bit address space, alongside other improvements.

IoT, MTC, M2M Internet of Things (IoT), Machine Type Communications (MTC), and Machine to Machine (M2M) all refer to concept where Internet-enabled nodes – which are not directly operated by humans – talk to other nodes. The IoT is used especially in the context of low-cost, low-power, smart objects. The M2M is usually used in the context where not so low-power and low-cost machines talk to other machines. The MTC is used especially in 3GPP to refer to M2M type of communications. These Internet-enabled objects can be smart electricity meters, heart rate sensors, vending machines, street lights, and alike.

IP mobility Technologies that allow a node to change its point of attachment inside a network (intra-network) or between networks (inter-network) without any change in IP address and without disconnection of transport layer sessions. The key protocols are host-based mobile IP and network-based proxy mobile IP.

IP Security (IPsec) IPsec refers to IPv6 protocol features providing authentication, integrity protection, anti-replay protection, access control, and encryption. These are provided with AH and ESP headers. The IPsec does not refer to the general security of the IPv6 protocol.

IPv6 transition The IPv6 should one day replace IPv4 in all Internet communications. The path from IPv4 Internet to IPv6 Internet is a rocky and very complex path requiring standards, implementation, verification, policies, education, governance, deployment, willingness, money, and effort. This path from IPv4 to IPv6 is generally referred as IPv6 transition.

IPv6 transition tools The protocols and practices that help nodes with IPv4-only, IPv6-only, and dual-stack access to communicate with each other by using tunneling, protocol translation, and proxying technologies.

Link layer Layer two in the OSI model. Responsible for transporting data over the physical layer, detecting errors, and providing physical layer addressing (such as MAC addresses on WLAN). Ethernet and PPP are examples of link layer protocols.

LTE *See 4G and LTE.*

M2M *See IoT, MTC, M2M.*

Mobile Management Entity (MME) Responsible for mobility management, authentication, and authorization of the UE in 4G networks. The MME is, in practice, a very similar element to the SGSN in UMTS. The difference between the 3G-SGSN and the

MME is that the MME is only a control plane element, and it does not perform any user-plane functions.

Mobile Switching Center (MSC) The local exchange responsible a set of Base Station Controllers (BSC), and Mobile Stations (MS) under those base stations in a geographical area. Performs authentication and mobility management, and exchanges the subscribers' calls. MSC is a Circuit Switched network entity.

MTC *See IoT, MTC, M2M.*

Multicast address An address that identifies multitude of interfaces on, usually, a multitude of nodes. An IPv6 packet sent to a multicast address is delivered to all receivers of the multicast address.

Neighbor Discovery The functionality that IPv6 nodes on the same link use to discover each other's presence, when applicable to determine each other's link-layer addresses, to find routers, and to maintain reachability information about the paths to active neighbors.

Network Address Translation (NAT) Process whereby one network address is changed to another. The most common use of NAT has been to translate private IPv4 addresses into public IPv4 addresses. The translation is also referred as NAPT, if transport protocol port translation is taking place. However, due to port translation being very common in IPv4, the term NAT usually also indicates port translation. With the emergence of IPv6 translation discussions and protocol translation, the term NAT44 is sometimes used to highlight that it is IPv4 address translation that is taking place, and similarly term NAT66 to indicate IPv6 address translation.

Network layer Layer three in the OSI model. Responsible for transporting data over a series of link layers, including routing, error detection, possibly fragmentation and reassembly, and providing logical addresses to the network (such as IP addresses). IPv4 and IPv6 are examples of network layer protocols.

NodeB 3G radio network element that directly communicates with the UE over the air interface.

Offloading Process for directing the host's traffic to go through some other network than the one over which the traffic would have otherwise been sent.

Packet Data Convergence Protocol (PDCP) Responsible for transferring the IP packets over the 3G and LTE radio interface, and also providing header compression. The segmentation and reassembly of the user packets, and mapping the PDP Contexts to the radio bearers is provided by the underlying RLC. In 3G, the functionality, which was in SNDCP in 2G, has been distributed into PDCP and to the RLC protocol layers. In 4G the PDCP is further extended to provide user traffic ciphering and integrity protection, and to deliver packets in sequence and without duplication.

Packet Data Network (PDN) PDN can be an internal walled garden like an IP network within the mobile operator network, or any IP network outside operator administration such as the Internet.

Packet Data Network Gateway (P-GW, PGW, or PDN-GW) The gateway between the EPS and the external IP networks, and the anchor point for EPS mobility management. From a UE's IP-stack point of view PGW is the first-hop router.

Path Maximum Transmission Unit (MTU) Discovery A process for finding the smallest MTU on the links on the path from sending node to receiving node.

Pointer (PTR) resource record A DNS resource record that points to a hostname. Used for reverse DNS queries – to find which name maps to an address.

Protocol translation An important IPv6 transition tool is protocol translation, which refers to the process of translating IPv6 packets into IPv4 (NAT64) or IPv4 packets into IPv6 (NAT46).

Provisioning Technologies and processes used for automatic configuration of hosts with information needed for accessing networks.

Radio Access Network The 'last mile' connectivity from core network to the edge, for example to hosts. In 3GPP 4G the access network includes eNodeB, in 3G it includes eNodeB and RNC, and in 2G it includes BTS and BSC. Referred in 2G also as GERAN, in 3G as UTRAN, and in 4G as E-UTRAN.

Radio Access Network Application Part (RANAP) UMTS protocol that carries the higher layer signaling. It runs on the Signaling System No. 7 (SS7) protocol suite. Originally the Iu-interface was specified to be transported only on ATM. However, the standard has been updated to enable any transport network technology to be used.

Radio Network Controller (RNC) Controls a group of NodeBs and handles mobility management between the NodeBs under its control.

Request For Comments (RFC) Permanent documents created by IETF. The RFCs that represent IETF's consensus are, for example, classified as Standards Track or Best Current Practice (BCP). A standards track RFC is usually a protocol specification or a protocol framework. BCPs describe the currently best found ways to implement or deploy something. Additionally, an RFC can be classified as experimental or informational. Experimental RFCs are meant only for experimenting, and informational RFCs range from specifications to jokes.

Roaming A mobile station is roaming when it is attached to a 3GPP radio access that is not part of the home network. Attachment to non-3GPP accesses, such as WLAN, is not referred to as roaming.

Router discovery A process for hosts to automatically find routers present on the link. The router discovery is essential for global communications, as without manually configured or automatically discovered routers, hosts can only send packets to other hosts on the same link. Hosts discover routers by sending Router Solicitations and by listening for Router Advertisements.

Routing Technologies that cover rules and decisions made by individual routers to decide where to forward IPv6 packets. It also refers to the protocols that are used by routers to share information about paths to different destinations (to different IPv6 networks).

S1-MME interface Control-plane interface connecting the eNodeB and the MME. The interface uses SCTP as the transport protocol.

S1-U interface User-plane interface connecting the eNodeB and SGW. The interface is based on GTP.

S4 Serving Gateway Support Node (S4-SGSN) A 3GPP Release-8 (and onwards) compliant SGSN that connects 2G/3G radio access networks to the EPC via new Release-8 GTP-based S3/S4 interfaces, and to an HSS using a Diameter-based S6d interface.

S5/S8 interface The interfaces connecting the SGW and the PGW. S5 is used within the home network, and S8 is the roaming interface when the UE is roaming. This is similar to the Gn- and Gp-interfaces in GPRS. The S5- and S8-interfaces consist of user and control planes.

S6a interface Interface connecting the MME and HSS. The interface is based on Diameter.

S6d interface Interface connecting 3GPP Release-8 S4-SGSN and an HSS. The interface is based on Diameter.

Serving Gateway (SGW, S-GW) Mobility anchor in 4G networks for inter-eNodeB mobility. It routes the user traffic between the radio network and the Packet Data Network Gateway (PGW).

Serving Gateway Support Node (SGSN) Responsible for authentication, authorization, and mobility management in GPRS. In addition, SGSN gathers charging information and terminates the radio protocols between UE and SGSN.

SGi interface Interface connecting the EPS to the external IP networks, such as the Internet.

Signaling Connection Control Part (SCCP) Protocol used in the SS7 protocol suite.

Split-UE A UE that is either split into completely separate MT and TE devices, such as a modem and a computer, or a single device that is internally architectured to separate the MT and TE components from each other.

Subnetwork Dependent Convergence Protocol (SNDCP) Protocol that transports IP packets over the 2G GSM radio network. The SNDCP is responsible for multiplexing multiple PDP Contexts between the UE and the SGSN, perform header compression, packet segmentation and reassembly, and content compression.

Tethering A process where a mobile station shares the cellular uplink Internet connectivity to devices in a local area network, typically using WLAN or USB technologies.

Transit network The part of the Internet that connects different Internet service operators together.

Transport Control Protocol (TCP) Transport layer protocol that transmits a stream of data in reliable and in-order form. TCP implements error detection, retransmission, flow control, and congestion control.

Transport layer Layer four in the OSI model. Responsible for transporting data over the network, providing, if needed: in-order delivery, flow control, retransmissions, and multiplexing (e.g., via port concept). TCP and UDP are the most common transport layer protocols. In the context of 3GPP the transport layer also refers to protocols used to transport both control and user-plane protocols.

Unicast address An address that is assigned to a single network interface. A packet sent to a unicast address is delivered to a single recipient.

Universal Mobile Telecommunications System (UMTS) The UMTS Terrestrial Radio Access Network (UTRAN) is a communications network, commonly referred to as 3G, and consists of NodeBs (3G base stations) and Radio Network Controllers (RNCs), which make up the UMTS radio access network. The UTRAN allows connectivity between the UE and the core network. The UTRAN is composed of WCDMA, HSPA, and HSPA+ radio technologies.

User Datagram Protocol (UDP) Simple transport layer protocol capable of transporting datagrams that provides ports for multiplexing different flows and checksum for error detection. UDP does not guarantee packet delivery or that packets are delivered in order.

User Equipment (UE) or Mobile Station (MS) Everything that is on the user side of the cellular network connection. Traditionally refers to a mobile handset, but nowadays also covers dongles providing cellular access to computers (e.g., via USB or WLAN), fixed home gateways using cellular access as uplink, or even a set of devices behind the single cellular connection.

User layer In the 3GPP domain this refers to the layer that consists of users' packets, including IP packets and all payload. The addressing, IP version, and the traffic on the user layer are completely separate from the 3GPP network's transport layer.

User plane Takes care of the end user packet transmissions through the network.

Voice over LTE (VoLTE) VoLTE is an IMS-based solution for providing IP-based voice and short message services over LTE access technology. It is also the solution from 3GPP for replacing circuit switched voice services in LTE networks.

Wireless Local Area Network (WLAN) Wireless short-range radio standard defined in the IEEE 802.11 standard suite. Very commonly used for Internet access in homes, enterprises, cities and such. Supported by all smart phones, and nowadays also by many feature phones.

1

Introduction

Recently, Internet access has been revolutionized by mobile broadband. However, mobile Internet access is not a new technology – it has been available since the beginning of the 2000s, but only during the past last few years has the growth of mobile usage of the Internet exploded. This explosion is due to the increased data speeds that have brought mobile Internet access speeds close to those of fixed broadband access, and the prices dropping to affordable and competitive ranges. In addition, the exploding usage is due in very large part to the introduction of the smartphone.

At the same time, and partly as a result, the Internet is facing its biggest change and its biggest challenge since its introduction. This is the transition to the new version of the Internet Protocol (IP) – IP version 6. The old version – IP version 4 (IPv4) – has been in use since 1983 when the ARPANET transitioned from Network Control Program (NCP) to the Internet Protocol. Now, the exhaustion of readily available Internet Protocol version 4 (IPv4) addresses at the beginning of 2011 puts the growth of the whole Internet at risk.

The ongoing transition of the Internet to the new version of IP will, obviously, have implications for mobile networks as well. We have written this book to look at these important two topics together – mobile broadband access to the Internet, and the transition to Internet Protocol version 6 (IPv6). In Chapter 1, we start with an overview of the Internet technologies, and the background and implications of the transition to IPv6 to the Internet. Chapter 2 explains the basics of the Third Generation Partnership Project (3GPP) specified mobile broadband technologies, and Chapter 3 examines the IPv6 technology, giving a good understanding of how IPv6 works. Chapter 4 goes through how IPv6 is intended to work in the 3GPP mobile broadband networks. Chapter 5 concentrates on giving an understanding of different transition strategies that can be used in 3GPP networks. Chapter 6 gives a forward-looking view by the authors of some areas relevant to the future of IPv6 in 3GPP networks.

We wish the reader interesting reading moments, and we hope that this book provides help to the reader, whether a student, operator, network vendor, application developer, or handset manufacturer, to learn about and navigate through the IPv6 transition in the 3GPP network ecosystem.

Deploying IPv6 in 3GPP Networks: Evolving Mobile Broadband from 2G to LTE and Beyond, First Edition.
Jouni Korhonen, Teemu Savolainen and Jonne Soininen.
© 2013 John Wiley & Sons, Ltd. Published 2013 by John Wiley & Sons, Ltd.

1.1   Introduction to Internet and the Internet Protocol

The Internet and the Internet Protocol creation were originally funded by the Defense Advanced Research Agency (DARPA) in the United States. Yet, today the Internet has become the global network of the whole world connecting all continents, virtually all of the countries, and already has significantly over two billion users. This path from a relatively small research project to the global information superhighway has been both fascinating and relatively quick. The DARPA project was started at the very end of the 1960s, the current version of the Internet Protocol was introduced in the early 1980s, and the first commercial Internet access providers came online in the end of 1980s or early 1990s depending on country and region. As late as 2006, one of the main topics of the Internet Governance Forum (IGF) – a United Nations (UN) organization discussing matters that concern the governance of the Internet, both technical and non-technical – was to connect the unconnected, that is how to get Internet access to the developing countries. Since that day, most of the developing countries have at least Internet access in the bigger cities, usually through mobile networks. The Internet has very quickly encompassed our lives, regardless where we live.

This chapter concentrates on explaining what are the guiding principles that led and enabled this evolution, and to describe what is the Internet's most important building block – the Internet Protocol. For the interested reader, at the end of the chapter there are additional reading materials for more information about the fascinating history of the Internet.

1.2   Internet Principles

Today the Internet is used for file transfer, email, voice, video, gaming, and many, many other applications. We have become dependent on the Internet starting from, the world economy, via businesses big and small, to the normal people who trust the Internet either to keep them connected to the artery of the economic world, or to keep up relationships with their loved ones. It is surprising how big and important the Internet has become in such a short time. However, the versatility of the Internet is not accidental. The reasons lie in the design of the Internet.

In the heart of the design of the Internet and the technology that powers it has certain principles. These principles have ensured that the Internet and the Internet technologies enable the current usage of the Internet, which is way beyond the usage and expectations envisioned by anybody at the dawn of the Internet. Let's look at the main principles:

- packet switched networking;
- the end-to-end principle;
- layered architecture;
- Postel's robustness principle; and
- creative anarchy.

Packet Switched Networking

This first principle is very widely used in modern communication networks. However, traditionally, the voice centric networks, such as the Public Switched Telephony Network

(PSTN), were based on a different technological principle – circuit switched networking. In circuit switched networking connections or calls are switched through the network by reserving a circuit from the caller to the callee. Each connection has its own circuit, and the same resource is reserved for the call regardless how much traffic is transferred through that circuit. For instance, a voice call uses exactly the same resources within the network whether the participants speak or are silent.

In contrast, the modern data networks are built based on Packet Switched (PS) networking. In packet switched networking, the sent data is divided into smaller packets that travel independently to their destinations, each transmitted through a route that seems best for a given packet at a given moment. In the Internet, the packets can travel different routes even to the same destination, can get out of order at transit, and even get completely lost, never arriving to their destination.

End-to-End Principle

The end-to-end principle is one of the most important principles of the Internet and of the Internet technology. It states that the network should not interfere with or alter traffic on layers that are above the network layer. Sometimes this principle is also known as *'Intelligent endpoints – dump network'* principle, but that name does not do justice to the concept. Basically, what the principle states is that the network should not make any assumptions about any particular service or characteristic of the data in transit. The network must concentrate on moving the data from its source to its destination. It is up to the end points to understand the traffic and its use. This principle allows the Internet to be used for many different services and applications. Even applications and services can be supported that were at the time of the design of the Internet either technically unfeasible, completely impossible, or even unimaginable. Thus, the principle enables us to create new services without changing the network. Only the end points have to be changed to support the new service or application.

Layered Architecture

The layered architecture principle is closely linked to the end-to-end principle. The layered architecture principle states that there are different protocol layers that talk to each other on the same level. Figure 1.1 shows the principle in a drawing. The main idea behind the principle is that each layer does its work – no more and no less. This strict separation allows the layers to be independent of each other, and make sure that layers can be changed without changing other layers. The Open System Interconnect (OSI) [1] model defines seven layers. The Internet model, however, only defines five. The Internet technologies range from the layer three (network layer) upwards.

Robustness Principle

The Postel's robustness principle, also known as Postel's law, has gotten its name from its inventor – Jon Postel. Jon Postel was one of the Internet pioneers who has had an enormous effect on the Internet and its design through his contribution in engineering and governance. The principle is quoted as *'Be liberal in what you accept, and conservative*

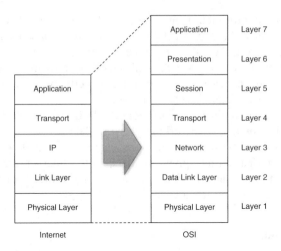

		Application	Layer 7
		Presentation	Layer 6
Application		Session	Layer 5
Transport		Transport	Layer 4
IP		Network	Layer 3
Link Layer		Data Link Layer	Layer 2
Physical Layer		Physical Layer	Layer 1

Internet   OSI

Figure 1.1   Networking layers.

in what you send.' [2] This principle means that the sender should be very conservative and strictly cohere with the standard, whereas the receiver should be able to receive even input that does not strictly conform to the standard. This is a very important principle for interpretability, and when making future extensions and improvements to the technologies.

Creative Anarchy

The final principle, creative anarchy, is something that seems to be at odds with proper telecommunications networking. However, this means that anyone (from a single individual to large enterprises or operators) can create and distribute new services and applications. This principle can be perhaps seen more as a symptom of the other principles, but comprehending it is still important to understand the Internet and its technologies. Without the creative anarchy principle, the Googles, Facebooks, and the like of this world would never have seen the light of day!

1.3   The Internet Protocol

The most important building block in the Internet is the Internet Protocol. IP is the networking protocol that ensures that packets travel from the source through the network and reach the right destination. Currently two versions of the Internet Protocol exist: IPv4 [3] and IPv6 [4]. At the time of writing, IPv4 is much more widely spread and used. IPv6 is a newer, upcoming protocol version which the world is transitioning towards. As you may have realised from this book's title and the number of pages we have been able to write, this transition is not quite straightforward. However, despite a few differences, the functionality of the two IP versions is very similar and in parts, the same. Therefore, the overview that this chapter gives to IP is applicable to the both versions.

The Internet Protocol has two parts: the header and the payload. The header consists of fixed length binary fields that show, among other information, where the packet has

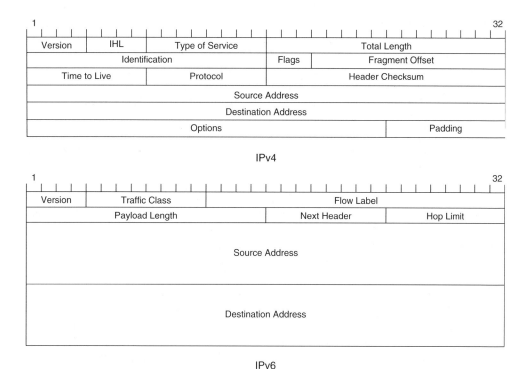

Figure 1.2   IPv4 header [3] and IPv6 header [4].

come from (source address), where it is going to (destination address), and what is in the payload – or better put, what is the upper layer protocol. The payload can be the transport layer protocol, and the protocols and data above it, or even an IP packet. The IPv4 and IPv6 headers are depicted in Figure 1.2.

IP provides an unreliable packet delivery service. Basically the main part of the service is addressing – the capability to tell where a packet should be delivered. In addition to the packet delivery, IP also performs other tasks. For instance it includes a Time To Live (TTL) field, which provides the mechanism to make sure that packets are removed from the network if they do not reach their destination due to, for instance, a loop in the network. Practically, this is achieved by decrementing the TTL at every hop in the network. When the TTL reaches zero the packet is discarded. There is also a simple per packet prioritization mechanism (Type of Service or Quality of Service). Everything needed for the communication, such as reliable transport, packet ordering, and identification of applications, is provided by upper layer protocols such as Transport Control Protocol (TCP) [5].

The nodes of an IP network are connected to the network via a network connection or an interface. A node can have one or more interfaces connected at a given time to the same, or different networks. Each interface has to have a unique IP address from the network to which they are connected. Network interfaces can be either physical or virtual. Physical interfaces are usually created by hardware based network technology support in the node. Physical network interfaces can be wired, wireless, or even inter- and intra-microchip connections within one physical device or even inside one microchip. Virtual interfaces are created usually by tunneling interfaces. Tunneling means transporting IP packets inside other IP packets. Practically the outer IP flow creates a 'tunnel' – a virtual link over a network. The inner IP packets then are transported via this tunnel to another network. The virtual interface is connected to this remote network, and the IP address comes from that network. This approach is used for instance in Virtual Private Networks (VPN) (see Section 3.8) or Mobile IP (see Section 3.7.2) technologies. In addition, one special kind of virtual interface is the loopback interface. This is a node internal interface that loops back, that is, sends back, all packets sent to it. The loopback interface is used for node internal communication.

Generally, the IP network nodes are divided into two categories: hosts and routers. Hosts are nodes that send and receive IP packets, but do not pass on packets for other nodes. A router is a device that connects two different networks, and forwards packets between them. Though this distinction is relatively straightforward, nodes can sometimes have both roles. A good example is the home routers that connect a home network to a wide area network. These usually function as routers, but also often implement, for instance, a web server for management purposes. As a web server the device is a host, but as a home gateway it is a router. It is important to note that IP does not have any special distinction between clients and servers. From IP's point of view, they are all hosts.

In addition to routers, hosts can also have multiple network interfaces connected to multiple IP networks at the same time. Hosts with multiple interfaces, and multiple IP addresses, are called multi-interfaced hosts. Logically, a host with just one interface is called a single-interfaced host or just a host. It is actually quite usual for hosts, such as personal computers, to be multi-interfaced. For instance, today it is normal for a laptop to be connected to a wireless and a wireline networks at the same time. In addition, when a host has a virtual interface, such as a VPN tunnel, it is always multi-interfaced: it has at least the one physical interface and the virtual interface.

The IP nodes are connected to each other by a network of lower layer network connections called links. These links can be either point-to-point links connecting two nodes, or complete layer 2 networks shared between multiple nodes. Different link-layers are used in modern networks. By far the most commonly used link-layer technology is Ethernet [6]. As Ethernet is a very common technology in IP, some parts of the IP technology expect Ethernet-like functionality as well as from other link-layer technologies, and even some non-Ethernet technologies simulate Ethernet for easier integration to their operating systems' IP implementations (which is not without issues, as we will see in Section 4.9.1).

1.3.1 Networks of Networks

When reading about IP, and the biggest IP network of all – the Internet – sometimes the usage of the word 'network' is very confusing. The IP networks come big and small. Some of them are connected to the Internet, and some of them are not. An example of the smallest possible IP network is two nodes directly connected to each other exchanging data. An example of such a small network may be just a single peripheral connected to a computer. As IP is a well standardized and supported networking protocol, sometimes even these simple connections are more favorable to do with IP rather than using an application specific protocol.

The Internet is said to be a network of networks. This description actually fits smaller IP networks, as well. Usually a section of a network, which is under one administrative control, is called a network. This can range from a peripheral connected to a single computer under the 'administrative control' of the computer's user, through a small home network, to large corporate or operator networks with thousands of nodes. These administrative domains can either be isolated or interconnected to other networks under a different administrative control. A good example is a home network; this has different nodes such as network attached storage, game consoles, different computers, etc. These create a network and communicate to each other sharing music, pictures, and other information. This home network may then be interconnected to an operator network such as an Internet Service Provider (ISP). The ISP is then connected to other Internet Service Providers. All these different ISPs (and the small home networks, and bigger enterprise networks) create the Internet.

IP itself does not have separate User-to-Network Interface (UNI) and Network-to-Network Interface (NNI) protocols, but IP handles the networking between the end-hosts, the network nodes, and the different networks in the same way. Hence, IP creates networks of networks.

1.3.2   Routing and Forwarding

The hosts in an IP network communicate between each other by sending IP packets. The source host has to know the IP address of the destination where it wants to send the traffic. The two hosts can be either directly connected to each other, or there can be one or more routers in the path between them. Directly connected hosts, such as hosts in the same Local Area Network (LAN), may be able to send packets directly between each other. Alternatively, a host can send packets to a router for further delivery.

Routers pass packets between different networks and different routers, as described in the previous section. Theoretically, this packet passing can be divided into two different processes: routing and forwarding. Routing involves selecting the route where the packet should be going next. The idea of routing is to find the shortest or fastest route to a destination, and then selecting the outgoing interface based on that information. Forwarding, on the other hand, is the process that actually moves the packet through the router to the intended direction. When a packet is forwarded by a router, the router decreases the TTL by one. If the TTL reaches zero, the packet is discarded.

Routing decisions are based on routes. Routes define which networks or destinations are reachable through which interface. In addition to routes, the routing decisions can be affected by routing policy. Routing policy is basically a rule that overrides the technical routing calculation. There are various reasons to set a routing policy. For instance one reason is cost – sending a packet through one operator may cheaper than a possibly shorter path through another operator. Routes can be either statically configured, or dynamically calculated. Statically configured routes are usually configured by a network administrator. These routes do not change until the network administrator changes the configuration, regardless of the network status. For dynamic route calculation, routers use dynamic routing protocols to exchange network topology information between each other. The most used routing protocols are Open Shortest Path First (OSPF) [7, 8], Intermediate System to Intermediate System (IS-IS) [9], and Border Gateway Protocol (BGP) [10]. Sometimes routers do not have the complete picture of the whole network, such as the whole Internet. A router might not know which of its interfaces would bring the packet to the right direction. Therefore, the router may have a default route. The default route basically states that if there is no more specific information – a specific route that fits the packet's destination address – by default the packet should be sent through a specific interface to a specific router.

Directly connected routers exchange information about the routes they have gotten from other routers to which they are connected. A router uses this information to create a view of the network topology. This network topology map is stored in an internal data structure called a routing table. When some new information about new routes or old routes disappearing comes to the router either by the routing protocol updates, or by the router noticing a neighboring router disappearing, the router updates the internal routing table. It also informs the other neighboring routers about the changes it has noticed.

Route information changes when new networks are connected or old networks are disconnected from the IP network. New networks appear when a network administrator somewhere installs a new router on the network, perhaps exposing a complete network behind that router. Networks can be detached either administratively when a network administrator takes a router away from the network, or because of a link failure – perhaps detaching a whole network behind it. To reduce the risk of a complete network being detached, networks are sometimes connected via multiple links and even through multiple routers. A good example is when an enterprise buys Internet services from two ISPs to protect against a case where one fails for some reason or other.

When an IP packet arrives at a router, generally the router looks at the destination address of the packet, and forwards the packet to one of its network interfaces. The forwarding decision is usually done by looking up the destination address from another data structure called a forwarding table. The forwarding table is created by the router using the routing table and possible routing policies existing in the router. The router then creates a table where it lists which networks are accessible through which of the router's network interfaces. An IP router looks at every IP packet individually and does the forwarding decision on a packet-by-packet basis. If something changes in the router's routing table, a packet can take a different route from a preceding packet even if it has the same destination.

In addition to routers, hosts make forwarding decisions. For instance, a host has to decide if it thinks the destination host is directly connected to it, or if it has to forward the packet to a router instead. This first router is often called the first-hop router. There may be multiple routers available in a network of which the host can choose. One of these routers is usually assigned as the default router for the host. If the host has no more specific information about the location of the destination host, the host sends the packet to the default router. Hence, the host's default route points to that default router. In the same way as routers may be configured with routing policy, the host might have a set routing or forwarding policy. The reasons for the policies include, security, some assumption about the cost or characteristics of a network connection, or just user preference. A corporate IT-department might configure the user's traffic to always go through an active VPN tunnel regardless of where the packet is going, in order to make sure that company secrets do not traverse through unknown networks. Some operating systems prefer a wireline connection over wireless because it is assumed to have more bandwidth and be more reliable, and some operating systems allow the user to define the preference order of the available interfaces.

1.4   Internet Protocol Addresses

Regardless of whether a network is big or small, the IP addresses have to be unique within the network. As described above, the IP packet's destination address is used to route the packet through the network to its final destination. If the addresses were not unique, there would be no way of knowing where a packet should be delivered to. In a simple network, which is not connected to the Internet, it is enough that the addresses are unique locally within that domain. On the Internet, the IP addresses have to be globally unique.

The IP address itself is a fixed length binary identifier. An IPv4 address is 32 bits, and an IPv6 address is 128 bits long. The IP address has two functions in an IP network: it uniquely identifies an interface or a node in the network, and it represents a location in the network topology.

The identity is related to the uniqueness of the IP addresses – no other node in the network should have the same address as the node that is identified by a certain address. That address uniquely identifies a single network interface of that node and no other.

The location means that the address uniquely identifies a network where the node is within the IP network. The IP address can be divided into two parts; the network prefix and the host identifier, which together make the complete IP address. In the following pages, we will describe in more detail what this means in practice for IPv4 and IPv6.

In this section we will have a more in-depth look at the IP addresses. First we will explain the IPv4 addresses, and then IPv6. The focus is on the addresses themselves: what are addresses, how do we represent them, and what kind of different address types exist?

1.4.1   IPv4 Addresses

As describe above, an IP address is a fixed length binary field. In IPv4, the IP address field is 32 bits long, thus giving a theoretical address space of $2^{32} = 4$ billion addresses. Because a binary bit sequence would be a bit difficult to remember, represent, and tell to a friend, a text representation is used to denote the address information. In IPv4, the 32 bits are broken into four 8-bit (one-byte) fields each separated by a dot. These eight bit fields are presented in decimal numbers, for instance, 198.51.123.234.

The network prefix length is variable in IPv4. The size of the prefix is denoted by putting a slash, '/', and the length of the prefix in bits to the end of an address. For example, 192.51.100.0/24 shows that the network prefix size is 24 bits. Blocks of addresses are presented in the same way, according to the prefix length. For example, a block of addresses from 198.51.100.0 to 198.51.100.255 is usually written as 198.51.100.0/24, or 198.51.100/24 (leaving the zero out at the end). This means that the prefix length is 24 bits, and the host part is the remaining 8 bits out of

Table 1.1   IPv4 address classes

Address Class	Prefix Length	Number of Addresses
Class A	/8	16,777,216
Class B	/16	65,536
Class C	/24	256

the full 32 bits. When a prefix has more bits, it is considered longer, and a prefix with fewer bits is considered to be shorter. The longer the prefix, the shorter the host part, and hence, fewer addresses are available for the hosts in that network.

Originally, the IPv4 address space was divided by the length of the network prefix into different classes. The class A address blocks were blocks of /8 – 16,777,216 addresses, class B address blocks were /16 – 65,536, and class C blocks of /24 – 256 addresses. Table 1.1 shows the address classes and their lengths. In the 1990s, the Internet moved to classless addressing – Classless Inter-Domain Routing (CIDR) [36, 37] – and the network prefix length is no longer dependent on the address class; the length can be different from the original address classes.

The addresses used for normal unicast communication between hosts are called unicast addresses. Quite obviously, the main part of the allocated address space is dedicated to unicast addresses. These addresses are sometimes also called public addresses or globally routable addresses. In addition, other address types also exist. The specification describing special use IPv4 addresses [11] lists 15 different special address blocks. We will here go through the most important ones.

As will be seen below, the IPv4 addresses are mainly used for the unicast communication between network nodes. In addition, however, there are special addresses with special meanings. Most of the IPv4 address space's 4 billion addresses are used for normal unicast addresses for the communication over the Internet, but there is quite a bit of address space that cannot be used for that purpose. This restricts the number of addresses in the Internet even further. We will discuss how IPv4 address scarcity is an issue later on.

Private Use Networks

10/8, 172.16/12 and 192.168/16 are reserved for private use networks [12]. Commonly this part of the address space is called private address space, and the addresses private addresses. These addresses can be used only within a private network. They cannot be used directly in the Internet, because the addresses given from a private address block are not unique over the whole Internet.

Shared Address Space

100.64/10 is reserved for shared address space [13]. Shared address space is similar to private addresses introduced above, but instead of general use, the shared address space is intended to be used within an operator network.

Loopback

127.0.0.0/8 is reserved to loopback addresses. These addresses are exclusively used for host internal communication. Packets with loopback addresses should not be seen on any network.

Link-local Addresses

169.254/16 address block is used for communication constrained to one link. These addresses are especially used before a host has gotten a real address, private or public. In some cases, such as with directly connected hosts, these may be the only addresses ever available on a link RFC 3927 [44].

Multicast

224.0.0.0/4 block is used for multicast services. How the multicast address space is used is specified in RFC 5771 [14].

Reserved for Future Use

The block 240.0.0.0/4 is marked as reserved for future use. There has been much controversy about the usage of this address block. It seems that some routers think it is an error if they see a packet from this address space. Therefore, the use of the address space is relatively difficult – at least at the level of the Internet. It is clear that there has been a lot of pressure and interest in doing something with this block as it is about 6 per cent of the overall address space. However, at the time of writing, there has been no use found for it.

Limited Broadcast Address

The address 255.255.255.255/32 is a special address for link-local broadcast. Packets with this destination address are not forwarded on the IP layer, but all the nodes within the link-scope get the packet. Some applications use this address as their destination for initial boot strapping.

1.4.2   IPv6 Addresses

The IPv6 address field is 128 bits long, and hence gives much bigger address space than IPv4. The theoretical maximum of the IPv6 address space is $2^{128} = 3.4 \times 10^{38}$. A number this big is rather difficult to understand, and people have tried to explain the number in various ways. One of the examples is that there are 6.5×10^{23} addresses per every square meter of the earth. It is not clear why anybody would want to allocate addresses per square meter, but the message is clear: the IPv6 address space is very, very big.

To distinguish IPv6 addresses from IPv4 addresses, and to have an address format that is at least vaguely readable, IPv6 has its own textual format. The textual format has been defined in RFC 4291 [15] and RFC 5952 [16]. The textual representation of an IPv6 address consists of eight 16-bit hexadecimal fields separated by colons – x:x:x:x:x:x:x:x. As the IPv6 address format is quite long, and rather complex, so are the rules on how to represent the addresses.

Below are examples of legitimate address representations:

```
2001:41d0:1:7827::1
2001:db8::a:0:0
2001:db8::a
::1
```

The 16 bit value of each x is presented without leading zeros. Furthermore, when the address has 32 or more consecutive zero bits, the zero bit sequence can be abbreviated with syntax ::. The :: can be used only once in an address. If an address has multiple series of zeros, the :: must be used where it makes the most difference. Here is an example of compressing zeros away from an IPv6 address: 2001:0db8:0000:0000:0000:0000:0000:000a can be presented in much more readable form as 2001:db8::a

A special textual addressing format has also been defined for IPv6 addresses that are carrying IPv4 address in the lowest 32 bits. The address format for such addresses is x:x:x:x:x:x:d.d.d.d, where ds represent an IPv4 address [15]. An example of such embedded address is: 64:ff9b::192.0.2.1 [17].

When an IPv6 address needs to be shown with a port number, square brackets, [and], should be used to make the address unambiguous. Without brackets it might be difficult to determine whether the last digit is a port number or part of an IPv6 address, as illustrated here: 2001:db8::a:80. A correct example, using brackets, makes the distinction clear: [2001:db8::a]:80. Many applications, such as web browsers, expect IPv6 addresses to be put in square brackets.

As in IPv4, so in IPv6, the address is divided into the network prefix and the host part. In IPv6, the host part is called the Interface IDentifier (IID). The IID for the unicast addresses is defined to be 64 bits long [15]. The IID must at least be unique in the scope of the link where the interface (and hence, the node) is located. However, the IID can also be unique in a larger scope as well. The network prefix length is noted in the same way in IPv6 as previously described for IPv4. For instance, 2001:41d0:1:7827::/64 means that the network prefix is 64 bits long, and in 2001:41d0:1::/48 the prefix length is 48 bits.

Consequently, the length of the IID to 64 bits also constrains the network prefix length at a maximum of 64 bits. Hence, allocating at least a network prefix of /64 for a single link is the norm, regardless of how many nodes are in that subnet. In IPv6, it is important not to count single addresses anymore, but count prefixes of /64 per link. This is quite a different philosophy from IPv4, and may be unintuitive for many. While in IPv4 the address policy was driven by conservation, in IPv6 the driver is network and numbering simplicity, with additional security considerations, as we will discuss in Chapter 3.

1.5 Transport Protocols

As described previously, the IP provides unreliable packet delivery over the IP network, such as the Internet. The only multiplexing between two end points that it provides is the distinction between the different protocols it is carrying. Thus, anything else has to be provided by the upper-layer protocols. These protocols are called transport protocols, as their job is to transport the actual application payload end-to-end. The two most important transport protocols used today are User Datagram Protocol (UDP) [18] and TCP [5]. Generally, the transport protocols are IP-address family agnostic. Thus, the same protocols can be used with IPv4 and with IPv6. Of course, in practice transport protocols themselves need to be able to work with different network layers, as we shall see in Chapter 3.

We will try to give a short overview of these transport protocols in this section.

1.5.1   User Datagram Protocol

UDP is a very simple protocol. It only provides two services: service and connection multiplexing, and a checksum for the receiving end point to check if bit-errors have been introduced during transport. The multiplexing is done by port number fields in the UDP header – the source and the destination ports. The quintuple – source address, destination address, and protocol number in the IP header – and the source and destination port numbers in the UDP header uniquely identify the connection for the end-host.

UDP does not guarantee that the packets are transported in order, or actually even that they are received by the other end at all. Thus, there is no reliability mechanism. Therefore, UDP is usually used for transporting application data, where occasional loss of packets is not fatal, and the transporting of the packets as fast as possible is more important than reliability. These are applications where it is better to forget the packet rather than transport it late. These applications include for instance Voice over IP (VoIP). If an application needs reliable packet delivery, it should either implement the reliability mechanism itself, or use TCP.

1.5.2   Transmission Control Protocol

TCP is one of the most important protocols in the IP protocol family. As a matter of fact, the IP protocol family is very often called TCP/IP. TCP provides reliable, ordered transport, with service and connection multiplexing, and with congestion control. Most Internet applications require reliable transport. Therefore, for example, file transfers, video steaming, and especially web traffic are transported over TCP. Hence, most applications use TCP for communicating over the Internet.

As in UDP, the application and service multiplexing is achieved by source and destination port numbers. The reliability is achieved by the receiving host acknowledging the packets it receives, and through this the sending host notices which packets have gone missing. These packets are then retransmitted. The TCP congestion control algorithms monitor this packet loss. TCP actually assumes packet loss to be always caused by congestion, and TCP will drop its transmission rate to adjust to this perceived congestion. Due to this behavior, TCP has received a bad reputation as an unfriendly protocol for wireless networks, because in wireless networks packet loss can be caused for many reasons other than just congestion. However, in recent years the new wireless access technologies are more TCP friendly, and new TCP extensions have taken wireless networks much more into account. Hence, TCP's bad reputation is mostly outdated.

TCP is relatively complex. Therefore, we will not try to describe it more than necessary in the book. However, an interested reader will find good suggestions for further reading in the additional reading section at the end of this chapter.

1.5.3 Port Numbers and Services

The port numbers perform an important function in the IP networks, and hence merit their own section. The port number field in UDP and in TCP is 16 bits long and therefore can carry numbers 0–65535. This port range is divided into different usages. The range of 0–1023 is called well known ports, or system ports, ports 1024–49151 are called registered ports or user ports, and the remaining 49152–65535 are called private, dynamic, or ephemeral ports [19].

The well known or registered ports generally identify a service, an application, or a usage. This means that a specific application or service is listening to that port, and is understood to be that service. For example, port 21 is the control part of the File Transfer Protocol (FTP) [20], port 23 is Telnet [21], and port 80 is HyperText Transfer Protocol (HTTP) [22, 23]. Therefore, a service listening and answering at port 80, for instance, is expected to be a web server.

1.6 Domain Name Service

IP addresses are used by computers – the hosts and routers that communicate to each other. However, the IP addresses are not the most intuitive identifier to be remembered by people. People remember names better than numbers. To address this, the Domain Name System (DNS) was developed RFC 1034 and RFC 1035 [43, 24]. DNS allows human-readable names to be used for protocols and user interfaces, though allowing the Internet still to function with numbers with better computational characteristics. Basically, DNS is a distributed database that provides name-to-address, and address-to-name mapping. So DNS is a bit like a big telephone book of the Internet.

1.6.1 DNS Structure

The DNS is a global database covering the whole Internet. Therefore, scalability is very important. Hence, DNS was designed to be a distributed, hierarchical database. Figure 1.3 shows the DNS structure. This structure is seen also in the DNS names – if we examine for instance the name 'www.example.com.'. A complete DNS name like this is called a Fully Qualified Domain Name (FQDN). It consists of the following parts, which can be found in Figure 1.3.

root is noted by the final dot ('.') at the end of an (FQDN). Root is the top of the hierarchy, and is the central point of the (DNS) database. For the user, this final dot is usually hidden, and the user usually sees 'www.example.com' rather than 'www.example.com.'.

Top Level Domain (TLD) is, as the name suggests, the highest point of the domain names. They are, for example, the .com, .org, and .de at the end of a domain name. There

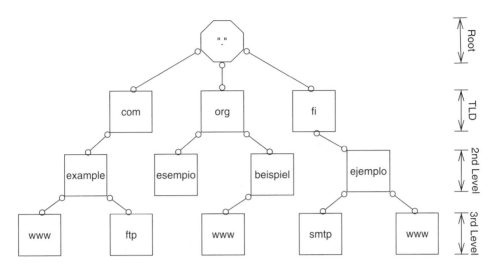

Figure 1.3   DNS structure.

are three categories of TLDs – generic Top Level Domains (gTLDs), country code Top Level Domains (ccTLDs), and an infrastructure TLD. TLDs are more closely examined in Section 1.6.3. An organization, which operates and owns a TLD is called a registrar. Registrars sell or distribute 2nd level domains to the organizations that need domain names.

2nd level domains are owned and controlled by registrars. Registrars include companies, organizations, and even private persons. A 2nd level domain owner can create 3rd level domain names, and distribute those in the way the 2nd level domain owner wishes.

3rd or lower level domain are domains created under the 2nd level. They can indicate a service (such as www, ftp, etc.) or they can be otherwise descriptive – for instance cs.helsinki.fi is the domain name of the computer science department of the University of Helsinki.

1.6.2   DNS Operation

The previous section explained the structure of the DNS name. This section will now examine how the actual DNS resolution – mapping a DNS name to an IP address – works actually. Figure 1.4 shows a simplified overview of the process, and in the following we examine the resolution, step by step:

1. The process starts when a host has an FQDN that it wishes to convert to an IP address. The host's DNS resolver sends a DNS query to a Recursive DNS Server (RDNSS) that has been configured in the host's DNS configuration (e.g., by means described in Section 3.10.2).
2. The DNS query for www.example.com is received by a RDNSS, which has to first find what server is responsible for the .com TLD. The RDNSS knows the root server

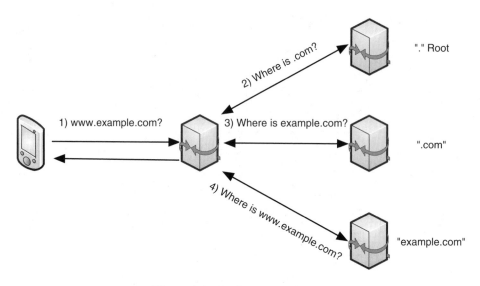

Figure 1.4 DNS name resolution.

addresses, and it sends the query to one of the root servers. The root server refers to the server responsible for `.com`.

3. The RDNSS sends the query to the authoritative server responsible for `.com`. This server responds by referring to the server responsible for `example.com`.
4. The RDNSS sends the query to DNS server authoritative for `example.com`. The DNS server answers with the DNS record that contains the IP address corresponding to `www.example.com`.
5. The RDNSS sends the record containing the IP address of `www.example.com` to the host.

The IP addresses are stored in DNS records. The IPv4 records are stored in A [24] records, and the IPv6 addresses in `AAAA` records [25] (also called quad-A records). Please see more detailed implications of IPv6 to DNS in Sections 3.9.4 and 3.10.2.

1.6.3 Top Level Domain

There are three categories of TLDs – gTLD, ccTLD, and infrastructure TLD. The Internet Assigned Number (IANA) Root Zone Database [26] lists the current TLDs. In the following, the different TLD categories are explained.

gTLD is a generic name TLD, which indicates something generic about the organizations or content on the next level. For instance, `.com` indicates commercial, `.org` organization (usually non-for-profit), `.net` network, and `.mobi` mobile communication. These indications are, however, not very strict in reality. A gTLD is always three or more latin characters long. The gTLDs are usually owned and operated by private organizations, which sell or distribute the second level names to other organizations somehow connected to that category. The gTLDs are administered by Internet Corporation for Assigned Names and Numbers (ICANN). At the time of writing, 21 unique gTLDs exist. However, ICANN has an ongoing new gTLD process in which new gTLDs are evaluated for approval. Almost two thousand unique applications were received by ICANN during the application window [27].

ccTLDs indicate a country, or a territory as defined in the ISO-3166 [28], and the ccTLDs consist of two-character strings defined in the same standard. The ccTLDs are usually operated and owned by the government of the country or territory, a government agency, or an organization appointed by the local government. (There are two-character TLDs that are used like gTLDs and might even be owned by companies. For instance, these include .tv. However, these are originally ccTLDs that have either been sold, or are just used in gTLD manner.)

Infrastructure (TLD) is the .arpa TLD. IANA administers arpa for the Internet Engineering Task Force (IETF) for technical purposes as a part of the Internet's infrastructure. The .arpa TLD is used, for example, for DNS reverse queries RFC 3172 [45] – mapping IP addresses to DNS names.

1.6.4 Internationalized Domain Names

Originally, the DNS supported only latin characters. This, obviously, had its origins in the birth of Internet in the USA, and in the challenges with computer systems of taking as input, and reproducing other character sets. However, modern computing can support many more scripts. In addition, the Internet has become a global phenomenon where most of its users are not native English speakers, or even use the latin script. Therefore, the IETF has specified the support for Internationalized Domain Names (IDNs) [29–32].

The IDNs started off in second level domains under different TLDs. In addition, the ICANN ccTLD Fast Track process enabled certain ccTLD IDN variants to be accepted to the root. It is expected that multiple IDN gTLDs will be introduced by the new gTLD process.

1.7 IPv4 Address Exhaustion

As explained earlier, the IPv4's 32-bit address field sets a theoretical maximum number of addresses to $2^{32} = 4,294,967,296$ addresses. To understand the address space usage more clearly, we can consider the IPv4 address space to consist of 256 address blocks of the size of /8. As described earlier, we have address space that cannot be used for normal communication purposes in the internet. Including all the different special use address spaces in RFC 5735 [11], we have 35.078 /8s (14%) that cannot be used on the Internet. This leaves 220.922 /8s, which can be used for Internet nodes – a range of 3,706,456,113 addresses – just under four billion.

As all nodes that are connected to the Internet need their own unique address, just under four billion individual nodes can be at the same time on the Internet. These nodes include the routers, servers, end-hosts – basically all nodes connected to the Internet, which have to be reachable from the Internet. At the time writing, the world population is around 7 billion, and there are already over two billion Internet users, and the number is growing fast. Introduction of the Internet of Things (see Section 6.5) creates significant additional pressure on the address space consumption. It is clear that the IPv4 address space is a seriously constrained resource. To clearly understand the IP address allocation problem, and the IPv4 address space exhaustion, we will now look at how IP addresses are allocated to the end users, what is the history of IPv4 address space exhaustion, and what has been done to mitigate the depletion of the IPv4 address space. Finally, we will look at the situation at the time of writing this book.

1.7.1 IP Address Allocation

The IP address allocation is a hierarchical system. Figure 1.5 shows how the IP address allocation hierarchy is set up. This is one representation of the hierarchy. You can certainly find other descriptions where the boxes are in a different order. The IP address allocation should be more of a technical, mechanical function. However, in the past few years the different interests around IP address allocation have politicized the process, and even the setup has been questioned. However, in our view, this picture makes sense when looking at the Internet from a technical perspective.

The highest level of the hierarchy is the IETF. The IETF is the standardization organization, which is responsible for specifying the IP technology – including the Internet Protocol itself. The IETF specifications specify how long is the address, what is the addressing architecture, and what are the special address ranges that are not allocated as unicast addresses.

The global allocation of IP unicast addresses is the responsibility of the IANA [33], which is operated by ICANN [34]. In IPv4, IANA allocated /8 address blocks for

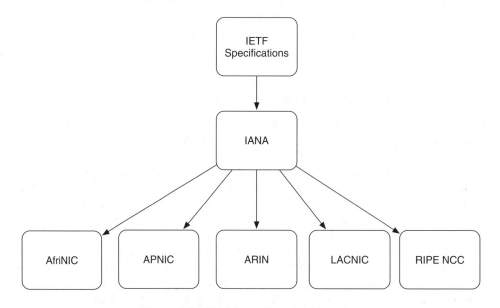

Figure 1.5 IP address allocation hierarchy.

the Regional Internet Registries (RIRs) [35]. The RIRs are responsible for allocating addresses regionally. Today five RIRs exist: African Network Information Center (AfriNIC) in Africa, Asia-Pacific Network Information Center (APNIC) in Asia Pacific, American Registry for Internet Numbers (ARIN), Latin America and Caribbean Network Information Center (LACNIC) in Latin America, and Réseaux IP Européens Network Coordination Centre (RIPE-NCC) in Europe.

The RIRs are responsible of allocating IP addresses to the Local Internet Registries (LIRs) in their region. The LIRs are generally operators, or larger organizations that allocate the IP addresses to the final users of the addresses – the hosts and routers that use the addresses. The allocation of IP addresses to the RIRs from IANA, and to the LIRs from the RIRs, are based on the global or RIR policies. The LIRs usually allocate addresses to the end users based on commercial or technical grounds. Basically, how many IP addresses are allocated to different computers behind a consumer broadband line is based on the contract that the end user has made with the operator.

Different urban legends exist about address allocation. Stories about how some North American universities have more addresses than some very large countries are used to show that the address allocation process is not fair, and favors certain geographic regions over others. Although there is some merit to these stories, the universities, organizations, corporations, and also countries, that participated in the Internet when it was still mostly a research network have gotten relatively large (IP) address allocations – such as class As in the beginning. However, as the Internet has grown, the address allocation policies and process have evolved to the one we have today, which is based on need. Hence, these stories are now mostly outdated.

1.7.2 History of IPv4 Address Exhaustion

On 3 February 2011, IANA stated that it had allocated the final /8 IPv4 address blocks to the RIRs. In April 2011, APNIC said it had reached its final block of /8, and RIPE NCC indicated the same situation in September 2012. Other RIRs are expected to follow suit very soon. Perhaps by the time you have picked up this book, others will have run out of IPv4 addresses as well. Hence, today easily available IPv4 addresses have been exhausted.

The exhaustion of the IPv4 address space, however, was not a surprise, and the technical community has been preparing for it for decades. The first time that the IPv4 addresses were about to be exhausted was at the beginning of 1990s. At that time, the issue was the address classes. As explained before, the addresses were divided in different classes, and the size of the IPv4 address allocation was dependent on the class. This system was very rigid, and it was hard to find a good fit per organization from the classes. Therefore, most bigger organizations needed at least a class B. Hence, the class B addresses were running out fast as the Internet started to grow at the beginning of the 1990s.

The Internet technical community at the IETF designed the CIDR [36, 37] to enable the allocation of address space in needed block sizes throughout the whole address space. This change prolonged IPv4's lifetime adequately for the community to start redesigning the Internet Protocol, and specifying a new IP protocol version that would have adequate address space for the growth of the Internet. CIDR addressed the efficiency of IPv4 address allocation well enough to accommodate the growth of the Internet throughout the 1990s and the beginning of 2000s.

The growth of broadband networking, and the proliferation of networked computers and other network devices, fueled the Internet's growth. Networks in enterprises, and even in homes grew. This introduced the need to have multiple addresses per subscriber both for enterprise connections, and for private subscribers. However, operators were conserving IP addresses allocated to the subscriber lines. Most operators gave just one address per subscriber, hence allowing just one computer to be connected to the network at a time. There was a need to multiplex multiple computers to a single public IP address. The introduction of private address space [12] and Network Address Translation (NAT) [38] made this possible. NAT is a technique that allows multiplexing of multiple private IP addresses to a single public IP address. The multiplexing is done by using upper layer protocol identifiers – for example the TCP and UDP port numbers. Figure 1.6 describes the NAT principle. Practically, the NAT follows the traffic coming from the inside the network (from the left in the picture) going to the Internet. The NAT rewrites the outgoing packet by replacing the source IP address with its external IP address, and the source port in the transport protocol with one of its available ports. When a downlink packet comes destined to the NAT's address, and the port, it knows to rewrite the packet to go to the host inside the private network. Hence, a NAT extends the IP address range using port numbers.

As the IPv4 address space approaches exhaustion, many operators deploy NATs in their networks, and give private addresses to their customers. Especially in mobile networks NATs have been widespread between the end user and the Internet. That fact combined with the NATs already existing at end user networks, means that usage of NATs can only increase in the future. In addition to CIDR, NAT is the main reason why IPv4 address space has lasted as long as it has.

Although these techniques increase the efficiency of IPv4 address space usage significantly, they do not change the fundamentals of the Internet Protocol and the need for globally routable IPv4 addresses. The IPv4 address space will be exhausted. It has just taken a longer time with these technologies.

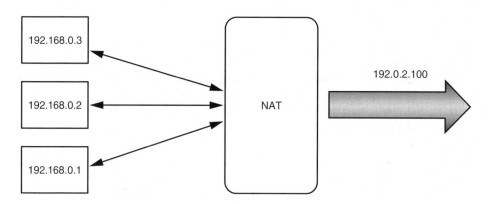

Figure 1.6 Network Address Translation.

1.8 IPv6 History Thus Far

Here we'll try to give a very short summary of the history of IPv6, and give a snapshot in time of the current IPv6 deployment. Writing down the current state of the IPv6 deployment is a calculated risk by the authors; by the time this book actually reaches its readers, the IPv6 deployment will have progressed significantly from what is written here. On the other hand, the authors have stated that the major IPv6 deployment will happen next year and we have said this now for a number of years. However, looking at the current IPv6 deployments seriously, it is clear that the technology maturity, and the overall deployment has taken major steps forward during the last few years. There is no reason to expect that this progress will stop. On the contrary, the deployment can only be expected to accelerate as the exhaustion of the IPv4 addresses progresses.

1.8.1 IPv6 Technology Maturity

At the beginning of 1990s, the IETF started to investigate new technology options for the technical evolution of the Internet Protocol. The consensus of the IETF decided on the new version of the Internet Protocol – the IPv6. By 1998 the IETF had finished the IPv6 base specification RFC 2460 [4]. At the same time of the standardization, multiple implementation projects also existed both in academia, and in the private sector. The best known IPv6 implementation has been done by the WIDE project in Japan [39]. The WIDE Project's IPv6 implementation KAME [40] is the basis of the IPv6 support of many widely used operating systems – Linux and FreeBSD, for instance.

Today all modern mainstream operating systems and many of their variants, including Windows, Mac OS X, Linux, FreeBSD, Symbian, and Android, support IPv6. Most of them also have IPv6 enabled by default. This means that if the local network to which the device is connected to supports IPv6, the operating system can use the capability without the user explicitly enabling IPv6. In addition, many of the mainstream applications have been enabled to support IPv6, including all modern web browsers. Yet, there are many applications that may not support IPv6, with no clear path to migrate to IPv6. Especially, older applications with limited to no support may never be updated.

The major switch, router, and other network equipment vendors providing equipment to operators and corporate networks have offered IPv6 capable devices for some time already. Initially, IPv6 support was included in the test or experimental branches of the software, and sometimes IPv6 was implemented only in software when the IPv4 support might have been hardware accelerated. In addition, initially the IPv6 support might have lacked features present for IPv4 in the same product – this is referred as feature imparity. Hence, the first IPv6 support in products might have been inferior to IPv4 in terms of features and performance. This has changed dramatically in the last few years, and most of the major vendors claim feature parity support with IPv4 and IPv6.

However, a segment that has been slower to adopt IPv6 is the home router vendor community. Only recently has IPv6 capable home network equipment become generally available. Sadly, the major part of the current, and only slowly being upgraded, installed base does not support IPv6 – at least not to the same level as IPv4.

At the beginning of 2000s, the introduction of 3rd Generation (3G) network services and mobile phones was seen as a major driver for IPv6 adoption. However, for a long time 3G phones were used mostly for voice calls, and only secondarily for Internet access. The introduction of mobile broadband services, and of smartphones, changed the course of events, and exactly what was expected happened – just about ten years later. Until recently basically only Nokia Symbian based smartphones supported IPv6. However, now IPv6 is slowly becoming a standard feature in mobile phones due to the requirements of the major mobile network operators.

1.8.2 IPv6 Network Deployments

The fist global IPv6 network was the IPv6 technology testbed 6bone (6bone) [41]. The 6bone was created by the IETF IPv6 community to provide a place where the IPv6 technology could be tested, connecting different IPv6 test projects, and for gathering IPv6 operational experience. The network itself was mostly a virtual network deployed over the global Internet infrastructure. It had its own network prefix 3FFE::/16 [42] that was temporarily assigned to the testbed. The testbed was operational from 1996 to 2006 when it was decommissioned on 6 June 2006 (note the date: 6.6.6) marking both the end of the 6bone and the first IPv6 day. Sometimes erroneously, 6bone has been said to be the Internet IPv6 network backbone. However, the intention was always for 6bone to be a temporary testbed in the test phase of the technology. As IPv6 technology matured, and the Internet itself became increasingly IPv6 capable, 6bone was no longer needed. The 6bone prefixes are not routed over the Internet anymore.

The definition of the Internet core is a difficult one. Due to the distributed nature of the Internet, it is difficult to state which operators provide the true core of the Internet. Often, so-called Internet Tier 1 operators are considered to be the core of the Internet. But this is not quite true. These operators are often the larger operators with global presence. The major operators providing Internet transit to other operators, including the Tier 1 operators, have provided IPv6 service both to their operator and business customers. In that sense, one can claim that the Internet core network has been IPv6 capable already for some time. This development has happened incrementally over many years, and has been mostly invisible to the end user. The operators have also had good transition tools, which have made it relatively easy to support IPv6 in the Internet core.

Though, some progressive access network operators have deployed IPv6 and provided it to the end users, most of the world's access operators do not provide IPv6, yet. There are many reasons for the slowness of IPv6 deployment in the Internet access networks. First of all, IPv6 is a not a feature that operators can sell to the end users. This absence of an evident business case has most probably been the number one reason for the slow adoption of IPv6 in access networks. This is not a feature that the users would currently demand, or for which they would be prepared to pay. Other important contributors are the relatively slow investing cycle – especially in the fixed access networks – and the result of the presence of very old equipment in the networks. Even though new equipment does support IPv6 well, access operators may have equipment that is as much as ten years old in their network, which would have to be replaced. Investment without a prospect of more revenue is never popular in the management of a company. In addition, as mentioned earlier, the lack of IPv6 capable home routers may have influenced the deployment plans. On the other hand, it is hard to say what is the cause and what is the symptom. Looking on the bright side, the number of IPv6 capable access networks is steadily growing, including big fixed access operators like Comcast providing commercial service already.

One reason stated for the slow uptake of the IPv6 access has been the lack of IPv6 services. For a long time, the major Internet services (like Google, Facebook, Yahoo and others) did not support IPv6, or supported it only partially. This was also recognized by the Internet community. The Internet Society started to organize World IPv6 Day events. The 2011 World IPv6 Day provided a day during which major Internet service providers turned IPv6 on their main websites for a day to operationally test IPv6. This was considered to be a great success. In 2012, the Internet Society coordinated the World IPv6 Launch, where many Internet service providers turned IPv6 on permanently. This has at least partly released the chicken-and-egg situation between the access network operators and the Internet content providers. It gives good proof of how positive coordination, or perhaps rather peer pressure, can help to bring the industry together.

1.9 Ongoing Cellular Deployments

The first 3GPP operator to deploy IPv6 in a commercial network was Sonera (nowadays TeliaSonera) in Finland. They started a trial of IPv6 in their commercial network back in 2004. The IPv6 support was not commercialized, and therefore not accessible to normal end users. Perhaps, the first commercial service deployments were in Slovenia where the national cellular network providers commercialized IPv6 at the same time in 2010, coordinated by the Slovenian GO6 institute – a not-for-profit industry association in Slovenia.

Recently major operators in North America have started open IPv6 trials, and some of them have commercialized IPv6 with the new Long Term Evolution (LTE) cellular networks. In addition, some Asian operators have launched their IPv6 service, together with the LTE network launch.

The major network equipment vendors have had at least basic support for IPv6 service for the end user for quite some time now. However, advanced services that the operators use in their current IPv4 Internet offering might not have been supported for IPv6. The same applies to mobile network equipment vendors as it does for fixed network equipment vendors – the support for IPv6 has been improved dramatically over the last few years. The network equipment, subscriber database solution, and network management solutions, are available with IPv6 support.

The deployment in operators' networks has been painfully slow. Though, in the early days of IPv6, 3G deployment was thought to be the driver of IPv6 deployment, the mobile broadband operators have been slow in adopting the new Internet technology. However, it seems that even the cellular deployment is now gaining momentum. All major cellular operators are either busy deploying IPv6 in their networks, or are planning the deployment. The mobile handset operating systems, the cellular chipsets, and the mobile applications increasingly support IPv6, eliminating one major problem from the IPv6 transition in the mobile broadband market. It should be expected that the cellular community will be very busy with IPv6 in the coming few years. A major reason that this book was written was to help people joining, or already participating in, the IPv6 deployment activities to get up to speed quicker.

1.10   Chapter Summary

In this chapter, we concentrated on giving an introduction to the Internet, Internet technology, address exhaustion, and the deployment status at the time this book was written. The Internet has been built on a set of principles, which are embedded in the technology and in the Internet itself. These principles are repeated in the following:

- packet switched networking;
- the end-to-end principle;
- layered architecture;
- Postel's robustness principle; and
- creative anarchy.

The IP is the technology that builds the Internet. It provides an unreliable packet delivery service where the packet delivery is based on the destination IP address. Currently, two versions of IP exist: IPv4 and IPv6. The IPv4 is the protocol that is predominantly used in the Internet. However, IPv4 addresses have been mostly depleted, posing a risk to the growth of the whole Internet. Therefore the Internet, including 3GPP cellular networks, is in the biggest technical transition of its existence – the transition to IPv6.

1.11   Suggested Reading

- Where Wizards Stay Up Late, The origins of the Internet, K. Hafner and M. Lyon
- Routing in the Internet, C. Huitema
- TCP/IP Illustrated, Volume 1: The Protocols (2nd Edition), K. Fall, W. R. Stevens
- Brief History of the Internet – Internet timeline http://www.internetsociety.org/internet/internet-51/history-internet/brief-history-internet
- DNS and BIND, 5th Edition, C. Liu, P. Albitz

References

1. ITU-T. Data Networks and Open System Communication Open System Interconnection – Model and Notation Information Technology – Open System Interconnection – Basic Reference Model: The Basic Model. Recommendation Q.700, International Telecom Union – Telecommunication Standardization Sector (ITU-T), July 1994.
2. Braden, R. *Requirements for Internet Hosts – Communication Layers*. RFC 1122, Internet Engineering Task Force, October 1989.
3. Postel, J. *Internet Protocol*. RFC 0791, Internet Engineering Task Force, September 1981.
4. Deering, S. and Hinden, R. *Internet Protocol, Version 6 (IPv6) Specification*. RFC 2460, Internet Engineering Task Force, December 1998.
5. Postel, J. *Transmission Control Protocol*. RFC 0793, Internet Engineering Task Force, September 1981.
6. IEEE Society Computer. *Part 3: IEEE Standard for Information technology – Specific requirements – Part 3: Carrier Sense Multiple Access with Collision Detection (CSMA/CD) Access Method and Physical Layer Specifications*. IEEE Standard for Information Technology 802.3, Institute of Electrical and Electronics Engineers Standards Association (IEEE-SA), December 2008.
7. Moy, J. *OSPF Version 2*. RFC 2328, Internet Engineering Task Force, April 1998.
8. Coltun, R., Ferguson, D., Moy, J., and Lindem, A. *OSPF for IPv6*. RFC 5340, Internet Engineering Task Force, July 2008.

9. ISO. *Information technology – Telecommunications and information exchange between systems – Intermediate System to Intermediate System intra-domain routing information exchange protocol for use in conjunction with the protocol for providing the connectionless-mode network service (ISO 8473)*. International Standard 10589, International Organization for Standardization (ISO), March 2008.

10. Rekhter, Y., Li, T., and Hares, S. *A Border Gateway Protocol 4 (BGP-4)*. RFC 4271, Internet Engineering Task Force, January 2006.

11. Cotton, M. and Vegoda, L. *Special Use IPv4 Addresses*. RFC 5735, Internet Engineering Task Force, January 2010.

12. Rekhter, Y., Moskowitz, B., Karrenberg, D., deGroot, G. J., and Lear, E. *Address Allocation for Private Internets*. RFC 1918, Internet Engineering Task Force, February 1996.

13. Weil, J., Kuarsingh, V., Donley, C., Liljenstolpe, C., and Azinger, M. *IANA-Reserved IPv4 Prefix for Shared Address Space*. RFC 6598, Internet Engineering Task Force, April 2012.

14. Cotton, M., Vegoda, L., and Meyer, D. *IANA Guidelines for IPv4 Multicast Address Assignments*. RFC 5771, Internet Engineering Task Force, March 2010.

15. Hinden, R. and Deering, S. *IP Version 6 Addressing Architecture*. RFC 4291, Internet Engineering Task Force, February 2006.

16. Kawamura, S. and Kawashima, M. *A Recommendation for IPv6 Address Text Representation*. RFC 5952, Internet Engineering Task Force, August 2010.

17. Bao, C., Huitema, C., Bagnulo, M., Boucadair, M., and Li, X. *IPv6 Addressing of IPv4/IPv6 Translators*. RFC 6052, Internet Engineering Task Force, October 2010.

18. Postel, J. *User Datagram Protocol*. RFC 0768, Internet Engineering Task Force, August 1980.

19. Cotton, M., Eggert, L., Touch, J., Westerlund, M., and ire, S. Cheshire. *Internet Assigned Numbers Authority (IANA) Procedures for the Management of the Service Name and Transport Protocol Port Number Registry*. RFC 6335, Internet Engineering Task Force, August 2011.

20. Postel, J. and Reynolds, J. *File Transfer Protocol*. RFC 0959, Internet Engineering Task Force, October 1985.

21. Postel, J. and Reynolds, J. K. *Telnet Protocol Specification*. RFC 0854, Internet Engineering Task Force, May 1983.

22. Fielding, R., Gettys, J., Mogul, J., Frystyk, H., Masinter, L., Leach, P., and Berners-Lee, T. *Hypertext Transfer Protocol – HTTP/1.1*. RFC 2616, Internet Engineering Task Force, June 1999.

23. IANA. Service Name and Transport Protocol Port Number Registry, http://www.iana.org//assignments/service-names-port-numbers/service-names-port-numbers.xml.

24. Mockapetris, P. V. *Domain names – implementation and specification*. RFC 1035, Internet Engineering Task Force, November 1987.

25. Thomson, S., Huitema, C., Ksinant, V., and Souissi, M. *DNS Extensions to Support IP Version 6*. RFC 3596, Internet Engineering Task Force, October 2003.

26. IANA. Root Zone Database, http://www.iana.org/domains/root/db/.

27. ICANN, New Generic Top-Level Domains, http://www.icann.org.

28. ISO. *Codes for the representation of names of countries and their subdivisions – Part 1: Country codes (ISO 3166-1:2006)*. International Standard 3166-1:2006, International Organization for Standardization (ISO), November 2006.

29. Klensin, J. *Internationalized Domain Names for Applications (IDNA): Definitions and Document Framework*. RFC 5890, Internet Engineering Task Force, August 2010.

30. Klensin, J. *Internationalized Domain Names in Applications (IDNA): Protocol*. RFC 5891, Internet Engineering Task Force, August 2010.

31. Faltstrom, P. *The Unicode Code Points and Internationalized Domain Names for Applications (IDNA)*. RFC 5892, Internet Engineering Task Force, August 2010.

32. Alvestrand, H. and Karp, C. *Right-to-Left Scripts for Internationalized Domain Names for Applications (IDNA)*. RFC 5893, Internet Engineering Task Force, August 2010.

33. IANA. Internet Assigned Numbers Authority(IANA), http://www.iana.org.

34. ICANN. Internet Corporation for Assigned Names and Numbers (ICANN), http://newgtlds.icann.org/en/.

35. Organization, Number Resource. Regional Internet Registries, http://www.nro.net/about-the-nro/regional-internet-registries.

36. Rekhter, Y. and Li, T. *An Architecture for IP Address Allocation with CIDR*. RFC 1518, Internet Engineering Task Force, September 1993.

37. Fuller, V., Li, T., Yu, J., and Varadhan, K. *Classless Inter-Domain Routing (CIDR): an Address Assignment and Aggregation Strategy*. RFC 1519, Internet Engineering Task Force, September 1993.

38. Egevang, K. and Francis, P. *The IP Network Address Translator (NAT)*. RFC 1631, Internet Engineering Task Force, May 1994.

39. Project, WIDE. The WIDE Project, http://www.wide.ad.jp/.

40. Project, WIDE. The KAME Project, http://www.kame.net/.

41. Fink, R. and Hinden, R. *6bone (IPv6 Testing Address Allocation) Phaseout*. RFC 3701, Internet Engineering Task Force, March 2004.

42. Hinden, R., Fink, R., and Postel, J. *IPv6 Testing Address Allocation*. RFC 2471, Internet Engineering Task Force, December 1998.

43. Mockapetris, P.V. *Domain names – concepts and facilities*. RFC 1034, Internet Engineering Task Force, November 1987.

44. Cheshire, S., Aboba, B., and Guttman, E. *Dynamic Configuration of IPv4 Link-Local Addresses*. RFC 3927, Internet Engineering Task Force, May 2005.

45. Huston, G. *Operational Requirements for the Address and Routing Parameter Area Domain (arpa)*. RFC 3172, Internet Engineering Task Force, September 2001.

2

Basics of the 3GPP Technologies

Mobile broadband has swept though the world during the past few years. The growth in the number of subscribers has been phenomenal, and it might seem as if the services have just appeared very quickly, but the systems, and the technology it relies on have been in the development for almost two decades. The key organization in the cellular, mobile broadband systems is the 3GPP [1]. 3GPP is a Standards Development Organization (SDO), which is a joint project between six regional and national SDOs. (The Association of Radio Industries and Businesses (ARIB) from Japan [2], the Alliance for Telecommunications Industry Solutions (ATIS) from USA [3], China Communications Standards Association (CCSA) from China [4], European Telecommunications Standards Institute (ETSI) from Europe [5], the Telecommunications Technology Association (TTA) from Korea [6], and the Telecommunication Technology Committee (TTC) from Japan [7].) The 3GPP is responsible for standardizing the requirements, the architecture, and most of the protocols used to create the mobile broadband system. Some of the protocols, however, are standardized by other organizations. The biggest set of protocols standardized elsewhere, but used in the 3GPP system, are the Internet Protocol (IP) protocols. Those are standardized by the Internet Engineering Task Force (IETF) [8].

In this chapter, we will look at the architecture, technology, and protocols used in 3GPP systems. In addition, we will briefly look at the standardization process used to specify the technology. We will start with the process, because understanding the process used to create the technology will help us to better understand the steps of the evolution of the system, and perhaps the technology itself.

2.1 Standardization and Specifications

A technology is strongly influenced by the process with which it has been created. A well-organized and well-defined process can create good results quickly. An inclusive, and less structured process can create innovative solutions. However, too strict or too loose processes can hamper the creation of technology – even to the point where a perfectly good technology does not come ready in time, or has inadequate or unsuitable content.

As with other technologies, 3GPP network technologies have also been influenced by the standardization process. Obviously, 3GPP is the main standardization organization for the technology. In addition, the IETF has been influential on the protocols used in

Deploying IPv6 in 3GPP Networks: Evolving Mobile Broadband from 2G to LTE and Beyond, First Edition.
Jouni Korhonen, Teemu Savolainen and Jonne Soininen.
© 2013 John Wiley & Sons, Ltd. Published 2013 by John Wiley & Sons, Ltd.

Table 2.1   3GPP standardization process stages

Stage	Purpose
Stage 1	Service description / requirements
Stage 2	Technical realization / architecture
Stage 3	Detailed protocol specifications

the 3GPP systems. In this section, we will first look at the 3GPP standards and the standardization process, after which we will briefly look at the IETF.

2.1.1   3GPP Standardization Process

3GPP was organized in 1998 to create a global 3rd Generation (3G) technology from the legacy of the successful Global System for Mobile Communications (GSM) 2nd Generation (2G) technology previously created by ETSI. Later 3GPP also got the responsibility to maintain and further develop the 2G technology from ETSI. As the 3G technology was a direct descendent from the 2G technology, it was only natural that the same process, and working procedures were kept. Hence, the 3GPP working procedures are more or less directly adopted from ETSI's processes.

As 3GPP is a partnership of many SDOs, the organizations participating in the 3GPP process are members of their respective SDOs, and participate through that membership in the 3GPP. The members are organizations such as network operators, network equipment vendors, handset Original Equipment Manufacturers (OEMs), and wireless chipset manufacturers. The process itself is contribution driven. Practically this means that the 3GPP organization drives the standards process, but the 3GPP members are responsible for the direction of the technology, and pushing the work forward by inputting, contributing to the process.

The 3GPP process has three stages – stages 1, 2, and 3. The three stages are described in Table 2.1. In layman's terms, the 3GPP first specifies the requirements for a feature or a complete new system, and then describes the system architecture. When there is reasonable consensus on the architecture, the detailed protocol specifications are written. In addition to requirements, architecture, and detailed protocol specifications, the 3GPP also performs feasibility studies, and also writes test specifications for its systems.

3GPP Organization and Work Procedures

The work in 3GPP is organized in Technical Specification Groups (TSGs), and then in Working Groups (WGs) within the TSGs. Figure 2.1 shows the 3GPP organizational structure. The Project Coordination Group (PCG) is an administrative group overseeing the 3GPP organization. The TSGs are the groups coordinating the WGs underneath each of the TSGs. The TSGs and their responsibilities are the following:

SA is the TSG responsible for services and system aspects. The system requirements, the system architecture, security, and operations and management issues are dealt in this TSG.

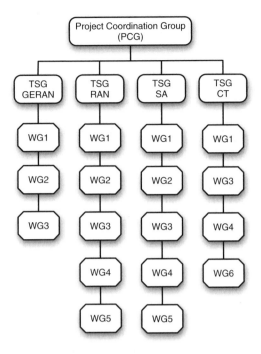

Figure 2.1   3GPP organization.

GERAN is the TSG responsible for the GSM radio maintenance and evolution.
RAN TSG handles 3G radio, and its evolution including the LTE.
CT works on stage 3 related to the core network and terminals.

The TSG SA is working mostly on the requirement and architecture aspects – hence, stages 1 and 2. The other TSGs are mostly working on stage 3. However, this division is not very clear cut. There is radio architecture related work in the Radio Access Network (RAN) and GSM/Edge Radio Access Network (GERAN) TSG as well.

In 3GPP, the bulk of the work is done in face-to-face meetings held by the WGs. The interested organizations send their representatives to the meetings to present their contributions, and to react to other organizations' contributions. The decisions on the direction of the technology is based on consensus on the technical merits of the different proposals. In difficult issues where the consensus is not very clear, the 3GPP relies on voting among the members to reach decisions.

The 3GPP produces specifications which it publishes in releases. We will the look at the different releases in the next section, followed by a view of how the 3GPP specifications are organized.

3GPP Standards Releases

The 3GPP inherited its release process from ETSI. ETSI used to produce yearly releases of the GSM standards named after the year that the release was produced, and 3GPP

Table 2.2   3GPP releases

Release	Description of the major 3GPP releases in the context of this book
Release-97	The first release of General Packet Radio Service (GPRS)
Release-98	
Release-99	The first release of the Universal Mobile Telecommunications System (UMTS)
Release-4	
Release-5	IP Multimedia Subsystem (IMS) [9], and High Speed Downlink Packet Access (HSDPA) [10] introduced
Release-6	High Speed Uplink Packet Access (HSUPA) [11] introduced
Release-7	HSPA+ introduced
Release-8	The first release of LTE
Release-9	Dual-Stack PDP Context introduced to 3G
Release-10	DHCPv6 Prefix Delegation (DHCPv6-PD) introduced
Release-11	
Release-12	

continued this approach for its first release. However, the second release – Release 2000 – was a bit too difficult to do in just one year, and 3GPP decided to move to releases that are not based on a particular year, and thus not named after a year either.

The ETSI releases defining the packet switched architecture, and hence, being of interest for the purposes of this book, were Releases 1997 and 1998. The first 3GPP release was Release-1999. After that the releases were named after the version number of the specifications as Release-4, -5, -6, etc. At the time of writing, 3GPP has just started working on Release-12. One might wonder why 3GPP started from number four instead of some more logical number such as one. However, the 3GPP specification version numbers also indicate the specification maturity, and the numbers from one to three were already taken. We will explain the specification maturity in the next section.

Table 2.2 lists the releases until today, and describes the major releases in the context of this book.

3GPP Specifications

The 3GPP specifications are usually identified by their numbers. Partly this is because the numbers are rather short, and can be more easily remembered than their sometimes quite obscure titles. Although this is relatively handy for the people who work with the 3GPP specifications every day, it might prove quite difficult for an outsider to find the right documents or understand the context of a particular document.

The 3GPP document number consists of two numbers – the document series number, and the document number. The document series number is a two-digit number, and describes which part of the system the document specifies. Table 2.3 shows the document series. The document number part is a three-digit number. For example, 23.060 is a document 60 from the document series 23. If we look at Table 2.3, we see it is a stage 2 document. We will refer to this particular document many more times in this chapter.

Table 2.3 3GPP specification structure

Subject	Specification number
21 Series	Requirements
22 Series	Service aspects (stage 1)
23 Series	Technical realization (stage 2)
24 Series	Signaling protocols (stage 3) – user equipment to network
25 Series	Radio aspects
26 Series	CODECs
27 Series	Data
28 Series	Signaling protocols (stage 3) – (RSS-CN) and OAM&P and charging
29 Series	Signaling protocols (stage 3) – intra-fixed-network
30 Series	Programme management
31 Series	Subscriber Identity Module (SIM / USIM), IC cards, test specs
32 Series	Operations, management and provisioning, and charging
33 Series	Security aspects
34 Series	UE and (U)SIM test specifications
35 Series	Security algorithms
36 Series	LTE (Evolved UTRA) and LTE-Advanced radio technology
37 Series	Multiple radio access technology aspects

Table 2.4 3GPP specification version numbers

Version	Description
0	Early draft
1	60% ready draft – ready to be presented to the TSG for information
2	Above 80% ready draft – ready to be approved by the relevant TSG when finalized
>= 3	Specification has been approved by the relevant TSG for the 3GPP Release indicated in the version number

We did briefly mention the document version numbers in Section 2.1.1. The meaning of the version numbers is shown in Table 2.4. Obviously, it is hard to know when something is 60% or 80% ready. Therefore, this practically means that in version 1 there is agreement on the way forward, and there is some feeling that the WG is getting its head around the concept. Version 2 means that the concept is all documented and there are just a few details to be worked out. However, as with all projects, those last details can take a surprisingly long time. When the specification has been approved by the relevant TSG the document version is set to the release in which it is to be published.

2.1.2 IETF Standardization Process

In Chapter 1 we described the IETF as the main SDO for the Internet protocols. Therefore, it is also a very important SDO for the 3GPP system. The IETF protocols are used within

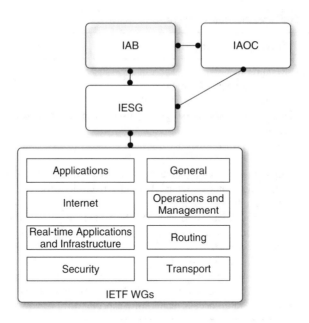

Figure 2.2   IETF organization.

the 3GPP system, and the service that the 3GPP system provides is the transport of the Internet protocol. Therefore, in our books, the IETF merits the ranking of second most important SDO for the 3GPP system.

The IETF standardization process is quite a bit different from the 3GPP process. Hence, they should not be compared directly – it would not do justice to either of these fine organizations. The work in the IETF is performed by WGs. These WGs are organized into areas based on the technology they are working on. The areas are managed by Area Directors (ADs). The IETF structure is depicted in Figure 2.2. The ADs form a group called the Internet Engineering Steering Group (IESG) headed by the IETF chair who is also the AD for the General Area. Currently, all other areas have two ADs. The General Area is a non-technical area responsible for the evolution of processes and the general structure of the IETF. All the other areas are technical.

The Internet Architecture Board (IAB) is responsible for the architectural guidance of the IETF. If the IESG is a very operational body, the IAB is very strategic with few short-term goals. The administrative part includes making sure the IETF organization has adequate funds and tools, and that it is capable to organizing its meetings; this is the responsibility of the IETF Administrative Oversight Committee (IAOC).

The interesting part of the IETF organization is that it has no membership. The idea is that the process is as inclusive as possible, and anybody and everybody can join to participate in the IETF process. The IETF participants are supposed to represent only themselves, and their best technical judgement. However, it is clear that most of the participants do work for organizations that have an interest for the IETF process.

To make sure that the barrier for entry to the IETF is as low as possible, most of the work and all of its decisions are conducted in the WG mailing lists. The IETF does hold three meetings a year, but according to the procedures it is not mandatory to participate in the physical meetings to participate in the technical process.

The decisions in the standards process are based on so-called 'rough consensus'. As the IETF has no membership, and the decisions are done on the mailing lists, the IETF cannot really vote. Instead, the WG chairperson or chairpersons try to understand using the mailing list discussions as their guide, which direction the consensus of the WG is going. The 'rough' part of the consensus means that the whole WG does not have to agree, but it is enough if most of the people voicing their opinion agree.

The IETF has two main types of documents: temporary document called Internet-Drafts, and permanent documents called Request For Comments (RFC). Internet-Drafts are documents that represent work in progress. An Internet-Draft has a lifetime of six months, during which it will have to be resubmitted again, approved to be an RFC, or it will cease to exist. There are two types of Internet-Drafts – individual submissions and WG Internet-Drafts. The individual submissions are, as the name suggests, something that somebody has submitted. There is no barrier to submitting one, and also no quality or content control. Thus, it mainly represents the submitter's opinion. The WG Internet-Drafts, however, have to be adopted by a WG. Therefore, they represent the interest of that working group to work on the topic.

RFCs come in many flavors as well. The real specifications that represent the IETF's consensus are either on the Standards Track or Best Current Practice (BCP). A standards track RFC is usually a protocol specification or a protocol framework. BCPs describe the currently best found way to implement or deploy something, which the IETF community thinks the industry and the Internet community should follow. In addition to standards track, and BCPs, IETF also produces Experimental, and Informational RFCs. An Experimental RFC describes a protocol which the IETF does not know, yet, if it is useful, works well, or might even be harmful in the Internet. However, the experimental status allows people to document the experiment.

Informational RFCs range from specifications that somebody wanted to publish, but for some reason or other did not go through the full IETF standardization process, over requirements documents, to even April fool jokes. If an Informational RFC has the publishing date of April 1st, one should be a bit careful. This has not stopped people from implementing such specifications. One of the classics is 'IP over avian carriers', RFC 1149 [12].

2.1.3 Other Important Organizations in the 3GPP-Ecosystem

In the past sections, we described two of the most important standardization groups that shape the 3GPP technology, and thus, are part of the 3GPP ecosystem. However, these are not the only SDO or industry organizations that shape the industry and the technology. In the following, we shall briefly describe a few organizations that are part of the same ecosystem.

GSM Association (GSMA) [13] is an organization that represents mobile operators globally. It has multiple roles, ranging from organizing the global roaming between operators, over creating technical industry specifications, to organizing the largest and most important of mobile communications events: the Mobile World Congress (MWC).

Open Mobile Alliance (OMA) [14] is an SDO concentrating on specifying standards for mobile services. 3GPP has had, and continues to have, close cooperation with OMA over such services as the Multimedia Messaging (MMS) [15] and IMS services.

ETSI is not only one of the 3GPP's founding SDOs, but 3GPP and ETSI continue to have close technical cooperation in multiple areas.

2.2   Introduction to 3GPP Network Architecture and Protocols

In this section we will start to get to know the different 3GPP architectures and radio accesses. This book will concentrate on the Packet Switched (PS) domain of the 3GPP architecture, as the Circuit Switched (CS) has little relevance when talking about IP and IPv6. However, to completely understand the 3GPP system, and its architecture, one must understand a bit of the technology it has been built on.

As described before, the 3GPP system has been specified in different releases where the newer ones build on the previous ones. There have been defining moments in the 3GPP release history where, in addition to adding new features, the architecture is significantly altered. Obviously, the first defining architectural event was the definition of the GSM architecture. However, the significant alterations to the architecture in the context of this book are related to the PS domain. These changes to the architecture were made when the PS domain was first introduced, and after that, when a new radio access technology was added. These bigger architecture steps have been the introduction of the PS domain – the addition of GPRS to GSM architecture, the move to the 3G radio system (and to 3GPP from ETSI) and hence the introduction of Universal Mobile Telecommunications System, and finally the latest new radio technology, Long Term Evolution. All of these three architectures are closely related, and they share many of the same technologies. However, GPRS and UMTS are so closely related that they are often now together described as 'GPRS', or sometimes 2G-GPRS and 3G-GPRS. Therefore in the section below, these two architectures are discussed together. But LTE is a more radical change to the 3GPP architecture and it is described separately.

We will now follow this evolution in the 3GPP system, starting with a high-level overview of the GSM architecture. After that we will move to a GPRS architecture overview covering both the 2G and the 3G variants. At the end of the section, we will cover the newest arrival to the 3GPP architecture family, LTE.

2.2.1   GSM System

As described earlier, the 3GPP system has evolved from the original European mobile telephony system GSM. The GSM system [16] was introduced in the early 1990s to provide digital mobile telephony service. It was the first digital, cellular mobile phone system to become widely used. The GSM used Time Division Multiple Access (TDMA) based radio technology. It was very successful, and the usage of mobile telephony started growing. Data was becoming more important for the users of telecommunication systems. These users included enterprise users, but also consumer interest was growing. At the time, IP and the Internet access had not become as dominant as it is today, and still another packet data networking technology called X.25 [17] was in wide use. At the time, the data

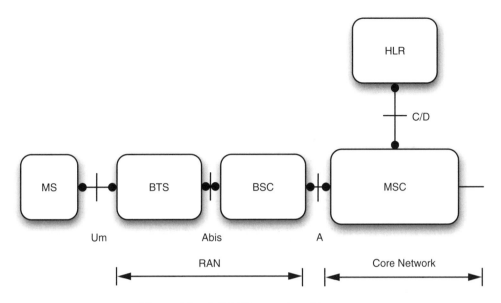

Figure 2.3 GSM CS network architecture.

service provided over the GSM system was purely CS based, as was the whole network architecture. However, there was an increasing interest in packet data.

The GSM architecture consists of network elements and the interfaces between those elements. (One has to be careful with the word 'interface'. In the 3GPP context, it means the whole vertical protocol stack between two network elements – sometimes also called a reference point, but in the IP language it means the network connection of an IP node.) You can see these described in the GSM network architecture in Figure 2.3. The network elements are the labeled boxes. In the following, we will briefly describe the main elements and the interfaces between them:

Mobile Station (MS) is the phone, handset, or other device that is used to connect to the network.

Base Transceiver Station (BTS) is also known as the base station. It is the radio network element that is directly connected to the MS over the air interface.

Base Station Controller (BSC) controls a group of base stations. It performs the mobility management between the BTSs under its control.

Home Location Register (HLR) is the database that contains the subscriber data. This subscriber data includes the authentication data used to authenticate the user, user profiles including what services the user is subscribed to, and the location where the user currently is.

Mobile-services Switching Centre (MSC) is the local exchange responsible for a set of BSCs, and the MSs under those base stations in a geographical area. It performs authentication, mobility management, and exchanges the subscribers' calls.

In Figure 2.3, the interfaces are shown as lines between the labeled network elements, looking from left to right, *Um* for the MS to BTS interface, *Abis* between BTS and the

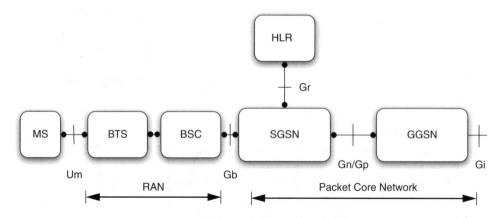

Figure 2.4   2G-GPRS network architecture.

BSC, and *A* between the BSC and the MSC. The interface between the MSC and the HLR is labeled with two names: *C* and *D*. The reason for this is the C-interface is used when the MS is in the home network, and the D-interface is used when the MS is visiting another network – roaming. Thus, this interface is between two operators when the subscriber is roaming. In the roaming situation, the HLR remains always in the home network (hence, the name), and the MSC is in the visited network.

2.2.2   General Packet Radio Service

GPRS was first introduced as the packet service for the GSM network. In addition, 3G is based on the same architecture. Therefore, the name 'GPRS' is used in both networks for the packet network. We will describe both flavors in this section.

2G-GPRS

As described earlier, GPRS was introduced to the circuit switched GSM system in ETSI Release-1997. Figure 2.4 shows the GSM-based GPRS architecture. When comparing Figure 2.4 with Figure 2.3 it is clear that the RAN (sometimes also called the Base Station System (BSS) as the architectural combination of BSC and BTS) are the same, but the Core Network (CN) differs. We will now focus on the parts that are different in the PS domain from the CS domain.

The GPRS uses many of the elements from the GSM telephony system. The BTS, BSC, and HLR are the same in both and shared between both of the domains. However, there are also two new elements that are specific to GPRS – the Serving Gateway Support Node (SGSN) and the Gateway GPRS Support Node (GGSN):

SGSN is mainly responsible for authentication, authorization, and mobility management. In addition, it gathers charging information. It is connected to the HLR for the subscriber profile information. The SGSN is an element responsible for both mobility management signaling, and user packet forwarding. Hence, SGSN is both a user-plane and a control-plane element. On the user plane, the SGSN terminates the radio protocols, and thus it is the first element to see the user IP packet when a packet comes from the MS, and the last element to see the IP packet in the downlink direction. The SGSN can also perform compression, such as IP header compression [18–21], and content compression (using V.42 bis). On the control plane, the SGSN handles mobility management by performing mobility management signaling towards the radio network and the GGSN. In addition, the SGSN performs signaling to other SGSNs during handovers. An inter-SGSN handover is performed when an MS moves to an area where the radio network is controlled by a different SGSN. Therefore, the MS can pass by multiple SGSNs while having an active connection open.

GGSN is the topological anchor point of the mobility management in the GPRS network. It is the gateway between the GPRS network and external networks, such as the Internet. Thus it terminates the GPRS protocols and mobility management functions. As the GGSN is the mobility management anchor point, unlike the SGSN, it does not change during the MS's connection. Like the SGSN, the GGSN also performs both control and user-plane duties. On the control plane, it receives and answers session initiation, handover, and session teardown signaling from the SGSN. On the user plane, it is responsible for the routing from and to the external IP-networks. In addition, it also gathers charging information, and is a central point in Legal Interception (LI). As pointed out above, the MS can go through many SGSN while moving in the network and having a connection open. Since the GGSN is the anchor point, one connection can be connected only to one GGSN. However, the MS can have multiple independent, concurrent connections to multiple GGSNs at the same time. We will describe these concepts later in this chapter.

New interfaces are specified to connect the new network elements to the existing network elements and to each other. These are the most important:

Gb-interface is the interface between the BSS and the core network – hence, between the BSC and the SGSN. This interface comprises both of the control and user-plane parts.

Gn/Gp-interface is the interface between the SGSN and the GGSN. The Gn-interface is used when the MS is in the home network, whereas the Gp-interface is used when the subscriber is roaming in another network. Thus, the Gp-interface is the interface between two network operators. The Gn-interface is also used between SGSN during a handover to exchange MS related information, and for forwarding user packets.

Gr-interface is the interface between the SGSN and the HLR.

Gi-interface connects the GPRS network to the external IP-networks.

There are elements in the GPRS architecture that pass both control and user data. These include the whole BSS, and the SGSN and the GGSN. The interfaces connecting those elements have two modes – the user-plane, and the control-plane modes. Figure 2.5 shows the 2G-GPRS user-plane protocols.

The protocols shown in the interface figure are described in more detail in Section 2.3. Here we will just give an overview.

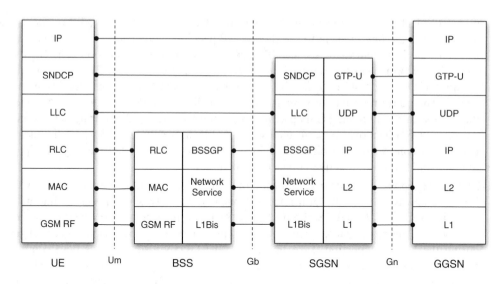

Figure 2.5   2G-GPRS user-plane protocols.

GSM radio protocols (GSM Radio Frequency (RF), Media Access Control (MAC) [22], Radio Link Control (RLC) [22], and Logical Link Control (LLC) [23]) provide the GSM radio functionality including ciphering, reliability, and other radio procedures.

Base Station System GPRS Protocol (BSSGP) provides routing-related and Quality of Service (QoS)-related information between the BSC and the SGSN [24]. The network service in the Gb-interface originally was defined as Frame Relay, but now also IP based configurations are possible [25].

SubNetwork Dependent Convergence Protocol (SNDCP) is a protocol that enables the transport of the multiple networking protocols over the GSM radio service [18].

GTP User Plane (GTP-U) is a tunneling protocol that allows the transportation of different networking protocols over the 3GPP core network.

Figure 2.6 shows the 2G-GPRS control-plane protocols. Many of the protocols are the same, but the functionality is different. These protocols carry signaling payload instead of the end-user traffic. The following protocols are different from the user plane protocols.

GPRS Mobility Management and Session Management (GMM/SM) performs mobility management functions and connection activation, modification, and deactivation.

GTP Control Plane (GTP-C) is the same for GPRS Tunneling Protocol (GTP) as the GMM/SM is for the GSM radio.

3G-GPRS

The radio technology used by GSM did not provide very high bandwidth, and the industry had been preparing for some time for the new generation of mobile phone radio interface. The new technology was called UMTS and it is based on a new radio

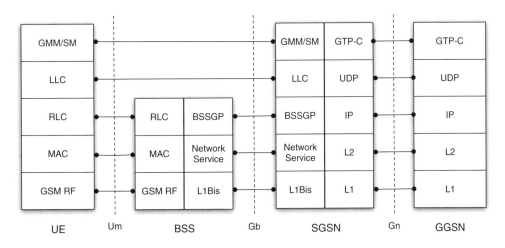

Figure 2.6 2G-GPRS control-plane protocols.

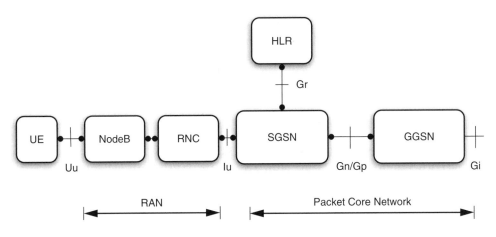

Figure 2.7 3G-GPRS network architecture.

technology – Wideband Code Division Multiple Access (WCDMA). The industry decided on specifying UMTS based on the GSM architectural principles, and reusing the GSM protocols. However, due to the transition to the WCDMA technology, the RAN architecture (more specifically called UMTS Terrestrial Radio Access Network (UTRAN) in UMTS) had to be changed to meet the new requirements. The UMTS PS domain architecture – the 3G GPRS – is illustrated in Figure 2.7 [26].

The similarity of the 2G and 3G architectures is relatively obvious. The difference is in the level of abstraction in the RAN architecture. The 2G radio network elements BTS and BSC have been replaced in the 3G-GPRS architecture with UMTS base station (NodeB) and Radio Network Controller (RNC) respectively. In addition, there have been changes to the functionality split between the elements. In 2G-GPRS, the SGSN terminated the

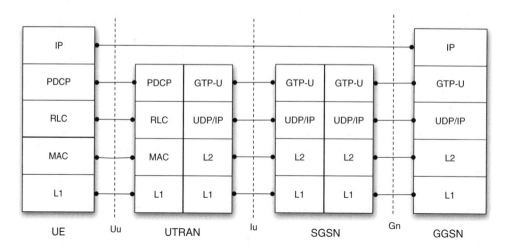

Figure 2.8   3G-GPRS user-plane protocols.

radio protocols. However, in UMTS the RNC terminates the radio protocols. In UMTS, the RNC is also responsible for header compression. Content compression is not supported in UMTS. The name of the MS has also changed to User Equipment (UE) in UMTS. However, the MS and the UE terminology is used somewhat interchangeably in literature and in the specifications. We will use the UE terminology in this book.

Together with the new radio access network, there is a new interface. Instead of Gb-interface, 3G-GPRS has the Iu-interface. Figure 2.8 shows the 3G-GPRS user plane protocols, and Figure 2.9 describes the control plane.

If compared to the 2G-GPRS interfaces in Figure 2.5, it can be seen that the Gn-interface has stayed the same. However, there is a significant difference in the Iu-interface compared to the Gb-interface. The Iu-interface is based on the same protocol as the Gn-interface – GTP. In addition, the transport in the Iu-interface has changed to IP from Frame Relay.

The 3G-GPRS control plane in Figure 2.9 shows that in addition to the radio interface the Iu-interface has replaced the Gn-Interface in the 2G-GPRS described in Figure 2.6. This change has introduced some new protocols.

GMM/SM is, like GMM/SM in the 2G-GPRS, the protocol that performs mobility management and session management functions.

Radio Access Network Application Part (RANAP) is a UMTS protocol that carries the higher layer signaling. It runs on the Signalling System No. 7 (SS7) protocol suite [27]. Originally the Iu-interface was specified to be transported only on Asynchronous Transfer Mode (ATM) [28]. However, the standard has been updated to enable any transport network technology to be used.

Signalling Connection Control Part (SCCP) is a protocol used in the SS7 protocol suite.

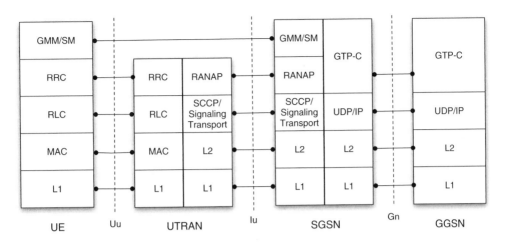

Figure 2.9   3G-GPRS control-plane protocols.

2.2.3   *Evolved Packet System*

The newest member of the 3GPP technology family is Evolved Packet System (EPS). As with GSM and UMTS, the architecture has two parts: the radio access technology, LTE, and the core network architecture – the Evolved Packet Core (EPC). As the saying goes, a beloved child has many names. So does the EPS. The terminology is slightly confusing. The whole system is named, depending on the source, 4th Generation (4G), EPS, or after the radio technology LTE. Especially in the popular press and marketing either 4G, or LTE is used. The EPS core network is called EPC, or also sometimes System Architecture Evolution (SAE). In this book, we will try to use relatively consistently the EPS terminology. When designing the EPS, 3GPP aimed to simplify the architecture, and make it more compatible with IP. Therefore, the new architecture is quite a bit different from GPRS architecture. The most important change in the new system is the complete lack of a traditional circuit switched domain. The telephony service is provided with a Voice over IP (VoIP) technology – the IP Multimedia Subsystem (IMS). In addition, the whole network architecture is also changed significantly. The EPS architecture is shown in Figure 2.10 [29]. If the *differences* between the 2G and 3G are in the details, between EPS and its predecessors it is the *similarities* that are in the details.

The EPS network elements and their functionalities are listed in the following:

Evolved Node B (eNodeB) is the LTE base station. However, it also includes some of the functionalities of the radio resource controller (BSC in 2G, and RNC in 3G). Therefore, there is no separate radio controller in the EPS network architecture.

Mobile Management Entity (MME) is responsible for mobility management, authentication, and authorization of the UE. The MME is practically a very similar element to the SGSN in UMTS. The difference between the 3G-SGSN and the MME is that the MME is only a control-plane element; it does not perform user-plane functions at all.

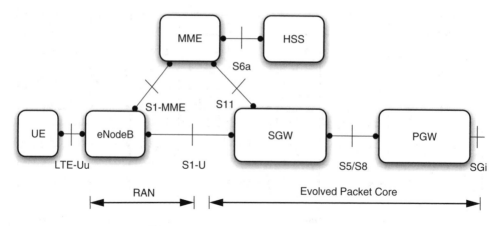

Figure 2.10   EPS network architecture.

Home Subscriber Server (HSS) is the database that holds the subscriber profile including the authentication and authorization information. It performs the same functions as the HLR in previous 3GPP network generations.

Serving Gateway (SGW) is the mobility anchor for inter-eNodeB mobility. It routes the user traffic between the radio network and the Packet Data Network Gateway (PGW). Practically it performs the same functionality as the 3G-SGSN performs on the user plane. Like the SGSN, SGW can change with the mobility management during a connection.

PGW or PDN-GW is the gateway between the EPS and the external IP-networks, and the anchor point for the EPS mobility management. Thus, it does not change during a connection as it is the topological anchor for the UE in the IP network. The PGW performs the same functions as the GGSN in GPRS. The PGW can also be used as a GGSN in a setup where both GPRS and EPS are present.

We described in Section 2.2.2, how the SGSN can change during a connection and when the UE is moving about, and how the GGSN as the anchor point does not. Similarly, the PGW stays the same during a connection, and the eNodeB, the MME, and the SGW can change during the connection.

The standards allow the SGW and PGW to be implemented in the same physical element. This kind of combined element is called System Architecture Evolution Gateway (SAE-GW). In this configuration, the functionality stays the same as in the separated elements case.

Obviously, as the network architecture has changed, so have the interfaces connecting them. The interfaces listed in Figure 2.10 are the following:

S1-MME is the control-plane interface connecting the eNodeB and the MME.
S1-U is the user-plane interface connecting the eNodeB and SGW.

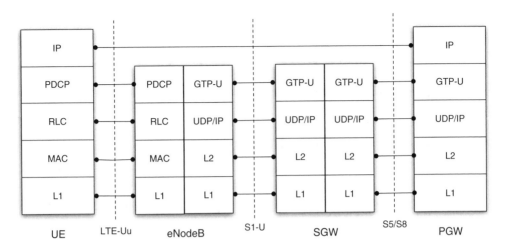

Figure 2.11   EPS user-plane protocols.

S5/S8 are the interfaces connecting the SGW and the PGW. S5 is used within the home network, and S8 interface is used when the UE is roaming. This is similar to the Gn- and Gp-interfaces in GPRS. The S5- and S8-interfaces comprise the user and control planes.

S6a interface connects the MME and the HSS.

S11 is a control-plane interface connecting the MME and SGW.

SGi is the interface connecting the EPS to the external IP networks, such as the Internet.

When looking at the EPS user plane in Figure 2.11, the legacy from the previous 3GPP architecture is noticeable. When comparing the EPS user plane to the 3G-GPRS user plane in Figure 2.8, the basic difference between the two can be found in the naming of the interfaces, and of the nodes. However, again the difference is in the details. The radio interface, LTE-Uu is completely new, and it is based on the Orthogonal Frequency-Division Multiple Access (OFDMA) technology instead of TDMA or WCDMA in the previous generations. In addition, we can see differences in the protocol versions when we take a closer look at the 3GPP protocols in the next section.

The differences between EPS and legacy are more apparent at the control plane. Figure 2.12 depicts the EPS control plane. Many of the interfaces are quite different from the previous generations. The main difference is that the control plane goes through some different elements from the user plane. In addition, there are no SS7 based interfaces anymore. All the interfaces are on IP.

The protocol architecture shown in Figures 2.12 and 2.11 is sometimes called GTP based architecture. This is to distinguish it from another architecture variant that 3GPP has defined for EPS – the Proxy Mobile IP (PMIP) based EPS architecture [30]. The difference between these two architectures, however, at this level is very small. Therefore, we will concentrate at this point on the GTP variant, and we will look at the PMIP variant later, when we look in more detail at the 3GPP protocols.

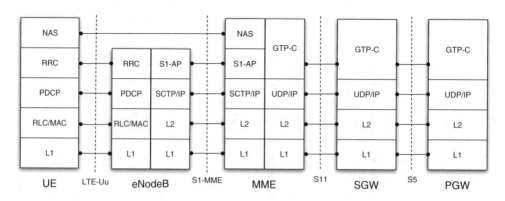

Figure 2.12 EPS control-plane protocols.

2.2.4 Control and User Planes, and Transport and User Layer Separation

One important part of the 3GPP architecture is the differences between the control and user planes, and transport and user layers. This terminology is sometimes quite confusing, but understanding this is an important part of understanding the 3GPP architecture, and how it works. We will first look at the user and control planes, and then look at the transport and user layers.

Control and User Planes

The 3GPP architecture is divided into the control and user planes. The control plane takes care of the signaling in the network. It transports the messages that are needed to attach a subscriber to the network, authenticate and authorize the user, open and close connections, and perform handover signaling. As the name implies, the control plane transports signals that control the network and the mobile station. The user plane, on the other hand, is responsible for transporting the user packets through the network.

Depending on the interface, the control and user plane can be realized with completely different protocols, or different parts of the same protocol. This is due to the very different nature of the control and user planes. The control plane messages are sent when there is something to signal, and to refresh the state in the network. It is under heavy load, especially when the network connection is set up and also during handovers. The user-plane load is dependent on the download and upload speed of the subscriber.

Transport and User Layers

The transport layer refers to the IP network and the related transport protocols used to transport the control and user planes in the 3GPP network. This is the network that connects the network elements to each other. End users do not have access to this network. The transport layer network of the operator is connected to other operators' networks to enable control- and user-plane connections during roaming.

The user layer consists of the users' packets. This includes the users' IP packets and their payload. The addressing, IP version, and the traffic are completely separate from the transport layer.

These two layers are completely separated, and thus independent of each other. This separation allows different IP versions to be used by the user from those used by the network. Hence, the operator can provide an IPv6 service to the user even if the operator's internal network is not yet fully IPv6 capable. The opposite is also possible: in the case of fully IPv6 capable network, the operator can provide IPv4 services for legacy devices.

2.3 3GPP Protocols

Above, we became familiar with the 3GPP architecture variants, and saw that they are a bit different, although there are similarities in all of them. We also learned the difference between the control and user planes. In this section, we will drill a bit deeper on what are the different functions that the control and user planes actually perform. In addition, we will look a bit more closely at the main protocols used in the 3GPP networks.

The different interfaces throughout the network have very different characteristics. The radio interface is a wireless interface, where it is important to make sure that the messages – network control signals or user packets – can get over that hop with adequate robustness. Some other interfaces within or between the radio access and packet core networks can actually be transported over a complete network with both wireline and wireless segments. The network topology of those networks does not necessarily reflect the clean architecture pictures shown in this book and in other literature. The different interfaces are not necessarily in their different networks, or segments, but in some cases are in the very same network segment. Depending on the network, the network elements can be either co-located in the very same data centers, or on very different locations. However, obviously, the base stations and (e)NodeBs have to be located relatively close to the subscriber to actually create a wide area wireless network.

The differences in interfaces mean that messages have to be carried by quite different protocols – depending on the characteristics of an interface. The radio interfaces, 2G, 3G, and LTE, are just a single hop long, and therefore they do not have to be routed over a network. However, errors or complete loss of a message can easily occur over the radio interface. Therefore, the protocols have to make sure that the messages actually get over the hop with a relatively high probability. In addition, the radio interface is out there in very public space. Anybody could try to listen or even attempt to inject traffic. Therefore, the protocols have to make sure that the integrity and confidentiality of the messages are protected.

On the other hand, the protocols within the network itself may have to be transported over the operator's underlying transport network. Sometimes these may be over long distances and over different kinds of transport technologies. Some of these links can be point-to-point (or made to look like point-to-point links by the underlying transport technology). However, they can also be networks of different links. Therefore, many of the protocols within the network are on top of IP. Non-IP-based networking protocols were also used in the networks, especially within the RAN. However, the 3GPP architectures have mostly either replaced these interfaces with IP-based transport, or at least provided an IP-based alternative.

In the following, we will be taking a closer look at the most relevant control- and user-plane protocols.

2.3.1   Control-Plane Protocols

As described earlier, the control-plane protocols are, as the name suggests, controlling the network. They are used to attach the subscriber to the network, authenticate and authorize the subscriber, and for session and mobility management. Excluding the actual moving of user packets, the whole functionality of the network is performed by the control-plane protocols. All this is extremely fascinating technology, and especially the mobility management, makes very interesting read. However, 3GPP technology is built to hide the mobility from the upper layers. Therefore, it has little effect on the IP layer. Hence in this book, we will concentrate on the network functions that are needed for the user to connect to the Internet – and eventually disconnect from the network.

The 2G and 3G GPRS variants work very much in a similar manner. Therefore, we are going to explore the control plane of the two GPRS variants together. However, EPS uses a slightly different approach, and we will study EPS separately in its own section.

The aim of this book is to understand the functionality of the protocols. There is not enough space to dive into the details of the different protocols. There is better literature and much better authors for that purpose. An interested reader can take a look at the suggested reading section at the end of the chapter. We already described the protocol architecture in Section 2.2.2 for GPRS, and the EPS protocol architecture was defined in Section 2.2.3. In the following, we will show how the connections between the UE and the network are managed, by explaining the used signaling. The protocol messages described in the following sections are messages of the upper protocol layers of the control plane part of an interface. The layers below the uppermost layer make sure that the message is transported correctly over the interface.

For 2G-GPRS, the control-plane protocols can be found in Figure 2.6, and for 3G-GPRS control plane protocols are in Figure 2.9. The EPS system control-plane protocols can be found in Figure 2.12.

GPRS

Before connecting to the Internet, a UE has to get access to the mobile network itself. This is done by performing a network attach function. The purpose of the attach function is for the UE and the network to get to know each other. The UE identifies itself to the network, and the subscriber database, the HLR, authenticates the user. During this authentication procedure the network and the UE also set up network radio interface ciphering to protect the confidentiality and authenticity of future exchanges over the radio interface. Some networks may also perform a check of the device identity by comparing the International Mobile Equipment Identity (IMEI) to the Equipment Identity Register (EIR) to see if the used device has been stolen, or for some other reason blacklisted from attaching to the network. After the subscriber has been authenticated, and the radio communication has been secured, the location of the subscriber is updated to the HLR.

In 2G/3G networks the UE is informed when the attach has been successful, after which the UE has finished the network attach procedure. The UE is attached to the network,

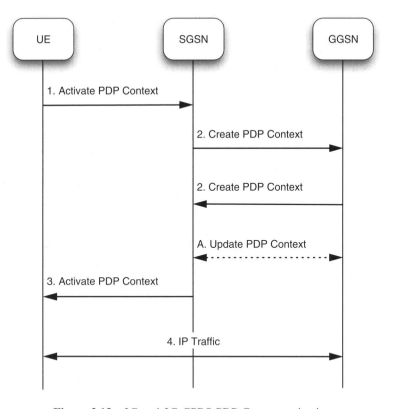

Figure 2.13   2G and 3G GPRS PDP Context activation.

but it does not yet have a connection to the Internet. The connection has to be separately activated with the Packet Data Protocol (PDP) Context Activation procedure.

In 2G and 3G-GPRS the establishment of the IP layer connectivity is not part of the network attach procedure, but has to be explicitly performed. This procedure is called PDP Context Activation. The PDP Context is the name of the connection, or session, which actually connects the UE to the external Packet Data Network (PDN) – such as the Internet. GPRS supports different types of PDP Contexts – hence, different PDP Types. The different PDP Types currently supported are IPv4, IPv6, Dual-Stack (IPv4v6), and Point to Point Protocol (PPP) [31]. However, practically speaking, PPP is not used anymore. Thus, we ignore the PDP Type PPP in this book.

Figure 2.13 shows PDP Context activation procedure, and the explanation can be found below:

1. *Activate PDP Context* request message is sent by the UE to the SGSN to indicate that the UE wishes to open a packet data connection, such as a connection to the Internet. The message indicates if the UE would like to have service on IPv4, IPv6, or both. The SGSN adds the PDP Context to its context table.
2. *Create PDP Context* request is sent by the SGSN to the GGSN. The GGSN adds the new PDP Context to its context data structure. The GGSN allocates the IP address or

addresses to the context, and sets the Protocol Configuration Option (PCO) information and sends them in the *Create PDP Context* response to the SGSN.

3. *Activate PDP Context* response message is sent by the SGSN with the IP address and PCO information received from the GGSN. This finishes the PDP Context activation procedure.

4. The IP traffic can then start flowing over the new PDP Context.

In GPRS, the procedure in Figure 2.13 is used for the first PDP Context activation, and for subsequent activations. There are two types of PDP Context activations: Primary PDP Context activations, and Secondary PDP Context activations. The same procedure described in Figure 2.13 is used for both Primary and Secondary PDP Context activations – the difference is the message '*Update PDP Context*'. It is used in Secondary PDP Context activation to update the actual received QoS in the radio bearer negotiation (not shown in the picture). The Primary PDP Context activation does not use this message. The difference between Primary and Secondary PDP Context negotiation is the Primary PDP Context activation allocates a new IP address for the PDP Context. Secondary PDP Context activation only opens a new connection related to an already active PDP Context using the IP address allocated to it. The Secondary PDP Contexts can be used for QoS differentiation where different traffic is transferred over a different PDP Context with a different QoS profile. Making an analogy to the physical world, if the Primary PDP Context is the highway, a Secondary PDP Context is a car pool lane (in the USA, this is a lane reserved for cars containing two or more people, and therefore being less congested than the other lanes). GPRS does not distinguish between the Primary and Secondary PDP Contexts after the activation.

As we have now learned, the UE starts the PDP Context activation process. 3GPP has also specified a network requested PDP Context activation procedure. In this procedure, the subscriber has to have a statically configured address. When the GGSN that has the given address receives an IP packet destined to this address, it will query the HLR for the location of the UE and send Protocol Data Unit (PDU) Notification Request to the SGSN that is handling the UE. The SGSN will then send a Request PDP Context activation message to the UE, which would trigger the normal PDP Context activation procedure. (A careful reader might now be a bit confused – neither Figure 2.4 nor Figure 2.7 shows an interface between the GGSN and the HLR. This has been left out of the drawing for simplicity. The interface between the GGSN and the HLR is the Gc-interface and it is specified only for the Network Requested Primary PDP Context Activation.) However, this network requested Primary PDP Context activation procedure has, to the authors' knowledge, never been implemented in any network. In addition to Primary PDP Context activation, a Network Requested Secondary PDP Context Request activation is also specified.

In addition to being activated, the PDP Contexts can be also be modified and deactivated. The modification and deactivation procedures can be activated by the UE, the SGSN, or the GGSN. The modifications usually relate to changing the QoS parameters of the PDP Context. As described earlier, the Primary and Secondary PDP Contexts are not distinguished after activation. This can be seen in the deactivation process. The IP address related to a bundle of PDP Contexts is deallocated from the UE when the last PDP Context related to that address is deactivated. The deactivation of the last PDP Context does not automatically detach the UE from the network. Obviously, the UE is then no

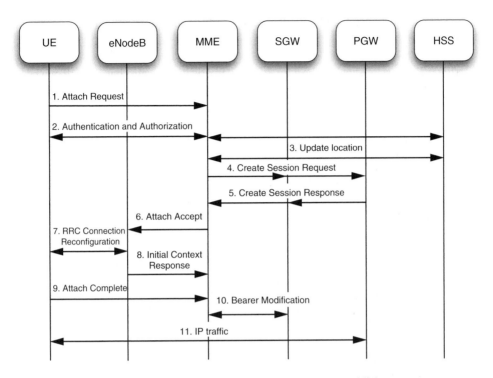

Figure 2.14 LTE attach and Default EPS Bearer establishment.

longer capable of packet data communication. However, it can still receive Short Message Services (SMSs) via GPRS signaling [32].

Traditionally, mobile phones activated a PDP Context when they needed one. The activation of an application, such as a web browser, also triggered the activation of the PDP Context, and shutting down that application triggered the deactivation of the PDP Context. Some devices have also been able to support multiple concurrent PDP Contexts. For instance, often different services have had their own PDP Context. Thus, Internet access, or access to an operator service such as MMS have had their own PDP Contexts. However, modern smartphones usually open one PDP Context after starting, and keep it active most of the time. This benefits applications and users by optimizing away the time required to set up a PDP Context on demand.

EPS

The EPS architecture was designed to be completely packet switched with no CS mode. Hence, it makes little sense in an EPS network to just attach to the network without service. Therefore, in the EPS networks, in the network attachment procedure also the packet data connection is activated. This is called Default EPS Bearer Establishment. A simplified version of the network attach procedure signaling with the Default EPS Bearer establishment is shown in Figure 2.14. (The packet data connection in LTE is called a 'bearer', and in GPRS a 'PDP Context'. The names are different, but conceptually they are very similar.)

The LTE network attach and default EPS Bearer establishment procedure is performed as follows:

1. *Attach Request* is sent by the UE to the network to indicate that the UE wishes to get service from this network.
2. *Authentication and security* functions are performed to authenticate the subscriber and to secure the communication channel.
3. *Location update* is done by the MME to the HSS to record the UE's location in the subscriber database.
4. *Create Session Request* message is sent from the MME to the SGW to begin the establishing of the default EPS Bearer. The SGW creates a new entry for this bearer in its EPS Bearer Table, and sends the message to PGW to establish the bearer at the PGW.
5. *Create Session Response* is sent as the response to the Create Session Request, indicating the success or failure of the bearer establishment. In case of success, the PGW creates an entry to its EPS Bearer table, and it allocates the IP address or addresses for the bearer. The PGW sends the message to the SGW, which sends it forward to the MME. The message includes the IPv4 and/or IPv6 address allocated to the UE, and the additional PCO needed to configure the IP stack of the UE.
6. *Attach Accept* message is sent to the eNodeB to indicate successful attachment to the network. This message includes the address or addresses and the PCO included in the previous step.
7. *Radio Resource Control (RRC) Connection Reconfiguration* procedure configures the Radio Resources and provides the IP address, and PCO information to the UE.
8. *Initial Context Response* is sent to the MME by the eNodeB to inform the MME of the eNodeB's address. This message and the Attach Complete message can also be sent in a different order to the MME.
9. *Attach Complete* message indicates that the UE has finished the attach and the default EPS Bearer setup. At this point, the UE can start sending user layer packets.
10. *Bearer Modification* procedure is performed between the MME and the SGW upon the MME receiving both Initial Context Response and the Attach Complete messages. The Bearer Modification procedure provides the eNodeB address to the SGW, which allows the SGW to start forwarding IP packets towards the UE.
11. The attach and default EPS Bearer establishment procedures are complete, and the IP packets can flow in both directions.

Comparing the two connection establishment procedures in Figure 2.14 and in Figure 2.13 may make the GPRS PDP Context activation procedure look simpler. However, this is not completely true as the GPRS PDP Context activation in Figure 2.13 does not include the network attach procedure. In GPRS, the network attachment also has to happen before PDP Context activation can be performed.

In addition to the default bearer establishment, additional connections can also be created. These are called dedicated bearers. The dedicated bearers are similar to the Secondary PDP Contexts in GPRS. They provide, for instance, additional QoS capabilities to the communication. The dedicated bearer activation procedure is shown in Figure 2.15, and each related message is described below.

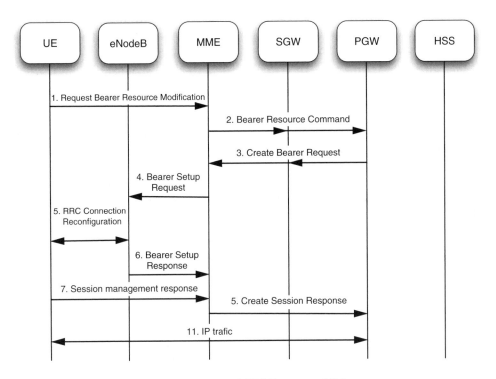

Figure 2.15 Dedicated EPS Bearer establishment.

1. *Request Bearer Resource Modification* is sent by the UE to the MME indicating the request for a dedicated bearer.
2. *Bearer Resource Command* is sent by the MME to the SGW. The SGW sends the Bearer Resource Command to the PGW; if it is accepted it triggers the Dedicated Bearer Activation process.
3. *Create Bearer Request* is sent from the PGW via the SGW to the MME, indicating the creation of a dedicated bearer. The message contains the information about the characteristics of the bearer including QoS and PCO parameters.
4. *Bearer Setup Request/Session Management Request* is created by the MME with the information received from the SGW in the Create Bearer Request. The MME sends to Bearer Setup Request to the eNodeB. The Bearer Setup Request is a message between the MME and the eNodeB. It contains information for creating the Session Management Request, which is sent from the MME and to the UE, but transferred within other messages and hence not shown as a separate message.
5. *RRC Connection Reconfiguration* procedure is initiated by the eNodeB and conducted with the UE. The Session Management Request is passed to the UE within the RRC Connection Reconfiguration message. The new radio bearer is created between the UE and the eNodeB.
6. *Bearer Setup Response* is sent by the eNodeB to the MME to indicate the radio bearers have been created.

7. *Session Management Response* is sent by the UE to the MME to indicate finalization of the dedicated bearer creation process. This is the answer to the Session Management Request received within the RRC Connection Reconfiguration.
8. The upper layer traffic can now flow between the UE and the network via the dedicated EPS Bearer.

The biggest difference between the EPS dedicated bearer and the GPRS Secondary PDP Context activation procedures is that the EPS dedicated bearer activation was designed to be network initiated, while the GPRS PDP Context activation is UE initiated. Both systems' specifications support both UE and network initiated connection establishment. However, in EPS the UE can request the network to perform dedicated bearer activation, and the network then performs the procedure. In GPRS, it is the other way around.

The dedicated bearers can be activated, modified, and deactivated. Modification means changing the characteristics of the bearer itself. For instance, by modifying the bearer, the QoS characteristics of the bearer can be changed. The bearer modification can be initialized by PGW, HSS, or the UE.

When a dedicated bearer is deactivated, the traffic that would have been transferred over the dedicated bearer moves back to the default bearer. Practically, this means that the traffic no longer enjoys the QoS improvements it had in the dedicated bearer. Coming back to our highway analogy, the car pool lane ends, and the cars – the packets – move among the normal traffic. Usually, however, the traffic that was moved on the dedicated bearer had already stopped, and hence the bearer was deactivated. The dedicated bearers can be deactivated for various reasons by the UE, the MME, and the PGW.

The EPS system also has what is called multiple PDN support. Practically, this means that a UE can be connected to multiple IP networks at the same time. Hence, it has multiple IP addresses. Figure 2.16 describes the UE requested PDN Connectivity procedure. The idea is the same as in GPRS with multiple Primary PDP Contexts.

1. *PDN Connectivity Request* is sent by the UE to the MME via the eNodeB. The PDN Connectivity Request indicates what kind of connectivity (IPv4, IPv6, or both) is requested to which network.
2. *Create Session Request* message is created by the MME and sent to the SGW. The SGW creates an entry for this new bearer in its EPS Bearer table, and forwards the message to the PGW. The PGW also generates a new entry for in its EPS Bearer table and allocates the IP address(es) for the bearer.
3. *Create Session Response* is created by the PGW including the needed configuration information such as the IP address, PCO, and QoS information. The message is sent to the SGW, which forwards it to the MME.
4. *Bearer Setup Request/PDN Connectivity Accept* message is sent to the eNodeB. The Bearer Setup Request indicates to the eNodeB to start the radio bearer negotiation with the UE. The PDN Connectivity Accept is a response to the UE to the PDN Connectivity Request. The PDN Connectivity Accept is sent within the Bearer Setup Request message.

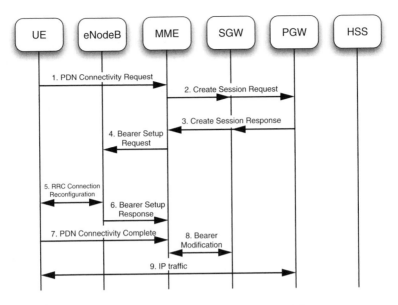

Figure 2.16   UE requested PDN connectivity.

5. *RRC Connection Reconfiguration* message exchange sets up the radio bearers between the UE and the eNodeB. The PDN Connectivity Accept message is delivered to the UE at this point.
6. *Bearer Setup Response* is sent by the eNodeB to the MME to indicate that the radio bearers have been set up.
7. *PDN Connectivity Complete* message is sent from the UE over the eNodeB to the MME to indicate the UE has finished the PDN Connectivity procedure.
8. *Bearer Modification* procedure between the MME and the SGW sets up the location of the eNodeB to the SGW.
9. The upper layer traffic can start to be transferred between the UE and the PGW.

2.3.2   User-Plane Protocols

We showed in the previous section how the connections between the UE and the network are managed. Now that we have successfully created a connection it is time to transmit some user-plane IP packets. The user-plane protocols make sure the user packets make it from one side of the network to the other. Conceptually, this is relatively simple: the user packet is either sent from the UE in the upstream direction, or the packet enters the network at the GGSN in 2G and 3G, or at the PGW in EPS. In order to make sure that a packet makes it safely through to the other side, the packet has to be wrapped in the right protocols, which carry the packet over the different interfaces. As in all IP networks, the IP packets can get lost in the process. However, the user-plane protocols make sure the user packets have the best possibility of getting through.

All three systems, 2G-GPRS, 3G-GPRS, and EPS are a slightly different. Therefore, we shall examine these separately:

2G-GPRS

In Figure 2.5 we learned the user plane protocol architecture for 2G-GPRS. We can see that there are three interfaces that the packet has to travel over – the GSM radio interface (Um), the radio network interface between the BSS and the SGSN (Gb), and the core network interface between the SGSN and the GGSN (Gn). However, the IP packet comes in contact only with two protocols – the SNDCP and GTP. The lower layer protocols provide services to the upper layer protocols including transport, over the radio interface retransmission, and ciphering. In the following, we shall only concentrate on the upper layer GPRS protocols. The other protocols are not that interesting in the context of this book. An interested reader can find literature that will cater to the hunger for knowledge of the other protocols. The SNDCP and GTP functionality are described below:

SNDCP is the protocol that transports IP packets over the GSM radio network. The SNDCP is responsible for multiplexing multiple PDP Contexts between the UE and the SGSN, performing header compression, packet segmentation and reassembly, and content compression. The header compression mechanisms supported by the specifications are RFC 1144 [19], Degermark [20], and Robust Header Compression (RoHC) [21]. Content compression, however, is not used at all in reality, and mostly not even implemented in the networks and terminals. Header compression currently shares the same fate. However, RoHC perhaps might be supported one day for VoIP. Currently, SNDCP carries IPv4, IPv6, and PPP [31]. Originally, it was also designed to carry X.25 [17]. However, X.25 never became a reality in the GPRS networks, and also later the support was dropped from the 3GPP specifications.

GTP is responsible for transporting the users' IP packets over the GPRS core network. The basic functionality of GTP is to uniquely identify the GTP-tunnel to which the user packets belong. The GTP-tunnel then maps directly to the PDP Context. As shown in Figure 2.5, GTP is transported over User Datagram Protocol (UDP) [62]. As we learned in Section 1.5.1, UDP only provides unreliable transport. Therefore, the transported packets can be delivered in a different order than sent, multiplied, or not even reach their destination. Originally, a Transport Control Protocol (TCP) [34] transport option was included for X.25 transport. However, this option was removed with the support for X.25. The GTP header holds a sequence number, which allows the reordering of out-of-order packets, if needed. IP itself does not need the packets to be delivered in order, but PPP expects the packets to be transmitted in order.

Although 3GPP removed X.25 support after there was clear consensus that it would not be needed anymore, sometimes reaching consensus on a feature being outdated can be difficult. PPP has been used in very few networks, and practically it was never supported by mainstream mobile phones. Still, the support for PPP is included in the specifications.

3G-GPRS

The 3G-GPRS user-plane protocol architecture is shown in Figure 2.8. Even at first glance it is clear that there are similarities with the 2G-GPRS architecture. However, there are also some obvious differences: the radio interface protocols are only between the UE and the NodeB, and GTP extends to the radio access network. In the following, we take a look at the two protocols that transport the users' packets over the 3G-GPRS network – the Packet Data Convergence Protocol (PDCP) and the GTP.

PDCP is responsible for transferring the packets over the radio interface, and providing header compression. The segmentation and reassembly of the user packets, and mapping the PDP Contexts to the radio bearers is provided by the underlying RLC [35]. In 3G-GPRS, the functionality that was in one protocol, in the SNDCP in 2G, has been distributed into PDCP and to the RLC protocol layers. In addition, PDCP and RLC are only used between the UE and the RNC, and they are not extended to the SGSN as SNDCP did in 2G. The header compression algorithms supported by PDCP are Degermark [20] and RoHC [21].

GTP in 3G-GPRS is the same protocol as in 2G-GPRS. Therefore the description of the protocol itself in the previous section applies also for 3G. However, in addition to the core network, GTP is also used inside the radio access network. In the radio access network, GTP is used in the user plane in the Iu-interface, which is an interface specific to the 3G-GPRS architecture. GTP is used in the core network in the Gn- and Gp-interfaces, which also exist in 2G-GPRS.

EPS

If the 2G and 3G-GPRS architectures are very similar with few differences, we have already learned that the EPS network architecture is very different, and the EPS procedures are sometimes quite different. However, in the EPS user plane we can see clearly the GPRS pedigree in the EPS architecture – although the interface names are different, the user-plane protocol architecture seems exactly the same in 3G-GPRS. On the other hand, when we look more closely the following differences become apparent.

PDCP is a new version in EPS [36]. The PDCP's responsibility in EPS includes transmission of the user packets over the radio interface, header compression, user traffic ciphering and integrity protection, and delivery of the packets in sequence and without duplicates. Hence, in EPS the PDCP's functionality has increased quite a bit from the 3G version of PDCP. In addition, the PDCP is terminated in the EPS architecture in the eNodeB.

GTP is responsible for transporting users' packets over the EPS network. This is the same GTP that is also used in the GPRS architectures.

2.3.3 GPRS Tunneling Protocol Versions

Although we promised not to dwell too deeply on the protocols themselves, it is good to understand a little about GTP. GTP is used in both control and user planes in all of the 3GPP architectures. Therefore, it is an important building block in the 3GPP system. Currently, three versions of GTP exist:

GTP version 0 was the first version of GTP. It is now deprecated, and not used in modern systems.

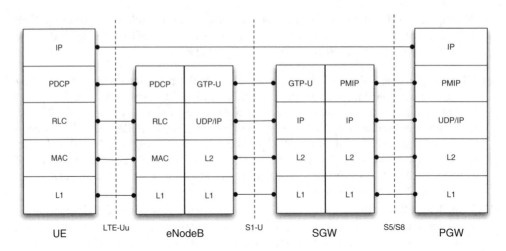

Figure 2.17   EPS control plane protocols in PMIP architecture variant.

GTP version 1 is used in GPRS in both control and in user planes, and in EPS in user
plane.

GTP version 2 is used in EPS in the control plane.

3GPP has defined fallback from version 2 to version 1 in nodes that support the fallback.
However, fallback to version 0 is not supported, and version 0 is compatible neither with
version 1 nor version 2.

2.3.4   PMIP Based EPS Architecture

As we briefly mentioned earlier, there are two different variants of the EPS
architectures – the GTP variant we described above, and the PMIP variant. We talk
about variants because the EPS architecture shown in Figure 2.10 is valid for the both
variants. The difference in the PMIP variant is in the S5/S8-interface, which uses PMIP
[37] instead of GTP. Figure 2.17 shows the control plane protocol architecture. When
comparing the PMIP-based control plane architecture with the GTP-based control plane
architecture in Figure 2.12, the difference is apparent. Instead of using GTP to create,
manage, and delete the tunnels between the SGW and the PGW, the PMIP protocol
is used. The PMIP protocol is a technology defined by the IETF, whereas GTP is a
technology specified by 3GPP.

As the control plane protocol architecture is different in the PMIP-based EPS architecture, the procedures used are a little different as well. An example can be seen in the LTE
attachment procedure described in Figure 2.18. When comparing to Figure 2.14, which
we studied earlier, we can see an extra step (marked as 4b and 5b). We do not want
to repeat the whole signaling flow explanation at this point, but the reader can refer to
the explanation connected to Figure 2.14 in Section 2.3.1. The added step is described
below.

4b *Proxy Binding Update* is created with the information that SGW obtained from the Create Session Request from the MME. The SGW creates a new entry in its bearer table, and sends the message to the PGW.

5b *Proxy Binding Acknowledgement* is a response to the Proxy Binding Update indicating the failure or success of the default bearer creation. In case of success, the PGW creates itself a bearer table entry and sends the Proxy Binding Acknowledgement message to the SGW.

As can be seen from Figure 2.18, at this level abstraction of the difference between the GTP and the PMIP EPS architecture variants is in the control plane and the names of the messages over the S5/S8-interface. Obviously, there is more than meets the eye. However, for the purpose of the scope of this book, the difference is small at this point.

In addition to the difference in the control plane, the user plane tunneling protocol is also different in the PMIP-based architecture. Figure 2.19 describes the EPS PMIP variant user-plane protocol architecture. Comparing it to the GTP-variant architecture in Figure 2.11, the difference is the GTP that has been changed to the Generic Routing Encapsulation (GRE) protocol [38].

The obvious difference in the PMIP based EPS user-plane architecture is very small at this level of abstraction – as is the difference in the control plane. However, though we will not see the difference at this level, there is a difference in the link model that is visible to IP. We will examine the link model differences in Section 4.1.3.

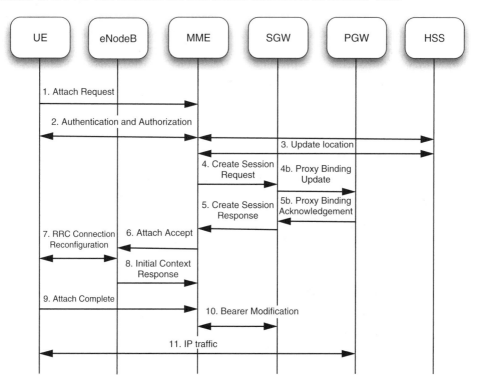

Figure 2.18 LTE attach and Default EPS Bearer establishment in PMIP-based architecture.

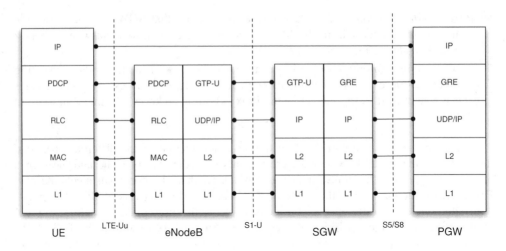

Figure 2.19   EPS user-plane protocols in PMIP architecture variant.

In addition to the technical differences of the two architecture variants, there is a difference between the protocol approaches of PMIP and GTP. The tradition in protocol design in the IETF is a little different from that in 3GPP. This is also visible when comparing the GTP to the PMIP. 3GPP has a tradition of specifing a single protocol to cover a protocol layer over one interface. Originally, GTP was designed for the Gn/Gp-interface to cover both control and user plane. PMIP, on the other hand, was designed as a protocol to provide network based mobility management. However, the IETF already had different generic tunneling protocols available, and PMIP can use many of those. 3GPP decided to use the GRE option because, with some extensions [39], it could basically support the same use cases as GTP did.

The original idea of the PMIP-based EPS architecture was to be simpler, better compatible with the IP and IETF concepts, and facilitate better interworking to other cellular wireless technologies such as technologies standardized by 3rd Generation Partnership Project 2 (3GPP2 [40] or the WIMAX Forum [41]. However, in the end, the PMIP technology was enhanced to support practically the same use cases as the GTP was [39, 42, 43]. In addition, at the time of writing, PMIP has had limited success in interworking to other mobile networking technologies. As of today, most EPS deployments use the GTP variant, though some PMIP based deployments do exist as well.

2.4   Mobility and Roaming

One of the main functions of the mobile, cellular networks is mobility management. Simply put, mobility management is making sure that the network always knows precisely enough where a UE is, in order to be able to send packets to it, or to receive packets from it. This is still relatively simple. On the other hand, managing mobility of potentially millions of UEs while ensuring the network does not become overloaded with signaling load, and the UE's battery is conserved, is not that simple anymore. However, it is not the aim of this book to go through the secrets of mobility management. Thus, we will not overload the reader with the details.

In this section, we try to give a very short overview of the mobility management. This should be enough to understand the tasks of the different nodes in the 3GPP network architecture, and understand how the UE's IP stack sees the 3GPP networks.

2.4.1 Mobility Management

Mobility management makes sure that a UE is connected to the mobile network while it moves within the network's coverage. To ensure this, the UE and the network exchange signaling messages including radio measurement information to decide when is the right time to move from one BTS, NodeB, or eNodeB to another. The UE's movement from one network element's control to another network element's control is called a handover. The handovers, and thus the mobility, are controlled by the 3GPP radio access specific handover procedures between the UE and the radio access network. Inside the network, the mobility management is provided by the mobility management protocols, GTP in GPRS and in the GTP variant of EPS architecture, and by a combination of GTP and PMIP in the PMIP variant. (During a heavy debate on the merits of PMIP versus GTP an intelligent man stated that he does not know of any case where GTP-based network would have lost a UE forever. This nicely crystalizes what the main purpose of a mobility management protocol is.)

When the UE moves far enough from the point where it originally attached to the network, the handover has to be performed between higher level network elements. For instance, in 3G-GPRS the next level handover would be between two RNCs. If the UE continues to move further, the next handover level is between two SGSNs. The only network element that does not change during a session is the network anchor point – the GGSN in GPRS architecture, and the PGW in the EPS architecture. These levels of mobility management create a hierarchy. The most handovers happen between the network elements closest to the UE, and the fewest the furthest from the UE. This hierarchy is also noticeable in the number of network elements in a deployed network. There can be thousands, or even tens of thousands of BTSs, NodeBs, or eNodeBs, hundreds of BSCs, and RNCs, and tens of SGSNs, GGSNs, MMEs, SGWs, and PGWs. (This is dependent on the network operator size, the geographic and population coverage, and the capacity needs of the network.)

Figure 2.20 depicts the mobility management hierarchy for GPRS, using 3G-GPRS as an example, and for EPS. A UE would be in the bottom of the picture moving horizontally. The levels of hierarchy are shown from top to bottom, where the anchor point is at the top, and the NodeBs and the eNodeBs are at the bottom.

In addition to the intra radio access technology handovers, the 3GPP system specifies handover between the different 3GPP radio access technologies. Thus, when a UE is capable of multiple radio technologies, it can also move between networks depending on the coverage.

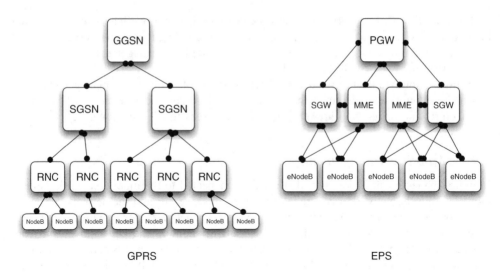

Figure 2.20 Mobility management hierarchy in GPRS and EPS architectures.

2.4.2 Roaming

In the section above, we discussed the mobility management within one operator's network. Additionally, the 3GPP system is capable of enabling a subscriber to move between networks – to roam. Roaming means that the UE moves to a different network from the one to which they subscribe. An operator's network is called a Public Land Mobile Network (PLMN). Consequently, the network to which the subscriber belongs is called a Home PLMN (HPLMN), and a network where the subscriber is roaming is a Visited PLMN (VPLMN).

There are two types of roaming: domestic and international. Domestic roaming can be used in a country where some operators cannot provide adequate national coverage, and therefore may augment their coverage through roaming partners' networks. The international roaming is used to provide network service to subscribers when they are in a different country. The operators have their own roaming partners governed by roaming contracts. Therefore, it is not always possible to roam to every network in a foreign country.

As the 3GPP network technology is the most used cellular network technology, and most major operators have practically global roaming partner networks, it is now possible to find a network that provides service almost anywhere in the world.

2.4.3 Mobility Management Beyond 3GPP

3GPP also specifies interworking and mobility management with other access technologies, such as Wireless Local Area Network (WLAN), or 3GPP2 networks. The mobility management provided can be either seamless or non-seamless. Non-seamless mobility management refers to mobility where the network attachment and authentication is performed automatically, but the UE is connected to a different IP network, and thus, its IP address changes. Practically, the handover is non-seamless, and the UE's IP stack has to reconfigure itself, and the applications running on the UE will notice the change. Seamless mobility management is either performed using GTP or PMIP network based mobility management technologies, or the mobility management can be host based using technologies such as Dual-Stack Mobile IPv6 (DSMIPv6 [44]).

2.5 Central Concepts for IP Connectivity

We have now illustrated what the different 3GPP specified architectures look like, and how they do session management – that is, manage the connections between the UE and the network – and what technologies are used to actually move the end user's packets back and forth over the network. However, it might not be quite clear how all of these work together, and how certain things are decided in the network. In this section, we try to add a bit of information to pull some of the concepts together.

2.5.1 PDP Contexts and EPS Bearers

We have gone through the different procedures of session management – on how to activate, modify, and deactivate PDP Contexts and EPS Bearers. We have said that these are the connections that connect the UE to the network. In this section, we will look at these connections, their characteristics, and the relationship of the different PDP Contexts and EPS Bearers to each other.

First of all, conceptually the PDP Contexts and the EPS Bearers are very similar, or even the same. They are connections that extend from the UE to the network gateway – the GGSN or the PGW. To send and receive packets, the UE needs a connection to the gateway. This gateway will then provide the UE its IP address. The UE will use this IP address to send and receive traffic on that particular connection. That connection is in GPRS a PDP Context, and in EPS architecture it is the EPS Bearer. The packets sent or received on this connection will all be treated the same way – they will all be put in the same packet queue and treated in first-in-first-out manner.

A PDN Connection can be used as a synonym for an EPS Bearer and PDP Context when there is no need to differentiate between dedicated bearers or Secondary contexts, just to refer to the IP level connectivity. The PDN Connection is the association between a UE and a PDN represented by an Access Point Name (APN) and the associated IP level configuration.

Multiple Connections to One Network

If there is a need to provide a different level of service for different IP packets, another connection needs to be opened – a Secondary PDP Context in GPRS as described in Figure 2.13, or a dedicated EPS Bearer in EPS as described in Figure 2.15. The traffic that needs different treatment can be moved to that new connection. As long as that connection is open, selected traffic will flow through that connection and be treated separately from the other traffic in the original connection. The new connection has its own packet queue, and the packets that flow into that queue are also treated in first-in-first-out manner,

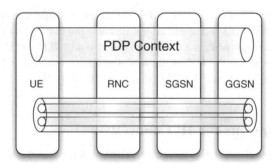

Single PDP Context, and two PDP Contexts associated as one connection

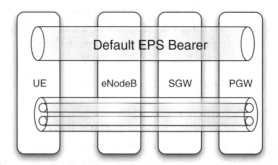

Default EPS Bearer, and Default EPS Bearer with two dedicated bearers

Figure 2.21   Relationship between PDP Contexts, and default and dedicated EPS Bearers.

but separately from the previous queue. The selection, which packets go down which connection, is done based on a set filter – the Traffic Flow Template (TFT). We will describe the TFT in more detail in Section 2.5.3. If there is a need to distinguish between multiple types of traffic, multiple Secondary PDP Contexts or dedicated EPS Bearers can be opened with their own TFTs. Figure 2.21 shows the relationship between PDP Contexts, and the default and dedicated EPS Bearers. Though the concept of the PDP Contexts, and the different EPS Bearers is very similar, the session management treatment is slightly different; after activation, all related PDP Contexts are equal irrespective of whether they have been activated as Primary or Secondary PDP Contexts. If one PDP Context is deactivated, the others will survive. However, in EPS, deactivation of the default EPS Bearer will result in the deactivation of the dedicated bearers as well.

Multiple Connections to Multiple Networks

The Secondary PDP Context activation, or the dedicated EPS Bearer activation, will create a new connection with the same IP address as the original connection possibly with different parameters for QoS. Thus, from the IP's point of view, it still was just one link. However, both GPRS and EPS are also capable of activating connections to networks other than the already ongoing connection. Hence, connections have their own

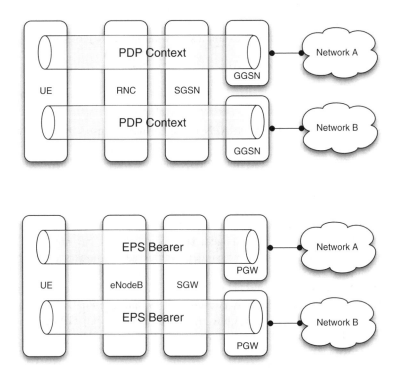

Figure 2.22   Multiple connections to different PDNs.

IP addresses from other IP-networks than the original connection. This can be useful, for instance, when the UE needs to be connected at the same time to the Internet, and to some operator-provided services that are not available through the public Internet. In order to create a new connection to a new IP-network, the UE will perform either Primary PDP Context activation as described in Figure 2.13, or a new default EPS Bearer activation as shown in Figure 2.16. The selection of the gateway that leads to the desired PDN is done by providing the Access Point Names (APNs) during the activation sequence. We will look at the APN in Section 2.5.2. Figure 2.22 depicts the multiple PDN concept.

2.5.2   Access Point Name

As described in Section 2.5.1, the selection as to which external network the UE will be connected is based on the APN [45]. An APN is an identifier, which points to an external network or service. The APN has two parts – the Network Identifier, and the Operator Identifier.

The Network Identifier uniquely identifies the network to which a GGSN or PGW is connected, or the service it provides. The Operator Identifier defines in which operator's network that GGSN or PGW is located. The overall structure of the APN follows the Domain Name System (DNS) format [46–48]. However, the operator identifier also has its own specified format. The format is `mnc<MNC>.mcc<MCC>.gprs`. Where the

MNC is the three-digit Mobile Network Code, and the MCC is the three-digit Mobile Country Code. Here are a couple of examples of what an APN looks like:

```
province1.mnc012.mcc345.gprs
ggsn-cluster-A.provinceB.mnc012.mcc345.gprs
internet.apn.epc.mnc015.mcc234.3gppnetwork.org
```

The DNS names resulting from the APN name format are used in the network inside the network operator's infrastructure, and in the roaming interface between the network operators. The operator identifier is not usually visible to the user. However, sometimes the user may have to manually configure the network identifier part of the APN to their connection settings on a UE.

An observant reader would notice a strange thing in the domain name that results from the APN: the Top Level Domain (TLD) is .gprs. We discussed the DNS in Section 1.6, and how the DNS is managed. However, .gprs has not been applied for through the Internet Corporation for Assigned Names and Numbers (ICANN) process, and it is managed by GSMA. Basically, the GSMA has created an alternate DNS root for the 3GPP network infrastructure. This .gprs TLD is, however, only used for GPRS networks. The .3gppnetwork.org is used in EPS, and eventually will supersede the .gprs TLD. However, it will take quite a long time before all GPRS networks have been replaced with EPS networks – if ever.

The 3GPP DNS naming is described in [45], and the EPS procedures in [49]. The GSMA equivalent DNS document for both naming and procedures is [50], which relies entirely on the 3GPP counterparts.

2.5.3 Traffic Flow Template

The concept of the TFT is relatively simple – it is a filter that identifies what kind of packets are supposed to go down a certain PDP Context or EPS Bearer. There has to be always one PDP Context or EPS Bearer without a TFT associated to it. Hence, there is always a default connection where packets can pass through. On the other hand, the other Secondary PDP Contexts and dedicated EPS Bearers do need a TFT associated with them. Again, this is to make sure that the intended traffic will go through the connection. There is a TFT for downlink and uplink traffic to ensure that the packets go through the right connection regardless of their direction. There are eight different parameters that can be set in a TFT:

- local address [33, 51]
- remote address
- protocol number (IPv4) / next header (IPv6)
- local port range
- remote port range
- IPsec Security Parameters Index (SPI) [52]
- Type of Service (TOS) (IPv4) / Traffic class (IPv6)
- Flow Label (IPv6).

The local address and port refer to the UE's address and port. Consequently, the remote address and port refer to the address of the remote host and its ports.

2.5.4   3GPP Link Model Principles

We have now described how 3GPP networks open, modify, and close sessions, the principles of mobility management, and how user-plane packets are mapped to the open sessions. However, we have not yet touched on how the IP stack sees the 3GPP access. We will next concentrate on the IP stack's view of the 3GPP networks. This section concentrates on the principles on which the 3GPP link model relies. We will look at the details in Section 4.1.

Compatibility with IP and the Internet

The 2G-GPRS was being specified during the 1990s, 3G-GPRS in the early 2000s, and EPS in the later part of the first decade of the 2nd century. However, mobile broadband only really started growing during the specification of the EPS architecture. Hence, it is obvious that during the design and specification time the equipment which is now used on these networks, was not available. Especially, while specifying GPRS, the capabilities of handsets and even computers were much more restricted than the smartphones of today. The phones were special devices with very restricted computing capabilities. In addition, the networks themselves were much more restricted in the terms of bandwidth and latency. Still, at the beginning of the 2000s, the norm for mobile data was circuit switched data with bandwidth between 9.6 kbps and 14.4 kbps. In addition, in the early stages of 3G standardization even WLAN was not widely used, and the mobile phones were strictly supporting only cellular access. Therefore, people who standardized the 3GPP system could not even envision what the future mobile phones would be, and for what the networks would in the end be used.

In the early days of GPRS standardization, IP stack was considered to be rather heavy-weight, and something more suitable for personal computers than for mobile phones. (At the time, personal computer operating systems only had IPv4 support. There was debate whether normal mobile phones could even support IPv6. Interestingly, these days work is ongoing to deploy Internet of Things (IoT) with coin cell battery operated IPv6 enabled smart objects, as we describe in Section 6.5.) Full Internet connectivity was considered to be much too heavy for mobile phones, and there were questions if phones could ever be connected to the Internet the same way the personal computers were. Thus, it is logical that the 3GPP engineers had a temptation to take shortcuts in the specification, and try to optimize the IP transport, and the IP stack itself to make IP more suitable for the mobile communication. When shortcuts are taken, and if protocols are strongly optimized, there is at least a risk that the optimization would have made cellular IP incompatible with wireline IP. This might have made 3GPP networks, at least in the early stages, incompatible with the Internet.

Luckily the 3GPP community had more vision than this. Perhaps the most important principle during the 3GPP IP service standardization has been strict compatibility with standard Internet Protocol. This means that the IP stacks in the UE behave in the standard way, the network expects standard behavior, and the IP packets are transported in the same way over the network as they would in any other network. This is one of the key factors that has made the 3GPP networks successful, and the Internet access service is usable for a range of devices not imaginable at the time of the standardization.

Configuration and Address Management

The IP stack, especially the IPv6 stack, has been designed to work well in an unmanaged environment. The IP stacks try to configure themselves as autonomically as possible. However, the principle of telecommunication networks, and especially mobile telecommunication networks, is that the network should be always in control. There are a multitude of reasons for this principle. The reason for relatively strict network control is better quality control (the network does not do anything unexpected), better scalability (the usage of the network resources for configuration can be minimized), and also control over charging and billing. However, as described earlier, compatibility for standard IP stacks, and the Internet has to be kept as well.

Therefore, the configuration and address allocation in the 3GPP networks is designed to be controlled by the network and the network operator. This means that the different configuration information, including UE's DNS server configuration and the IP addresses, are configured by the network. In IPv4 this was easy, as IPv4 was anyways lacking self-configuration mechanisms that later became available for IPv6. Without other configuration schemes in place, the network configures one single IPv4 address to the UE per Primary PDP Context, or default EPS Bearer. However, for IPv6 the network has to give more freedom and assign a range of addresses to the UE. We will describe this in more detail in Section 4.4. However, despite this, the network always knows which UE has what IP address, and therefore can always map the IP address to a subscriber.

Mobility Management and Network Infrastructure

As discussed in Section 2.4, the purpose of mobility management is to make sure that the UE stays connected to the network and maintains service even when the UE is moving. The IP stacks, and IP networks, do not understand mobility very well. IP assumes that network links are relatively stable, and if they fail or change, the reasons are usually link failure, network detachment (as in cable detachment), or eventual network attachment to a different network. Therefore, the 3GPP technology almost completely hides mobility from the IP stack. This means that the IP address is kept the same during a session, and therefore, the UE is connected to the same IP network, and stays in the same topological location within it. Hence, the 3GPP link emulates a wired access link. 3GPP mobility management does not only hide mobility within one 3GPP access technology, but also during handovers between the different 3GPP radio technologies.

In addition the mobility being invisible for the IP stack, the radio network and the whole network infrastructure is invisible for the UE's IP stack. (This is true for GPRS and EPS in GTP mode. However, the PMIP variant has a small deviation from this, which is described in more detail when describing bearers in Section 4.1.1.) A PDP Context, or an EPS Bearer seems like a single-hop link. In addition, the UE is alone on that link as the PDP Contexts and bearers of one UE are separated from other UE's PDP Contexts and bearers. The mobile gateway, GGSN or PGW, is the first-hop router for the UE. Practically this means that the IP packet does not see all the infrastructure and network elements during transit over the network, and therefore cannot address the network elements in the infrastructure either. Hence, also communication between the UEs in the 3GPP network has to always go through the network, and the GGSN or PGW. Note that within the scope of this book we do not consider or cover Local IP Access (LIPA) [29] or Selective IP Traffic Offload (SIPTO) [29] 3GPP system architectural extensions that may change the earlier stated communication assertion.

Special Link Characteristics

Although, as we explained above, the 3GPP access technologies emulates wireline IP-connectivity, it is still really a wireless system with different characteristics. Whereas fixed network links provide stable conditions with relatively stable bandwidth and latency characteristics, the mobile networks characteristics can significantly change due to mobility, load, and changing radio conditions.

In mobile networks, the radio conditions can change drastically very quickly. This can even cause considerable packet loss for extended periods, and as quickly as problems come, they can also go. In addition, due to radio conditions or intra-system handovers between the 3GPP network technologies, the bandwidth and latency characteristics can change in either direction. Hence, the bandwidth available for a UE can suddenly either decrease or increase. In addition, the network latency can vary greatly. Where in the best case the network the Round Trip Time (RTT) can be few milliseconds, in bad conditions RTT of multiple seconds is not uncommon.

An interesting characteristic of the 3GPP network is the network Maximum Transmission Unit (MTU) size. MTU is the maximum size of a single packet that the network can carry without the packet having to be fragmented that is, split into two or more IP packets. The standardized MTU of 3GPP networks is 1500 bytes. This is to be similar with other networking technologies, such as the Ethernet [53], as they use the same MTU size. As a matter of fact, 1500 bytes is the default MTU for most hosts.

However, the 3GPP mobility management is provided by tunneling in the network. Let's consider this for a moment. The tunneling providing the transport for the PDP Contexts and EPS Bearers creates overhead. In GTP, it is the size of the IP, the UDP, and the GTP headers. At minimum, with IPv4 based transport this is 20 bytes for the IPv4 header, 8 bytes of the UDP header, and another 8 bytes minimum for the GTP. Hence, we are talking about a minimum of 36 bytes of overhead. If we now add a user-plane packet with an MTU of 1500 bytes we have $36 + 1500 = 1536$ bytes. The probable MTU of the transmission technology that connects the mobile network elements to each other is 1500 bytes. The technology mostly used, at least in the core network, is Ethernet. Now we see the dilemma: the transmission path's MTU is 1500 bytes, but the packet with all its tunneling overhead is 1536 bytes. Either it will not fit, or it has to be fragmented into two separate packets. Dropping the packet in the network is of course not very nice, but fragmentation and reassembly of packets is resource intensive, and increases latency and network load. Thus, despite the specified 1500 byte MTU, many commercial UEs use smaller MTU sizes by default. The network can also communicate more optimal MTUs to UEs, as we will see in Section 4.2.2.

2.5.5 Multiple Packet Data Network Connections

3GPP networks support multiple PDN Connections – Primary PDP Contexts, or default EPS Bearers, hence, making the UE a multihomed host. However, 3GPP does not provide a mechanism to convey the routing information or policy. Thus, the routing policy has to be preconfigured in the host, or it is a bit random through which session the packets end up going. The routing policy can be application specific. For instance, some mobile operating systems allow application specific routing policy to be configured. Practically this means that the user interface asks which connection should be used, or the information is statically configured in the application settings. Some mobile phone systems may have inbuilt routing policy for specific applications such as MMS or IMS.

The operators use multiple PDN Connections for various reasons. One reason is to differentiate billing and different services. For instance, for MMS the subscriber might pay per message, but not for data transfer, but for other services there is some kind of volume based charging. Using a different PDN Connection for the MMS traffic makes sure that data packets related to MMS are not charged, but the network is not connected to the Internet. The Internet connection is on another PDN Connection, and so it is accounted for on data volume. In addition, operators may want to segregate important traffic from less important traffic. For this reason, the telephony traffic for IMS might have its own PDN Connection separated from the Internet traffic.

We do not want to judge if this approach is the best possible way to achieve the target, but it is not necessarily very IP friendly either. As the modern smartphones are now based on general purpose operating systems, they have a bit of an issue with the multiple PDN Connections, and customization work is needed to support this concept. We believe that improvements are possible, as we will later discuss in Section 6.1.

2.6   User Equipment

In the traditional 3GPP architecture, everything that is on the end user side is generally referred as a UE (or as an MS in the case of GPRS) as shown in Figure 2.7. The UE then consists of two main domains called the Universal Integrated Circuit Card (UICC) and the Mobile Equipment (ME), of which the ME can be further split down into the Mobile Terminal (MT) and the Terminal Equipment (TE) [54, 55]. The actual shapes and sizes for implementations of the abstract concept of UE, including those of MT and TE, can however vary significantly. In this section we will look at how the traditional UE model works, and how that has evolved into various Split-UE implementations.

The first thing to understand is how responsibilities between MT and TE are split. The MT is the part that is always 3GPP aware, and its responsibilities include the following [54]:

1. Radio transmission termination and channel management.
2. Speech encoding and decoding.
3. Error protection for all information sent over the radio.
4. Flow control of signaling and user data.
5. Rate adaptation of user data, and data formatting for the transmission.
6. Support for multiple terminals.
7. Mobility management (of the part covered by 3GPP, not the IP layer mobility).

The responsibilities left for the TE are then everything else that is protocol-wise 'above' the MT's responsibilities: starting from the IP stack itself. So to be clear, and for the following sections, the first significant part of the protocol stack that is on the TE is the IP stack, including the IPv6 that this book is about.

2.6.1   Traditional 3GPP UE Model

For a long time a UE, and in particular a 2G MS, was simply a single device consisting of an MT and an ME tightly integrated into each other, as illustrated in Figure 2.23, case 'A'. This was and is the typical approach in data-enabled mobile handsets. In these type of device the interface between MT and ME can be slightly 'looser' and utilize standardized channels and ATtention (AT) command sets – very much like traditional dial-up modems did [56]. The interface can also be 'tighter' and fully closed and/or using proprietary methods. An important property of case 'A' is that the TE is typically very aware of the MT's characteristics.

Some UEs have evolved a bit from the original model and have switched to using a more distinct interface between TE and MT, for example, presenting TE to MT as an Ethernet interface (see more about Ethernet in Section 3.6.2). This approach is illustrated in Figure 2.23 case 'B' with interface I_i (these interfaces have been given illustrative names that are not used elsewhere). What is noteworthy in case 'B' is that it may be easier to support the MT in the operating system running on the TE. However, this approach causes TE to lose some visibility into MT, unless quite a bit of additional control software is used to bind the MT and the TE together. In this approach, differences in link models used between the TE and the MT and between the MT and the 3GPP network can cause issues, as we will further discuss in Section 4.9.1. However, the model of case 'B' can be helpful if the same type of interface is also used for interfacing external TEs.

2.6.2   Split-UE

With the emergence of fast data connections on 3GPP accesses, it has become popular to share the cellular data connection with other devices via some local connectivity technology. We call this activity *tethering* throughout this book. The local connectivity method is typically WLAN or Universal Serial Bus (USB), but it can also be infrared, *Bluetooth*, low-power radio technology, or anything human imagination can produce. With Figure 2.23's case 'C' interfaces I_{e1} and I_{e2} we illustrate the tethering scenario with a

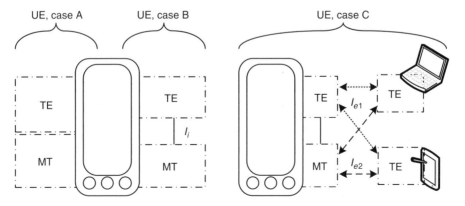

Figure 2.23   UE with tight integration (A), loose integration (B), and Split-UE (C) – in the split case the tethering is performed by the MT or the TE on the mobile phone.

mobile phone that shares its cellular connectivity over WLAN to a laptop and a tablet device. As shown in the figure, the MT function resides in the device providing access to the 3GPP network. The TE function, on the other hand, not only resides inside the handset, but logically also on the other devices, such as on the illustrated laptop and tablet. A significant factor in the Split-UE case is that the other devices are not necessarily aware of the fact that they are utilizing 3GPP access and having the role of 'TE'. As in the case 'B', some issues may arise due to differences in link types between TEs and the MT and from the MT to the 3GPP network. As shown in the figure, the whole set of handset and tethered devices is considered as an abstract UE from the cellular network point of view – the network serves the set of devices as if it were one traditional UE.

IP address wise the handset is often, but not necessarily, performing IPv4 NAT functions and some variation of IPv6 Neighbor Discovery Proxy (see more in Section 3.4.10). When the tethered devices' data passes through handset's IP stack pointed by interface I_{e1}, the tethering can be considered to be done by the TE on a handset and not by the MT (indicated by interface I_{e2}). When the tethering is performed via I_{e2} the interface is often implemented in a dial-up fashion; that is, the data channel may present serial communications and the AT-command set may be in use. As is often the case with tricks doable inside 'black boxes', several different implementation variations of the tethering feature are possible.

The way of sharing connectivity described above and specified in pre-Release-10 3GPP standards does not require, or provide, any explicit support from the 3GPP network. Nevertheless, sometimes the tethering is (dis)allowed via administrative means: by policing the data users are sending, placing data transmission gaps, or even delivering to users' handsets that do not allow tethering without additional subscription to the service.

The 3GPP in the Release-10 version has defined advanced and explicit features for IPv6 prefix delegation (see Section 4.4.6). When the prefix delegation is in use, the 3GPP network also 'acknowledges' the fact that the UE is providing routing services, and thus connectivity services, for other devices. Calling that setup Split-UE no longer makes much sense (as the UE is no longer 'split'), but an official term for the UE that supports Dynamic Host Configuration Protocol version 6 (DHCPv6) Requesting Router (RR) role has not yet been decided.

2.7 Subscription Management Databases and Other Backend Systems

We have now examined almost all of the network elements in the GPRS and EPS network architectures. It is time for an overview of the different registers in the 3GPP system. These include the HLR and Authentication Center (AuC), the HSS, and the EIR. Finally, we will have a quick look at the other backend systems not shown in the 3GPP architecture pictures, but that are nevertheless vital for a 3GPP network deployment to work.

2.7.1 Home Location Register and Authentication Center

The Home Location Register (HLR) is a GPRS network element, which is in the GPRS architecture upto 3GPP Release-4, when it is replaced by the HSS. However, the HLR is a subset of the HSS, and sometimes, confusingly, also referred to in the specifications also beyond the Release-4. However, it can also be seen as the part of the HSS that is responsible for the data network connectivity services in GPRS and EPS.

The HLR is the database that stores information about the subscriber. It includes information that is relatively static, including security, authentication, authorization, and service provisioning information. In addition, the HLR stores more dynamic information related to, for example, mobility management – for instance, HLR stores information about the SGSN serving the subscriber.

The AuC is basically a subset of the HLR that holds the security information including the keys for mutual authentication, integrity protection, and the ciphering of the radio interface for a given subscriber.

2.7.2   Home Subscriber Server

The Home Subscriber Server (HSS) is a combination of the AuC, HLR, and functionality to support the operator service infrastructure, the IMS. HSS replaced the HLR from the 3GPP Release-5 onwards with the introduction of the IMS. However, as stated above, the specifications and the literature are not always consistent in the use of terms. However this book uses HLR for GPRS, and HSS for EPS, as this is in line with most specifications and literature. Anyway, the difference is just a terminology issue as the HSS is a superset of the HLR, and the difference is only in the IMS functionality. Hence, in the case of GPRS and EPS network service only the HLR functionality applies.

2.7.3   Equipment Identity Register

The Equipment Identity Register (EIR) is a database that stores International Mobile Equipment Identitys (IMEIs) information. IMEI is an identity hard coded into the UE that uniquely and globally identifies the equipment. The EIR contains three lists, white, gray, and black, for IMEI [57]. The white list contains UE that are permitted to use the network, and the black list contains equipment that is barred from the network. The gray list can contain UE that are tracked by the network for some reason.

2.7.4   Other Backend Systems

Other backend systems used in the 3GPP network deployments include Authentication, Authorization and Accounting (AAA) servers, and Dynamic Host Configuration Protocol (DHCP) servers [58, 59] for subscriber and address management. The AAA technologies used are Remote Authentication Dial In User Service (RADIUS) [60] and Diameter [61]. RADIUS is a technology standardized and designed by the IETF originally for dial-up networking, but is currently a very widely used AAA technology in many different network environments. Diameter is a newer IETF technology aimed to supersede RADIUS and which tries to address some of its shortcomings. Alternatively, DHCP can be used for external address management. Section 4.4 describes the usage of the AAA, and the DHCP technologies for address management.

2.8 End-to-end View from the User Equipment to the Internet

We have now learned the different building blocks that make up the 3GPP system. It is time to put this together, and to give an overview of how your packets actually get to the Internet and back. We will look at GPRS and EPS separately as the process of getting online is slightly different.

2.8.1 GPRS

We will first look at how the UE gets connected to the Internet. Then we will show how the packets are transmitted over the network. Finally, we describe how the connection is torn down, and the communication is ended.

Getting Connected

When we turn on a UE such as a mobile phone, the first thing the UE must do is to attach to the network. During the network attachment, the UE authenticates itself to the mobile network, and the mobile network authorizes the network attachment as described in Section 2.3.1. However, at this point the UE does not have an IP address, and it is not connected to the Internet, or to any other IP network for that matter. Therefore, it has to open a PDP Context before it can send or receive packets.

The PDP Context activation procedure is shown in Figure 2.13. The UE can indicate either the APN it wishes to connect to, or leave the APN field empty in the Activate PDP Context message. If the APN is not provided, the network will use a default APN. The APN settings are usually operator dependent. As we want Internet access, the APN network identifier could be for instance `Internet`. The network would then find, based on the APN, where the GGSN is that provides Internet access, and send the Create PDP Context to it. The GGSN would allocate the IP address (IPv4, IPv6, or both, depending on the PDP Type that the UE indicated). The GGSN also provides additional information to configure the IP stack in the UE. This information includes for instance the DNS server information. The UE configures and brings up a network interface in its operating system. (Interface here refers to an IP-interface as described in Chapter 1, not a 3GPP interface, as the connections between 3GPP network elements are called in this chapter. The Internet technical community and the 3GPP community use the same word for two different definitions. It is confusing, we know.)

After the PDP Context is up and the IP stack is configured, the UE is ready and connected to the Internet.

Packet Delivery

After PDP Context activation, the UE can start communicating to the Internet. The IP stack in the UE creates a packet with the above allocated IP address as the source address, and sends it to the GPRS interface. There the GPRS modem sends it first over the radio interface to the BTS or the NodeB. Before sending, the packet might be header compressed. In 2G-GPRS, the packet is decompressed back into its original form in the

SGSN, whereas in 3G-GPRS the packet is put back together in the RNC. The RNC encapsulates the user's packet into a GTP tunnel (the packet is put inside a GTP packet). The tunnel is between this RNC, and the SGSN serving the UE. The SGSN receives the GTP encapsulated packet, and decapsulated the IP packet sent by the UE. The SGSN's context table and the GTP tunnel identifier allows the SGSN to find the upstream GTP-tunnel. The SGSN re-encapsulates the IP packet to the GTP tunnel going to the GGSN where the PDP Context is anchored. The GGSN then liberates the packet and sends it off further into the Internet. The first TTL-field decrease is done at the GGSN. Even though the packet has hopped over multiple network elements, and the connections between them, from the point of view of the IP packet that the UE sent, the packet has just done one hop. The network has been completely transparent to it.

In the downlink direction, the packet destination is set to the IP address allocated to the UE. As the UE's IP address has been allocated from the address pool of the GGSN, the packet travels the networks of the Internet, and arrives at the GGSN. The GGSN has a context table – practically a forwarding table that maps an IP address to the GTP tunnel that makes the PDP Context. The GGSN encapsulates the IP packet going downstream towards the UE, and sends it to the SGSN serving the UE. As for the packets coming upstream, the SGSN decapsulates the packet and in 3G case re-encapsulates the packet into the GTP tunnel, which is between the GGSN and the RNC serving the UE. The RNC de-encapsulates the packet. The packet is then mapped by the RNC to the radio bearers and sent to the UE. Again, all these steps inside the GPRS network are completely transparent to the IP packet. The IP packet then arrives at the UE's IP stack.

Ending the Connection

Modern smartphones activate PDP Contexts automatically. The PDP Context is often activated all the time, unless a preferred alternative is available such as WLAN. Thus, the PDP Context is activated as the phone is turned on, and is deactivated when the mobile phone is turned off. The PDP Context may be deactivated if the phone finds a preferred WLAN network, and re-actives when the connection to the WLAN is lost. However, some mobile phones keep a PDP Context activated only when there is an application that needs the connection to be active.

However, as described in Section 2.3.1, the PDP Context deactivation can be done independently of the network attachment. Thus, the PDP Context can be deactivated, and the UE can stay attached to the network.

The network detachment is usually only done when a phone is switched off, or switched to offline state such as 'flight mode'.

2.8.2 EPS

As with GPRS above, in this section we will go through the life cycle of the network connection. This section will first go through the network attachment with the default bearer establishment, over the IP communication over the network, to the closing of the connection.

Getting Connected

We described in Section 2.3.1 that in the EPS system the default EPS Bearer is activated as a part of the network attachment procedure. Therefore, after the network attachment the UE has a working network connection. Similarly to the GPRS, the IP stack configuration, and IP address allocation is done during the default EPS Bearer establishment. The UE is provided by the IP addresses – one IPv4 address, a range of IPv6 addresses, or both. As in GPRS, the UE indicates which kind of connectivity it wishes to have (IPv4, IPv6, or both), and it may also provide the APN. The MME may adjust the requested IP version and the APN according to the subscription information it obtains from the HSS. The PGW allocates the IP addresses and provides additional configuration information, for instance the DNS server information.

After the UE has configured its IP stack with the address, and additional configuration information provided, the UE is ready to start communicating to the Internet.

Packet Delivery

Perhaps not unexpectedly, the journey of a packet through the EPS system is quite similar to the journey of a packet through GPRS. There is a slight difference between the GTP and PMIP EPS variants due a difference in the link model. However, the difference is explained in detail in Section 4.1.1.

Starting from the UE, the IP packet with the UE's IP address in the source address field starts its journey off by being sent by the UE's LTE modem over the LTE radio interface. As there is no radio controller element in the EPS architecture, the eNodeB is responsible for header compression, and terminating all radio protocols. The eNodeB has a mapping from the radio bearers to the equivalent GTP tunnels. The packet gets encapsulated by the eNodeB and sent to the SGW. The MME is not involved in the user-plane forwarding. Similarly to the SGSN in GPRS, the SGW has a bearer table mapping the tunnels on the S1-U to the equivalent tunnels in the S5/S8-interface. Thus, the SGW decapsulates the packet, and then re-encapsulates the packet. As described previously, in the GTP variant the tunneling between the SGW and the PGW is, as already the name says, using GTP. In the PMIP variant, the tunneling is provided by GRE. The packet is sent of by the PGW to further into the Internet.

In the opposite direction, the packet arrives at the PGW. The UE's address is allocated from the PGW's address space, and therefore the packet is delivered to the PGW by normal IP routing. The PGW looks up the IP packet's destination address in its bearer table, and forwards the packet downstream on the right GTP or GRE tunnel depending on the EPS variant. The packet is then forwarded to the SGW, which is selected based on the downstream GTP tunnel. The SGW decapsulates the packet coming downstream over the S5/S8-interface, and re-encapsulates the packet to the S1-U interface. The packet is transmitted over the S1-U interface to the eNodeB. The eNodeB selects the right radio bearer based on the tunnel identifier of the GTP tunnel on S1-U, and forwards the packet over the radio interface to the UE. The packet in the UE is received by the LTE modem and forwarded to the IP stack. The packet has arrived.

Ending the Connection

The EPS default EPS Bearer is always on. Hence, as long as the UE is connected to the network, it has a valid IP address, and a working connection to the network. The connection is only deactivated at phone power off, or when transferring the phone to an offline state, such as 'flight mode'.

2.9   Chapter Summary

In this chapter, we have examined the packet switched 3GPP network architectures – GPRS and EPS. The purpose was to provide an overview for the reader of the 3GPP network architectures, and the network connectivity services they provide. This should provide an adequate basis of the 3GPP concepts for the reader to understand the rest of the book.

There are two GPRS variants, the 2G-GPRS with GERAN radio interface, and the 3G-GPRS with WCDMA radio interface. Although architectural differences between these two systems exist, the procedures and concepts are very similar.

The EPS network architecture is the newest 3GPP network architecture family member with the new LTE radio interface. 3GPP has specified two variants of the EPS architecture – the GTP, and the PMIP variant. EPS uses a new network architecture with significant changes from the GPRS architecture. However, due to the common pedigree similarities between the architectures also exist.

The main difference in the philosophy between the GPRS and EPS is that the EPS does not have the CS mode and the default bearer establishment is part of the EPS network attachment procedure, whereas in GPRS the PDP Context has to be activated separately.

Both architectures support primary network connections – Primary PDP Context in GPRS, and default EPS Bearer in EPS. In addition, both architectures support secondary connections, for instance for QoS reasons – Secondary PDP Context in GPRS, and dedicated EPS Bearer in EPS.

APNs are used for selecting the PDN where the sessions (PDP Context, or EPS Bearer) are connected. The TFT is used to decide into which Secondary PDP Context or Dedicated EPS Bearer the traffic is directed.

2.10   Suggested Reading

- WCDMA for UMTS: HSPA Evolution and LTE, H. Holma and A. Toskala.
- LTE for UMTS: Evolution to LTE-Advanced, H. Holma and A. Toskala
- Voice over LTE (VoLTE), M. Poikselkä, H. Holma, J. Hongisto and J. Kallio
- SAE and the Evolved Packet Core: Driving the Mobile Broadband Revolution, M. Olsson, S. Sultana, S. Rommer, L. Frid, C. Mulligan
- The IMS: IP Multimedia Concepts and Services, M. Poikselkä and G. Mayer
- The 3G IP Multimedia Subsystem (IMS): Merging the Internet and the Cellular Worlds, G. Camarillo and M-A. Garcia-Martin

References

1. 3GPP. 3rd Generation Partnership Project, http://www.3gpp.org/.
2. ARIB. Association of Radio Industries and Businesses, http://www.arib.or.jp/english/.
3. ATIS. Alliance for Telecommunications Industry Solutions, http://www.atis.org/.
4. CCSA. China Communications Standards Association, http://www.ccsa.org.cn/english/.
5. ETSI. European Telecommunications Standards Institute, http://www.etsi.org/.
6. TTA. Telecommunications Technology Association, http://www.tta.or.kr/English/.
7. TTC. Telecommunication Technology Committee, http://www.ttc.or.jp/e/.
8. IETF. Internet Engineering Task Force, http://www.ietf.org/.
9. 3GPP. IP Multimedia Subsystem (IMS); Stage 2. TS 23.228, 3rd Generation Partnership Project (3GPP), September 2010.
10. 3GPP. High Speed Downlink Packet Access (HSDPA); Overall description; Stage 2. TS 25.308, 3rd Generation Partnership Project (3GPP), December 2011.
11. 3GPP. Enhanced uplink; Overall description; Stage 2. TS 25.319, 3rd Generation Partnership Project (3GPP), December 2011.
12. Waitzman, D. *Standard for the transmission of IP datagrams on avian carriers*. RFC 1149, Internet Engineering Task Force, April 1990.
13. GSMA. GSM Association, http://gsm.org/.
14. OMA. Open Mobile Alliance, http://www.openmobilealliance.org/.
15. 3GPP. Multimedia Messaging Service (MMS); Functional description; Stage 2. TS 23.140, 3rd Generation Partnership Project (3GPP), March 2000.
16. ETSI. European digital cellular telecommunication system (phase 1); Network Architecture. GSM 03.02, European Telecommunications Standards Institute (ETSI), February 1992.
17. ITU-T. SERIES X: DATA NETWORKS AND OPEN SYSTEM COMMUNICATION Public data networks – Interfaces Interface between Data Terminal Equipment (DTE) and Data Circuit-terminating Equipment (DCE) for terminals operating in the packet mode and connected to public data networks by dedicated circuit. Recommendation X.25, International Telecom Union – Telecommunication Standardization Sector (ITU-T), October 1996.
18. 3GPP. Mobile Station (MS) – Serving GPRS Support Node (SGSN); Subnetwork Dependent Convergence Protocol (SNDCP). TS 44.065, 3rd Generation Partnership Project (3GPP), December 2009.
19. Jacobson, V. *Compressing TCP/IP Headers for Low-Speed Serial Links*. RFC 1144, Internet Engineering Task Force, February 1990.
20. Degermark, M., Nordgren, B., and Pink, S. *IP Header Compression*. RFC 2507, Internet Engineering Task Force, February 1999.
21. Bormann, C., Burmeister, C., Degermark, M., Fukushima, H., Hannu, H., Jonsson, L-E., Hakenberg, R., Koren, T., Le, K., Liu, Z., Martensson, A., Miyazaki, A., Svanbro, K., Wiebke, T., Yoshimura, T., and Zheng, H. *RObust Header Compression (ROHC): Framework and four profiles: RTP, UDP, ESP, and uncompressed*. RFC 3095, Internet Engineering Task Force, July 2001.
22. 3GPP. General Packet Radio Service (GPRS); Mobile Station (MS) – Base Station System (BSS) interface; Radio Link Control/Medium Access Control (RLC/MAC) protocol. TS 44.060, 3rd Generation Partnership Project (3GPP), March 2012.
23. 3GPP. Mobile Station – Serving GPRS Support Node (MS-SGSN); Logical Link Control (LLC) Layer Specification. TS 44.064, 3rd Generation Partnership Project (3GPP), December 2011.
24. 3GPP. General Packet Radio Service (GPRS); Base Station System (BSS) – Serving GPRS Support Node (SGSN); BSS GPRS protocol (BSSGP). TS 48.018, 3rd Generation Partnership Project (3GPP), March 2012.
25. 3GPP. General Packet Radio Service (GPRS); Base Station System (BSS) – Serving GPRS Support Node (SGSN) interface; Network service. TS 48.016, 3rd Generation Partnership Project (3GPP), December 2009.
26. 3GPP. General Packet Radio Service (GPRS); Service description; Stage 2. TS 23.060, 3rd Generation Partnership Project (3GPP), March 2012.
27. ITU-T. SPECIFICATIONS OF SIGNALLING SYSTEM No. 7. Recommendation Q.700, International Telecom Union – Telecommunication Standardization Sector (ITU-T), March 1993.

28. ITU-T. Series I: Integrated Services Digital Network General structure – General description of asynchronous transfer mode. Recommendation I.150, International Telecom Union – Telecommunication Standardization Sector (ITU-T), February 1999.

29. 3GPP. General Packet Radio Service (GPRS) enhancements for Evolved Universal Terrestrial Radio Access Network (E-UTRAN) access. TS 23.401, 3rd Generation Partnership Project (3GPP), March 2012.

30. 3GPP. Architecture enhancements for non-3GPP accesses. TS 23.402, 3rd Generation Partnership Project (3GPP), March 2012.

31. Simpson, W. *The Point-to-Point Protocol (PPP)*. RFC 1548, Internet Engineering Task Force, December 1993.

32. 3GPP. Technical realization of the Short Message Service (SMS). TS 23.040, 3rd Generation Partnership Project (3GPP), September 2010.

33. Postel, J. *Internet Protocol*. RFC 0791, Internet Engineering Task Force, September 1981.

34. Postel, J. *Transmission Control Protocol*. RFC 0793, Internet Engineering Task Force, September 1981.

35. 3GPP. Radio Link Control (RLC) protocol specification. TS 25.322, 3rd Generation Partnership Project (3GPP), June 2011.

36. 3GPP. Evolved Universal Terrestrial Radio Access (E-UTRA); Packet Data Convergence Protocol (PDCP) specification. TS 36.323, 3rd Generation Partnership Project (3GPP), January 2010.

37. Gundavelli, S., Leung, K., Devarapalli, V., Chowdhury, K., and Patil, B. *Proxy Mobile IPv6*. RFC 5213, Internet Engineering Task Force, August 2008.

38. Farinacci, D., Li, T., Hanks, S., Meyer, D., and Traina, P. *Generic Routing Encapsulation (GRE)*. RFC 2784, Internet Engineering Task Force, March 2000.

39. Muhanna, A., Khalil, M., Gundavelli, S., and Leung, K. *Generic Routing Encapsulation (GRE) Key Option for Proxy Mobile IPv6*. RFC 5845, Internet Engineering Task Force, June 2010.

40. 3GPP2. 3rd Generation Partnership Project2 (3GPP2), http://www.3gpp2.org/.

41. Forum, W. WIMAX Forum, http://www.wimaxforum.org/.

42. Muhanna, A., Khalil, M., Gundavelli, S., Chowdhury, K., and egani, P. Y. *Binding Revocation for IPv6 Mobility*. RFC 5846, Internet Engineering Task Force, June 2010.

43. Devarapalli, V., Koodli, R., Lim, H., Kant, N., Krishnan, S., and Laganier, J. *Heartbeat Mechanism for Proxy Mobile IPv6*. RFC 5847, Internet Engineering Task Force, June 2010.

44. Soliman, H. *Mobile IPv6 Support for Dual Stack Hosts and Routers*. RFC 5555, Internet Engineering Task Force, June 2009.

45. 3GPP. Numbering, addressing and identification. TS 23.003, 3rd Generation Partnership Project (3GPP), March 2012.

46. Mockapetris, P. V. *Domain names – implementation and specification*. RFC 1035, Internet Engineering Task Force, November 1987.

47. Braden, R. *Requirements for Internet Hosts – Application and Support*. RFC 1123, Internet Engineering Task Force, October 1989.

48. Elz, R. and Bush, R. *Clarifications to the DNS Specification*. RFC 2181, Internet Engineering Task Force, July 1997.

49. 3GPP. Domain Name System Procedures; Stage 3. TS 29.303, 3rd Generation Partnership Project (3GPP), June 2011.

50. GSMA. GSMA PRD IR.40 'DNS/ENUM Guidelines for Service Providers and GRX/IPX Providers'. PRD IR.40 7.0, GSM Association (GSMA), May 2012.

51. Deering, S. and Hinden, R. *Internet Protocol, Version 6 (IPv6) Specification*. RFC 2460, Internet Engineering Task Force, December 1998.

52. Kent, S. and Seo, K. *Security Architecture for the Internet Protocol*. RFC 4301, Internet Engineering Task Force, December 2005.

53. Society, I.C. Part 3: IEEE Standard for Information technology-Specific requirements – Part 3: Carrier Sense Multiple Access with Collision Detection (CSMA/CD) Access Method and Physical Layer Specifications. IEEE Standard for Information Technology 802, Institute of Electrical and Electronics Engineers Standards Association (IEEE-SA).

54. 3GPP. GSM – UMTS Public Land Mobile Network (PLMN) Access Reference Configuration. TS 24.002, 3rd Generation Partnership Project (3GPP), December 2009.

55. 3GPP. Network architecture. TS 23.002, 3rd Generation Partnership Project (3GPP), September 2011.

56. 3GPP. AT command set for User Equipment (UE). TS 27.007, 3rd Generation Partnership Project (3GPP), December 2011.

57. 3GPP. International Mobile station Equipment Identities (IMEI). TS 22.016, 3rd Generation Partnership Project (3GPP), April 2010.

58. Droms, R. *Dynamic Host Configuration Protocol*. RFC 2131, Internet Engineering Task Force, March 1997.

59. Droms, R., Bound, J., Volz, B., Lemon, T., Perkins, C., and Carney, M. *Dynamic Host Configuration Protocol for IPv6 (DHCPv6)*. RFC 3315, Internet Engineering Task Force, July 2003.

60. Rigney, C., Willens, S., Rubens, A., and Simpson, W. *Remote Authentication Dial In User Service (RADIUS)*. RFC 2865, Internet Engineering Task Force, June 2000.

61. Calhoun, P., Loughney, J., Guttman, E., Zorn, G., and Arkko, J. *Diameter Base Protocol*. RFC 3588, Internet Engineering Task Force, September 2003.

62. Postel, J. *User Datagram Protocol*. RFC 0768, Internet Engineering Task Force, August 1980.

3

Introduction to IPv6

The 128-bit address space, which provides approximately 3.4×10^{38} unique addresses, is the main reason for Internet Protocol version 6 (IPv6)'s existence and why it is being deployed. However, the next generation Internet Protocol (IP) is much more than the enormous address space it provides. In this section we will introduce the core features of IPv6, as well as looking at the most relevant additional features, especially from 3rd Generation Partnership Project (3GPP) networks' point of view. These features include, for example, IPv6 packet header structures, address autoconfiguration mechanisms, network and host based IPv6 mobility, different link models, and so forth. In addition, at the end of the chapter, we provide detailed examples related to IPv6 address configuration, Domain Name System (DNS) use, and Transport Control Protocol (TCP) session establishment.

IPv6 is described in over a hundred Internet Engineering Task Force (IETF) RFCs, of which a small minority are essential for an IPv6 enabled node's interoperability with other nodes. Most of the RFCs are optional and need only to be supported by certain kinds of nodes, or in certain deployment scenarios. A node can be a host, a router, or in some cases both. Whether the node is a router or a host is perhaps the biggest single factor that defines what kind of protocol features the node needs. IETF has worked on and published the 'IPv6 Node Requirements' document that lists the main requirements for the IPv6 enabled nodes – be they hosts or routers [1]. IETF RFC 6204 defines further requirements for customer edge routers (but does not cover core Internet routers) [2]. Also other organizations have defined sets of required RFCs, for example, RIPE has published 'Requirements for IPv6 in ICT Equipment', RIPE-554 available at *http://www.ripe.net/ripe/docs/current-ripe-documents/ripe-554*. In the end, despite the requirements list or profiles, each individual implementation needs to choose the exact set of protocol features deemed necessary for the use cases of each implementation. IETF has also documented the basic IPv6 components needed on 3GPP accesses in RFC 3316 [3], which is rather old but is being updated. In this book, we list 3GPP User Equipment (UE) specific requirements in more detail in Section 4.7.

When IPv6 was designed, it was planned to be deployed in parallel with its predecessor Internet Protocol version 4 (IPv4) and therefore is not directly compatible with IPv4. A set of tools has been developed to provide interoperability between IPv4 and IPv6, which

Deploying IPv6 in 3GPP Networks: Evolving Mobile Broadband from 2G to LTE and Beyond, First Edition.
Jouni Korhonen, Teemu Savolainen and Jonne Soininen.
© 2013 John Wiley & Sons, Ltd. Published 2013 by John Wiley & Sons, Ltd.

are described in Section 5.3.1 of this book. In this section we will solely focus on the IPv6 core protocol features.

3.1 IPv6 Addressing Architecture

An important benefit of the large IPv6 address space is that it allows versatile addressing architecture. The 'IP Version 6 Addressing Architecture' RFC 4291 [4] and 'IPv6 Scoped Address Architecture' RFC 4007 [5] describe the main architecture, which is then further defined in various documents that make smart use of the bits provided by the IPv6 address. In this section we will provide an overview of the addressing architecture.

3.1.1 IPv6 Address Format

Figure 3.1 shows how a 128-bit IPv6 address is generally formatted, as of today. The first three bits have special meaning for addresses: the 000 indicates no constraint on the Interface IDentifier (IID) structure, 001 indicates currently assigned global prefixes, and 111 is for other addresses such as multicast. As it is visible from the current allocation of the first three bits, the main part of the IPv6 address space remains reserved for the future. This significant reservation allows totally different addressing schemes to be invented, and hence helps in creation of backwards compatible extensions.

An IPv6 address is split into three different sections: a global prefix, a subnet identifier, and an IID. The global prefix refers to a prefix that a network has obtained for its use. These prefixes typically range from 32-bits to 56-bits (/32 to /56), but any other length is of course possible. The subnet identifier includes the bits after global prefix that are used within a network to identify a subnet, which is typically a link. The final bits hold an IID, which identifies a particular network interface (typically a host), inside the subnet.

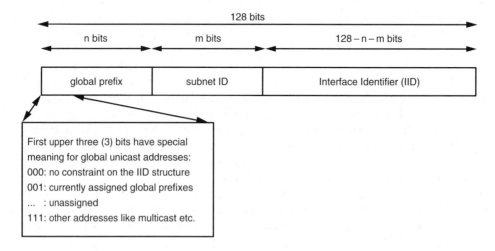

Figure 3.1 Generic IPv6 address architecture.

3.1.2   IPv6 Address Types

An IPv6 address can be of three different types: *unicast, anycast,* or *multicast.* The unicast type of address identifies a single network interface in a single node. The anycast address type identifies the set of interfaces typically on different nodes. An IPv6 packet sent to an anycast address is delivered by the routing system to the nearest instance of an anycast address. The multicast type of address identifies a multitude of interfaces on, usually, a multitude of hosts. An IPv6 packet sent to multicast address is delivered to all receivers of the multicast address (see more in Section 3.2.4). Notably, IPv6 does not have a concept of broadcast like IPv4. The multicast address type can be used to implement all broadcast use cases.

The address of all zeros, : :, is called the unspecified address. The unspecified address is used to indicate a lack of address. This address is never used as a destination. It's only use as a source address takes place during initial source address configuration (see Section 3.4.7).

3.1.3   IPv6 Address Scopes

The concept of address scopes is inbuilt in IPv6, as described by RFC 4007 [5]. The unspecified address, : :, has no scope. Addresses with *link-local scope* can be used only for communication within a single link, even when the link has no routers on it. Every network interface, with few exceptions, automatically creates a link-local address (see more in Section 3.5.1). A special case of link-local scope is the loopback address, : :1, which is considered to be an address in virtual interface inside a node, even if the address does not have the link-local prefix. Figure 3.2 illustrates how a link-local address is constructed: the first 10 bits are fe80, the following 54 bits are all zeros, and the last 64 bits contain an IID. The link-local addresses are useful for communications that are intended to stay on a link, and in cases where routers are not present or global addresses are not (yet) configured for a host.

The other major address scope used in IPv6 is *global scope*; addresses of global scope can be used for local or global – off-link – communications. IPv6 used to have the concept

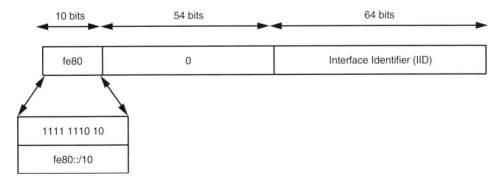

Figure 3.2   IPv6 link-local address format.

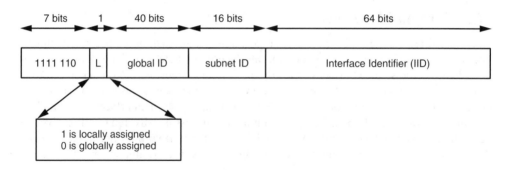

Figure 3.3   Generic IPv6 Unique Local Address format.

of *site-local* scope, but that was deprecated by RFC 3879 [6], and partially replaced by Unique Local Address (ULA) [7]. The ULA addresses can be distinguished by specific prefix of `fc00::/7`, as illustrated in Figure 3.3. At the time of writing, RFC 4193 specifies how ULAs are generated for local assignment, and hence the 'L'-bit is set to '1' (and hence making the value of the first 8 bits `fd`). The locally allocated ULA prefixes use a pseudo random global identifier (global ID). The subnet identifier (subnet ID) is then assigned administratively in a network as required. IPv6 implementations need to be able to differentiate ULA addresses from other global scope addresses for address selection purposes, as is described in Section 3.5.4.

For scopes of multicast addresses, please see Section 3.2.4.

3.1.4   IPv6 Addressing Zones

As stated in RFC 4007 [5]: 'a zone is a connected region of topology of given scope'. For example, for globally scoped addresses there is one zone, Internet, while for link-local scoped addresses there are as many zones as there are links. If an application is communicating to a non-global zone, it may need to indicate the zone identifier to lower layers, as the zone is not determinable from the IPv6 address itself – for example, all link-local addresses start with `fe80::` and that gives no hint about what link, or interface, it belongs to. Please see Section 3.1.9 about textual presentations of zones.

The zone identifiers are not typically communicated between peers, as the identifiers are implementation specific and generally have no meaning for other nodes than the one that has assigned an identifier to a zone.

3.1.5   IPv6 Addresses on Network Interfaces

In IPv6 every address is assigned to an interface. Each interface also usually has a multitude of addresses, of the same and different scopes. In addition to addresses directly allocated to interfaces, nodes are listening to a set of multicast and, possibly, anycast addresses. The exact addresses that nodes listen to depend on the role of the node. For example, every node listens to *All Nodes multicast addresses*, but only routers listen to *All Routers multicast addresses*. Please see more details in RFC4291 [4].

One important aspect where the subnet model defined by IPv6 differs from IPv4 is that the IPv6 prefix – except the link-local prefix when the interface is known – does not automatically mean that nodes with corresponding IPv6 addresses are on the same link [8]. This means that nodes cannot assume that they can directly send packets to an IPv6 address that has matching prefix with nodes' configured addresses. At the same time this means that even if the destination address does not match any of the IPv6 prefixes that the host sees in Router Advertisements, it does not mean that the destination would not be on the same link as the sender. In the case of communications with link-local addresses, a node needs to know a network interface that a destination link-local address is related to, as by definition all interfaces have the same link-local prefix.

3.1.6 Interface Identifier and the Modified EUI-64

The IPv6 address architecture defines that any IPv6 unicast address whose upper three bits are not zeros (i.e. a binary value `000`) must have a 64-bit Interface IDentifier constructed using the *modified EUI-64 format* [4]. Sections 2.5.1 and 2.5.4 of RFC 4291 are very clear on this.

Modified EUI-64 format-based Interface Identifiers may be derived from other unique tokens such as Institute of Electrical and Electronics Engineers (IEEE) 802 48-bit Media Access Control (MAC) identifier [9] or Non-Broadcast Multiple Access (NBMA) interface tokens [10]. See Figure 3.4 for an illustration of what the modified EUI-64 format looks like. In the figure 'C' forms the bits of the assigned *company_id* and 'M' forms the bits of the manufacturer selected extension identifier. To differentiate from the original IEEE managed EUI-64s and to make the system administration slightly easier, the modified EUI-64 has the universal/local 'U' bit inverted, that is the bit 70 of an IPv6 address. When the 'U' is set to 0 (zero), then the EUI-64 is non-global and managed locally. The 'G' is the individual/group bit. When creating interface identifiers using the modified EUI-64 format, care should be taken not to override the 'U' and 'G' bits' predefined meaning described in RFC 4291 and in [9]. Summarizing, when the 'U' is set to 0 (zero) the remaining 63 bits ('G' bit also included) can be of any value. For example, privacy addresses [11] makes use of this.

Figure 3.5 shows how an IEEE 802 48-bit MAC identifier is 'expanded' into a modified EUI-64 format. Two octets, `0xff` and `0xfe`, are inserted into the middle of the modified EUI-64. More specifically the two octets are placed in bits 24 to 39. Note that in the figure the 'U' bit is set to 1 (one), which means that the 48-bit MAC identifier had its corresponding 'U' bit cleared, indicating that the MAC identifier was managed and (supposedly) globally unique.

It is worth noting that for privacy and security reasons a node should never generate IID from sensitive yet unique values, such as from International Mobile Equipment

```
0                   1 1                 3 3                 4 4                 6
0                   5 6                 1 2                 7 8                 3
CCCCCCUGCCCCCCCC CCCCCCCMMMMMMMM MMMMMMMMMMMMMMMM MMMMMMMMMMMMMMMM
```

Figure 3.4 A modified EUI-64 format for IPv6 IID purposes.

```
0                      1 1                3 3                4 4                        6
0                      5 6                1 2                7 8                        3
CCCCCC1GCCCCCCCC CCCCCCCC11111111  11111110MMMMMMMM  MMMMMMMMMMMMMMMM
```

Figure 3.5 A modified EUI-64 derived from an IEEE 802 48-bit MAC identifier.

Identity (IMEI), Mobile Station International Subscriber Directory Number (MSISDN), or International Mobile Subscriber Identity (IMSI).

3.1.7 IPv6 Address Space Allocations

History has proven that no matter the size of the resource, it is always possible to exhaust it with mismanagement and wasteful use. IPv6 provides enormous address space that will be plentiful and will last for a long time, but only if used wisely. At the highest level the Internet Assigned Number (IANA) manages the use of the address pool. As of today, the IPv6 address space is divided into sections listed in Figure 3.6, which also shows a pie chart illustration of the part of the IPv6 address space that is currently reserved for the unicast address space, reserved for the future, and used for various IPv6 uses such as multicast, link-local addressing, ULA, deprecated site-local use, loopback, and so forth.

3.1.8 Special IPv6 Address Formats

The 128 bits used for IPv6 address is long enough to allow for embedding additional information into the address. Perhaps the most popular piece of information that is embedded into an IPv6 address is an IPv4 address. In particular, various IPv6 transition mechanisms

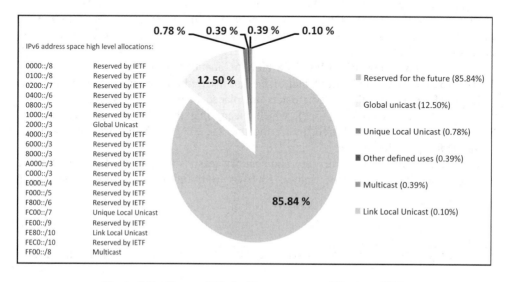

Figure 3.6 Usage of IPv6 address space as of January 2013.

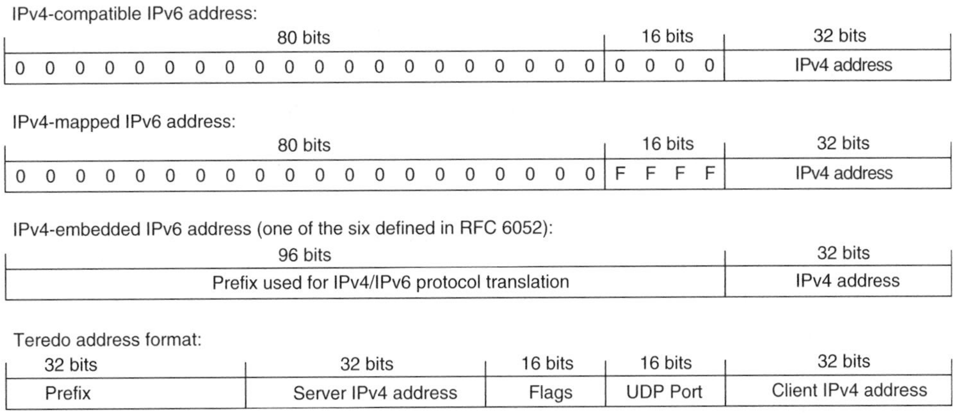

IPv4-compatible IPv6 address:

80 bits		16 bits	32 bits
0 0 0 0 0 0 0 0 0 0 0 0 0 0 0 0 0 0 0 0 0 0		0 0 0 0	IPv4 address

IPv4-mapped IPv6 address:

80 bits		16 bits	32 bits
0 0 0 0 0 0 0 0 0 0 0 0 0 0 0 0 0 0 0 0 0 0		F F F F	IPv4 address

IPv4-embedded IPv6 address (one of the six defined in RFC 6052):

96 bits	32 bits
Prefix used for IPv4/IPv6 protocol translation	IPv4 address

Teredo address format:

32 bits	32 bits	16 bits	16 bits	32 bits
Prefix	Server IPv4 address	Flags	UDP Port	Client IPv4 address

Figure 3.7 Examples of IPv4 addresses embedded within IPv6 addresses.

utilize inclusion of IPv4 addresses within IPv6 addresses (see more about transition mechanisms in Section 5.3, and in particular IPv4-Embedded IPv6 address format in Figure 5.3). Examples of such address uses include, but by no means are limited to, the following examples and the illustration in Figure 3.7:

IPv4-compatible IPv6 address: This deprecated address format was designed to help in IPv6 transition.

IPv4-mapped IPv6 address: This address format is to represent IPv4 addresses as IPv6 addresses. The format is used, for example, to simplify Application Programming Interfaces (APIs) by allowing both address families to be presented by the same fields, parameters, and variables.

IPv4-embedded IPv6 address: Six IPv4-embedded address formats have been defined for IPv4/IPv6 Network Address Translation (NAT64) use in RFC 6052 [13]. Figure 3.7 illustrates the simplest of these formats. For more details of possible NAT64 address formats please refer to Section 5.3.3.

Teredo address: The Teredo address format illustrates well how lots of information can be encoded in an IPv6 address [14]. The Teredo address consist of a 32-bit Teredo service prefix, 32-bit IPv4 address of a Teredo server, flag-bits for additional information, 16 bits for the obfuscated User Datagram Protocol (UDP) port of the Teredo service on a client, and 32 bits for the obfuscated IPv4 address of a client.

IP Addresses for Documentation Purposes

A special unicast IPv6 prefix of `2001:db8::/32` has been reserved for IPv6 documentation purposes in 'IPv6 Address Prefix Reserved for Documentation' RFC 3849 [15]. Similarly, special IPv4 subnets of `192.0.2.0/24`, `198.51.100.0/24`, and `203.0.113.0/24` have been defined for IPv4 unicast address documentation purposes in 'IPv4 Address Blocks Reserved for Documentation' RFC 5737 [16]. For documentation requiring multicast addresses, RFC 6676 'Multicast Addresses for

Documentation' reserves address blocks for both IPv6 and IPv4 purposes [17]. The multicast address block for use with IPv6 any-source multicasting documentation is ff0x::db8:0:0/96 (the 'x' standing for scope, see Section 3.2.4). For source-specific multicast cases the addresses allocated for unicast IPv6 documentation purposes are used.

In this book we will be using the addresses reserved for documentation in all cases where we are not either referring to IP addresses assigned by IANA for specific purposes, or looking at real world IP packet capture examples.

The addresses that are allocated for documentation purposes should never be used in real implementations. Furthermore, IP stacks and networks can and should filter traffic possibly sent to or from the addresses reserved for documentation. Even for experimentation purposes it is better to use ULA or private IPv4 addresses.

3.1.9   Textual Presentations of IPv6 Addresses

As the 128-bit IPv6 addresses in their standard hexadecimal textual presentation form may be challenging to handle, e.g. 2001:0db8:0123:4567:890a:bcde:f000:0000, a set of recommendations for textual presentation has been defined in RFC 4291 [4] and RFC 5952 [18]. The uniform textual representations are also needed to simplify machine-based IPv6 textual address parsing functions, such as those that work with log files.

The preferred format for presenting an IPv6 is x:x:x:x:x:x:x:x, where each x represents a 16-bit value in hexadecimal format using lowercase characters. The 16 bit value of each x is presented without leading zeros. Furthermore, in the case of the address having 32 or more consecutive zero bits, the zero bit sequence can be abbreviated with a syntax ::. In the case of an address having multiple series of zeros, the :: must be used where it makes the most difference (compresses the address the most).

Below are examples of legitimate address presentations:

```
2001:db8::a
2001:db8::a:0:0
::1
```

Here we list a few examples of address presentations that are not legitimate according to RFC 5952, even if they are otherwise unambiguous:

2001:0db8::1 – leading zeroes must not be shown

2001:db8:0:0:0:a:: – the compression syntax of :: should be used to compress the longest sequence of 16-bit fields of zeros

2001:DB::A – uppercase letters should not be used

A special textual addressing format has been defined for IPv6 addresses that are carrying IPv4 address in the lowest 32 bits. The address format for such addresses is x:x:x:x:x:x:d.d.d.d, where d's are presenting an IPv4 address [4]. An example address could be: 64:ff9b::192.0.2.1 [13].

When an IPv6 address needs to be shown with a port number, brackets '[' and ']' should be used to make the address unambiguous. Without brackets it would be impossible to determine whether the last digit is a port number or part of the IPv6 address, as illustrated here: `2001:db8::a:80`. A correct use of brackets makes the distinction clear: `[2001:db8::a]:80`.

Textual Presentation of IPv6 Prefix

Each IPv6 address logically consists of a prefix and an IID, as described in Section 3.1. The prefix length is presented in the notation familiar to the IPv4 Classless Inter-Domain Routing (CIDR) system [19], which means the prefix length is appended to the end of an IPv6 address using a slash-character. For example, if the length of the prefix is 64 bits, the address is presented as `2001:db8::a/64`.

Zone Identifiers

On some occasions, it is necessary to include a zone identifier into the Uniform Resource Identifier (URI) or Uniform Resource Locator (URL). This is needed, for example, if a link-local IPv6 address literal needs to be used by the web browser, and a node has multiple interfaces in use. Due to link-local scope of link-local addresses, it is impossible to determine the network interface that a link-local address is meant to be using. To solve this problem, IETF has defined the syntax for presenting a zone identifier in RFC 4007 [5] and is currently improving the syntax to be more exact [20]. In URIs the defined % symbol for presenting the zone identifier needs to be "escaped", and hence syntax containing an escaped %, that is %25, is needed. The following examples illustrate valid IPv6 address and URL formats for presenting a zone identifier *eth0*, and its use:

```
2001:db8::42%eth0
2001:db8::42%25eth0
http://[2001:db8::42%eth0]/
http://[2001:db8::42%25eth0]/
```

The exact values for zone identifiers are implementation specific and are different in each operating system.

3.2 IPv6 Packet Header Structure and Extensibility

The IPv6 packet header shown in Figure 3.8 was designed to be as compact as possible to minimize increased overhead of IPv6 addresses that take four times the space of IPv4 addresses. The first 8 bytes of the IPv6 header include fields for Version information, Traffic Class, Flow Label, Payload Length, Next Header type, and Hop Limit. The source and destination IPv6 addresses take 16 bytes each and hence the length of the IPv6 header is 40 bytes [21].

The 4-bit Version-field always contains decimal six, indicating that the IP version is IPv6. The contents of Traffic Class and Flow Label are described in Section 3.2.1.

People familiar with IPv4 will notice the lack of header length field. In IPv6 the header is always fixed at 40 bytes. The Payload Length field includes the length of any possible IPv6 extension headers, as well as any other payload packet being carried. The 16 bits reserved for the length therefore limits IPv6 packet payload size to 65535 octets.

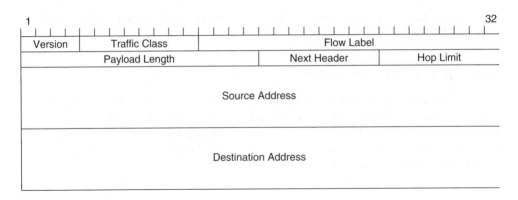

Figure 3.8   IPv6 header.

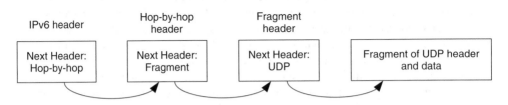

Figure 3.9   Chaining of IPv6 headers.

The Next Header field indicates what header follows the IPv6 header. These headers include transport protocol headers such as TCP and UDP, but also IPv6 extension headers. The Next Header field is also present in all IPv6 extension headers, thus creating a system that allows chaining of headers is illustrated in Figure 3.9.

The Hop Limit field is essentially the same as IPv4's Time To Live, and is decremented by one every time an IPv6 packet is forwarded. If the decrement causes the Hop Limit to reach zero, the packet is dropped and an Internet Control Message Protocol version 6 (ICMPv6) Time Exceeded message is sent to the originator of the packet.

The IPv6 header does not contain any fields for checksum, as the protocol suite design is such that relevant parts of the IPv6 header are included in the transport layer checksum (see more on this in Section 3.10.1).

3.2.1   Traffic Class and Flow Label

RFC 2460 defines Traffic Class and Flow Label fields [21] for distinguishing packets belonging to the same categories, as described below.

Traffic Class

IPv6 header includes a Traffic Class field to allow the distinguishing of different IPv6 packets as different classes [21]. The semantics of the Traffic Class field are defined in 'Definition of the Differentiated Services Field in the IPv4 and IPv6 Headers' RFC

2474 [22] and 'The Addition of Explicit Congestion Notification (ECN) to IP' RFC 3168 [23]. These documents define that the six leftmost bits of the Traffic Class field indicate Differentiated Services Code Point (DSCP) and the two rightmost bits are for ECN.

The DSCP bits of the Traffic Class field define Per-Hop Behavior (PHB) that a packet should receive at each node in its path through the Internet. The 6-bit pattern of `000000` is reserved for signaling default PHB. The IP packet forwarders aware of differentiated services, look at the 6-bit value and based on that can apply different treatment and different PHB for each packet. Hosts may be configured to mark selected packets they create with non-default DSCP values. However, the DSCP field values may be rewritten by forwarders in the Internet, as packets pass through different network boundaries where differentiated services are utilized. Due to intermediate nodes possibly modifying the DSCP field value, the field is not usable for end-to-end communications. A number of RFCs exist that define meanings for the DSCP bits. These include RFC 2474 that defines Class Selector Codepoints [22], RFC 2597 that defines DSCPs for assured forwarding [24] and expedited forwarding [25], and RFC 5865 for capacity-admitted traffic [26].

The two ECN bits of the Traffic Class field have been defined for four different meanings: '00' to indicate that ECN is not used, '10' and '01' indicate that endpoints are capable of understanding the ECN, and '11' to indicate that congestion is experienced in the network. The end points of a transport protocol that is capable of using ECN will set the ECN bits to either '10' or '01'. When congestion is experienced by a network node, but packets are not (yet) dropped, the node changes the ECN field bits to '11' in order to indicate the packet's receiver about network congestion taking place. The RFC 3168 does not take a stand on when nodes should set bits to '10' or '01'. When a node that learns about network congestion, the receiver of a packet that has ECN bits '11', has to take action at the transport protocol level in order to help decrease traffic volumes on the congested network path.

Flow Label

The Flow Label field was left experimental by RFC 2460 [21], which caused confusion and effectively left the field unused. Most of the implementations simply set the field always to zero, as instructed by RFC 2460 for nodes that do not support the function. The issues related to the field are explained in depth in RFC 6436 'Rationale for Update to the IPv6 Flow Label Specification' [27]. The updated functionality is normatively defined in RFC 6437 'IPv6 Flow Label Specification' [28]. The function that is defined for the Flow Label field is to help networks to perform efficiently various forms of load distribution functions, while helping to ensure that packets belonging to same flow always traverse the same path in order to avoid issues such as packet reordering.

The RFC 6437 specification for the Flow Label field recommends hosts to mark each packet belonging to a specific flow, such as to a transport layer session, with a same value. Furthermore, the hosts should select each value pseudorandomly so that the value for subsequent flows would be difficult to guess. For backwards compatibility reasons the Flow Label value of zero is still supported.

While the Flow Label field is not part of the transport layer checksum calculation, nor covered by Internet Protocol security (IPsec) Authentication Header (AH), the specification forbids networks rewriting the field except in two cases: host has set the value originally to zero, or value rewrite is required for security reasons described in depth in RFC 6437 [28]. In any case, a packet receiver cannot assume that the field value has remained unchanged in its path through the Internet. Therefore, applications should not assume that the Flow Label field can be used for end-to-end signaling purposes.

3.2.2   IPv6 Extension Headers

As space in the core IPv6 header is precious, features that did not have to be in the header were excluded and are provided in the form of extensions headers. The following list summarizes the main extension headers and their purposes. For the latest list of extensions headers please consult IANA's registry at *http://www.iana.org/ assignments/protocol-numbers/protocol-numbers.xml*.

Destination Options and Hop-by-hop Options Header

The Hop-by-hop and Destination Options headers are the only options containing additional options that are being processed at least by the destination [21]. The Hop-by-hop Options header is the only extension header that is processed by all nodes on a packet's path.

The option number space is shared by both Hop-by-hop and Destination options, and with exceptions the same options can appear on both. All of the options share the same Type-Length-Value format, and furthermore are embed into the two most significant bits of the 8-bit type-value field information about what node handling the packet must do if it does not recognize the option type. The four rules are: 'skip', 'discard', 'discard and send ICMPv6 Parameter Problem', and 'discard and send ICMPv6 Parameter Problem if the destination was not a multicast address'. This framework allows the introduction of new options in a backwards compatible manner. A third most significant bit of the type value indicates whether or not the option is allowed to change in transit, and hence whether the option content is to be included in the AH protection or not.

Currently defined options are the following. The list notes whether an option is only allowed to be present in the Hop-by-hop or Destination Options header:

Pad1 Option: A special case option, which has a total length of one and content of zero. This can be used when padding of just a single byte is required [21].

PadN Option: An option that allows padding of two or more bytes [21].

Jumbo Payload Option: This option, which is only allowed to be present in the Hop-by-hop Options Header, allows increasing the IPv6 packet's maximum length from $2^{16} - 1$ (65535) bytes all the way up-to $2^{32} - 1$ (4,294,967,295) bytes [29]. This option is still waiting for the future, as there are very few link types that can transport packets larger than 65535 bytes and through the Internet it is impossible to find paths that would allow Maximum Transmission Units (MTUs) larger than 65535 bytes (see Section 3.2.3 for Path MTU discussion). In practice the Path MTU is at maximum 1500 bytes and is often less.

Tunnel Encapsulation Limit Option: This option, defined only for the Destination options header, is inserted by a node performing tunneling to indicate to following nodes how many additional encapsulations are permitted for this packet. In the special case of 'zero' additional allowed encapsulations a tunneling node can prevent further tunneling taking place. If a node that would like to encapsulate the packet it receives finds it disallowed by this option, it will generate an ICMPv6 Parameter Problem message. Detailed description of this option's use is described in RFC 2473 [30].

Router Alert Option: The Router Alert option, which is only defined for the Hop-by-hop Options header, can be used to alert routers on the packet's path to examine the packet more closely [31]. The reason for alerting routers can be, for example, to require special processing such as for quality of service reasons or to get a router to pay attention to multicast messages that the router otherwise would have no interest in. The Router Alert option is specifically needed for the Multicast Listener Discovery (MLD) feature described in Section 3.2.4.

Quick-Start Option (experimental): RFC 4782 defines an experimental option for the Hop-by-hop Options header that allows transport protocols, especially TCP, to determine the allowed sending rate at the beginning of a transport session and after idle periods of time [32]. We will not discussing this option any further in this book.

Common Architecture Label IPv6 Security Option (CALIPSO) (informational): A Hop-by-hop option for including explicit sensitivity labels for IPv6 packets in Multi-Level Security networking environments is defined in the informational RFC 5570 [33]. This option comes with an Internet Engineering Steering Group (IESG) warning to not use it in the global Internet, but only in closed networks. This is not used in 3GPP networks, and hence we are not going to look at it in any more depth.

Simplified Multicast Forwarding Duplicate Packet Detection Option (experimental): This Hop-by-hop option is used in a simplified multicast forwarding mechanism defined in experimental RFC 6621 for mobile ad hoc and wireless mesh networking use [34]. Since such networks are not the scope of this book, we are not looking at this in any more depth either.

Home Address (HoA) Option: The Mobile IPv6 (MIPv6) technology uses a destination option for a Mobile Node (MN) away from its home link to use in its IPv6 packets for informing recipients about the MN's home IPv6 address, as the packet's source IPv6 address is MN's Care-of Address (CoA) [35]. See more in Section 3.7.2.

Routing Protocol for Low-power and Lossy networks (RPL) Option: The RPL Hop-by-hop option is used in low-power networks to convey routing information in every packet that a router forwards [159]. These low-power networks are not the focus of this book, and hence we are not looking at the RPL protocol in depth.

Options for experimentation: Eight option types are reserved for experimentation purposes by RFC 4727 [36]. These experimentation options can be used with both Hop-by-hop and Destination option headers. We will not look at these in any more depth other than to list them here for the sake of completeness.

For the changes since the printing of this book, please see the IANA's list of 'IPv6 Parameters' available at *http://www.iana.org/assignments/ipv6-parameters/ipv6-parameters.xml*.

Routing Header

The idea of the Routing Header is that the sender can list intermediate nodes through which the packet should traverse on its way to the destination.

The Type 0 Routing Header [21] was deprecated by the IETF [37], as significant vulnerability was found that allowed traffic amplification for Denial-of-Service attacks. Due to deprecation and vulnerabilities, the Type 0 Routing Header feature is expected to be removed from IPv6 stacks and is also filtered away by some networks in order to stop possible attacks.

MIPv6 uses another type of Routing Header, Type 2, which is used to contain the HoA of the destination MN. This routing header is used when a so-called Correspondent Node is sending packets directly to the MN. The packet sent by the Correspondent Node will have the MN's CoA as the destination address of the IPv6 header, and the HoA then inside the Type 2 Routing Header. We describe MIPv6 briefly in Section 3.7.2, but for readers deeply interested in MIPv6 details, we recommend studying RFC 6275 [35].

Fragment Header

The Fragment header is used when the originating node is sending a packet larger than the MTU of the path towards the destination node. This header is always inserted by the originating node, as in IPv6 fragmentation does not occur on the path. For more details and usage see next Section 3.2.3.

The Fragment header essentially contains identification of the packet it belongs to, the next header field informing what is the first header of the fragmented part, the offset indicating which part of the original packet a particular fragment contains, and the fragment data itself.

No Next Header

This header type indicates that there are no more headers in the IPv6 packet.

Authentication and Encapsulating Security Payload Headers

AH can be used to provide integrity protection and authenticity of origin of a transmitted IPv6 packet [38].

The Encapsulating Security Payload (ESP) header can be used to provide confidentiality, authenticity of origin, integrity protection, and an anti-replay service [39].

In this book we will not address features of IPsec in any detail, as it is a wide subject, for which dedicated books are available. However, we will give an overview in Section 3.8.

3.2.3 MTU and Fragmentation

The path from one node to another always includes at least one hop, and through the Internet usually many more. Each one of these hops, or links, is able to transport packets at some maximum size, called the MTU, as illustrated in Figure 3.10. A node must adjust the size of the packets it sends to accommodate the hop that has the smallest MTU in the packet's path from the node to its peer [40]. If a node sends too large a packet, it will be dropped by the router prior to the link whose MTU is not large enough to transport the packet. When a packet is dropped, an ICMPv6 Packet Too Big (see Section 3.3.1)

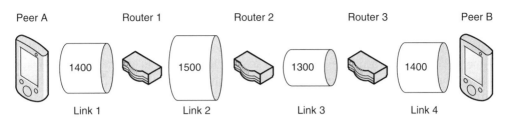

Figure 3.10 Illustration of links with different MTUs. Path MTU would be 1300 bytes – the smallest of MTUs on the path.

message is generated. IPv6 mandates 1280 byte minimum MTU that must be provided by every link on the Internet [21]. If a host intents to use larger than a 1280 byte packet size, it must perform a process called Path MTU Discovery [40].

The MTU of the first-hop link can have a default value depending on the link type, such as commonly 1500 bytes for Ethernet links, but it should also be dynamically learned via MTU Option on Router Advertisement [41].

In the hypothetical example scenario presented in Figure 3.10 two peers, A and B, are communicating over a route consisting of three routers and four links with different MTUs. When peer A sends a packet towards peer B, it initially uses a MTU of 1400 bytes, because that is the MTU of the first-hop link in this illustration. However, when a packet sent by peer A towards peer B reaches router 2, a problem occurs: the packet cannot be sent over link 3, which has an MTU of 1300 bytes – 100 bytes less than the packet sent by peer A. The router 2 will send ICMPv6 Packet Too Big message containing the MTU of the next-hop link, in this case 1300 byes, to peer A. Peer A, having received this message, learns what is the MTU of the link that could not transport peer A's packet. Subsequently, peer A reduces its Path MTU estimation and also reduces the size of the following packets that the node emits towards the same destination [40]. This process may be repeated, if there are other links on the path to peer B that have even smaller MTU (not illustrated in the figure).

A problem with RFC 1981 [40] based Path MTU discovery is that it uses unreliable ICMPv6 messages. An ICMPv6 packet may be lost on the network by accident, or on purpose by a firewall. When large packets do not go through the network and when ICMPv6 Packet Too Big messages are not received, a situation called ICMPv6 'black hole' is present. A sender can detect the black hole based on timers, which is a slow way, or more efficiently by performing Path MTU discovery with the help of the packetization layer as described in RFC 4821 [42]. The packetization layer Path MTU detection is performed by TCP, or some other transport layer protocol. A simplified description of the packetization layer Path MTU detection is as follows: a sending node initially sends smaller packets, such as smaller than the minimum MTU of 1280 bytes, to verify connection through the network in general. As receipts for successful packet reception are received, packet size is increased until the first-hop link's MTU is reached, ICMPv6 Packet Too Big is received, no acknowledgement is received (in which case the network path has a 'black hole') and sent packet was too large, or the functioning Path MTU estimation is good enough to not justify further probing (with possible cost of losing packets and hence time). The packetization layer discovery approach has the significant benefit of performing probing while transmitting useful data, and after the successful initial transmissions having an idea of when responses should arrive (estimation of Round Trip Time – RTT), and hence being faster and more efficient way to detect ICMPv6 'black holes'. For details and full description of the sophisticated algorithm, please see RFC 4821 [42].

Fragmentation is required if a host has to transmit packets larger than the Path MTU. In IPv6 the fragmentation is always performed by the sending host. Effectively, a host splits the transport layer packet into pieces that with IPv6 headers do not exceed the Path MTU. A receiving node will then reassemble the fragments before passing the transport protocol payload to the upper layers. In some cases a network node such as firewall may perform the reassembly as well, for example, for Deep Packet Inspection (DPI) reasons.

The original IPv6 specification allowed fragments to overlap, which introduced security issues. To fix these issues IETF decided to explicitly forbid overlapping fragments and standardized this in RFC 5722 [43].

While today link MTUs in the Internet are rarely larger than 1500 bytes, the IPv6 Jumbogram standard RFC 2675 [29] allows the creation of link types with MTUs up to 4,294,967,295 octets. The extended packet size is communicated in a Hop-by-hop Jumbo Payload option. The Jumbo Payload option is very rarely used and unlikely to be seen in the Internet and in the 3GPP networks.

3.2.4 Multicast

While in the Internet-scale multicast use for applications, or applications protocols, has not taken off too well, multicast functionality plays a key role in IPv6. Specifically, many of IPv6's major functions rely on multicast: Neighbor Discovery (see Section 3.4), Stateless Address Autoconfiguration (SLAAC) (see Section 3.5.1), and Dynamic Host Configuration Protocol version 6 (DHCPv6) (see Section 3.5.2).

Multicast Listener Discovery

Protocol for multicast functionality is described in 'Multicast Listener Discovery (MLD) for IPv6' [44], updated in 'Multicast Listener Discovery version 2 (MLDv2)' [45], and further tuned up in RFC 3590 [46], RFC 4604 [47], and RFC 4607 [48]. The MLD protocol takes place only on the directly connected a link, where nodes tell the router which multicast addresses they are listening to. The router then uses any multicast routing protocol (such as RFC 4601 [49], but multicast routing protocols are not included in this book) to ensure that multicast packets related to multicast addresses that nodes listen to are delivered to the router, and furthermore on the link nodes are attached to. The major difference between MLD and Multicast Listener Discovery version 2 (MLDv2) is that the latter allows nodes to specify from which source addresses they want ('include mode'), or do not want ('exclude mode'), to receive multicast traffic. However, as the 'include' and 'exclude' are not that commonly used, a lightweight MLDv2 has been defined in RFC 5790 [50] that supports the most commonly needed feature subset of the full MLDv2.

The MLD protocol makes use of three distinct messages. These messages are specified to work on top of ICMPv6 and are part of ICMPv6 message types as listed in Table 3.2.

Multicast Listener Query: A router in a network uses this query to find out for which multicast addresses the link has listeners. This message is further split into general query and multicast-address-specific query, which allow querying of all addresses that nodes are listening and for specifically finding out if a link has listeners for a particular address.

Multicast Listener Report: The nodes on a link utilize this report – both unsolicited and solicited as responses to queries – to indicate which multicast addresses they are listening.

Multicast Listener Done: A node that ceases to listen to a multicast address issues this done to indicate it to a router. Once there are no listeners for an address on a link, the router can use the multicast routing protocol it uses to stop receiving multicast traffic destined for the multicast address.

A node does not need to use MLD for joining 'All Nodes Multicast Address', as listening for that address is always mandatory.

Multicast Addresses

IPv6 multicast addresses are allocated by following the guidelines of RFC 3307 'Allocation Guidelines for IPv6 Multicast Addresses' [51]. Special purpose address blocks are used for documentation as described in Section 3.1.8. The multicast addresses can be recognized from the leading 8 bits of an IPv6 address, which are always `ff` as listed in Figure 3.6 and defined in RFC 4291 [4].

The flag bits, the 4 bits following the leading `ff` as illustrated in Figure 3.11, tell about the properties of the address. The 'T'-bit indicates whether the address is permanently assigned by IANA (value '0') or dynamically assigned address (value '1'). Bit 'P', however, has a more significant meaning: whether the address is based on a network prefix (value '1') or address is not based on a network prefix (value '0') [52]. As shown in Table 3.1, the IANA assigned addresses have all flag bits as zero, hence indicating permanently assigned and not network prefix based addresses. The dynamic addresses can be created by an end host or by an allocation server, using the prefixes they have at their disposal, based on a need for a temporary (but possibly long-lived) multicast group.

If the 'P' bit has a value of '1', the 'group identifier', which identifies a multicast group and for which the basic address format reserves 112 bits, is given more detailed structure as is illustrated in Figure 3.12. The group identifier field is split into 8 bits of reserved space, 8 bits for prefix length, 64 bits for the network prefix, and the 32 remaining bits for the actual group identifier. For example, a unicast address prefix, such as `2001:db8:cafe::/48`, is mapped into the multicast prefix by setting the 'plen' field to '0x30' (decimal 48) and the network prefix to `2001:db8:cafe:0000`, thus resulting in a multicast IPv6 address prefix of `ff3x:0030:2001:db8:cafe::/96`, where the 'x' denotes a selected scope. If the R bit is set to one, the multicast address contains an address of a rendezvous point. This is realized by using four bits of the reserved-field shown in Figure 3.12 to transport rendezvous point's interface identifier [161].

Multicast addressing utilizes scoping system to limit the distribution of packets sent to multicast addresses. The scope is presented in a 4-bit value inside a multicast address, as illustrated in Figure 3.11, and hence has 16 distinct possibilities. Most of the values

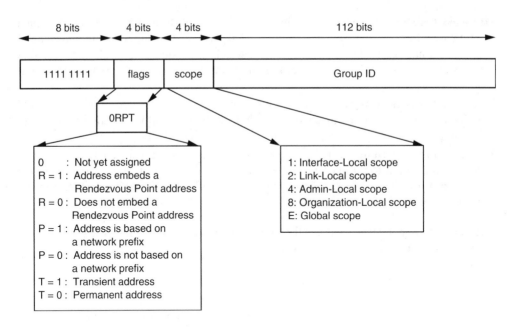

Figure 3.11   Generic IPv6 multicast address format.

Table 3.1   Subset of reserved IPv6 multicast addresses. 'X' in the multicast address indicates a variable scope address, as illustrated herein with the documentation address

Multicast address	Scope	Description	Reference
ff01:0:0:0:0:0:0:1	Interface-Local	All Nodes Address	[4]
ff01:0:0:0:0:0:0:2	Interface-Local	All Routers Address	[4]
ff02:0:0:0:0:0:0:1	Link-Local	All Nodes Address	[4]
ff02:0:0:0:0:0:0:2	Link-Local	All Routers Address	[4]
ff02:0:0:0:0:0:0:16	Link-Local	All MLDv2-capable Routers	[45]
ff02::1:ff00:0000/104	Link-Local	Solicited-Node Address	[4]
ff02:0:0:0:0:2:ff00::/104	Link-Local	Node Information Queries	[53]
ff05:0:0:0:0:0:0:2	Site-Local	All Routers Address	[4]
ff05:0:0:0:0:0:1:3	Site-Local	All DHCPv6 servers	[54]
ff0x::db8:0:0/96	Variable Scope	Documentation Addresses	[17]

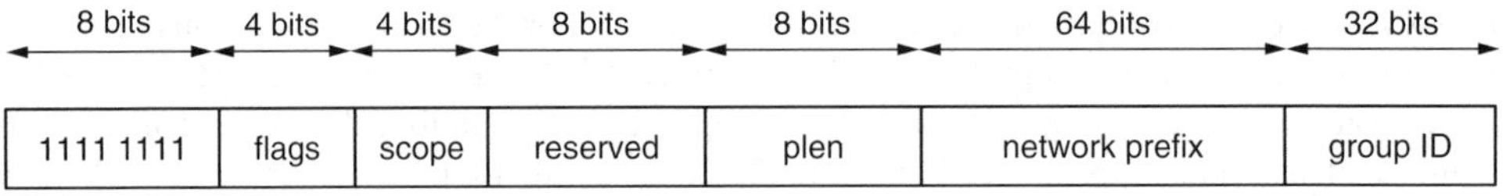

8 bits	4 bits	4 bits	8 bits	8 bits	64 bits	32 bits
1111 1111	flags	scope	reserved	plen	network prefix	group ID

Figure 3.12   Unicast Prefix based IPv6 Multicast Address Format.

are unassigned or reserved (scopes 0 and F). The allocated scopes, and related numerical values, are the following: *Interface-Local (1)*, *Link-Local (2)*, *Admin-Local (4)*, *Site-Local (5)*, *Organization-Local (8)*, and *Global (E)* [4].

Table 3.1 illustrates some of the IANA allocated, reserved, and permanent, IPv6 multicast addresses, most of which are referenced in this book. It is important to realize IANA has allocated many more addresses for various protocols than those listed here, and it is not feasible to include a full list here. The list of IANA's allocations is available at *http://www.iana.org/assignments/ipv6-multicast-addresses/ipv6-multicast-addresses.xml*.

Multicast Router Discovery

In some network deployments, layer two multicast-aware switches need to know link-layer addresses of multicast routers. They can find them via utilization of layer-two Multicast Router Solicitation and Multicast Router Advertisement messages [55]. This technology is not relevant in 3GPP networks and hence we do not look into that in any more detail.

Solicited-node Multicast Address Formation

Solicited-node multicast address is used when a node performs Duplicate Address Detection (DAD) (see Section 3.4.7) and link-layer address resolution (see Section 3.4.4). Figure 3.13 illustrates how a Solicited-node multicast address is created from a 128-bit IPv6 address by picking the 24 lowest bits and appending them to the `ff02::1:ff00::/104` prefix.

3.3   Internet Control Message Protocol Version 6

One of the most important components of the IPv6 protocol suite is ICMPv6 [56]. The ICMPv6 protocol is maybe best known for its use for 'pinging' other nodes for reachability by using ICMPv6 Echo Request and Reply messages, and for its ability to indicate failures in packet delivery to the originating node, such as 'destination unreachable'.

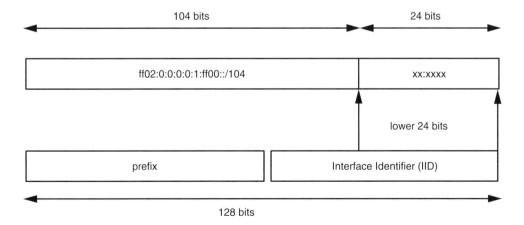

Figure 3.13   Solicited-Node IPv6 multicast address format.

However, in IPv6 ICMPv6 plays a much more significant role than Internet Control Message Protocol version 4 (ICMPv4) does in IPv4. This is mostly due to the use of ICMPv6 for the IPv6 Neighbor Discovery protocol (see Section 3.4).

ICMPv6 does not have any automated retransmission mechanisms in case of lost packets, and hence, for example, ICMPv6 errors may not always be received even if they are sent. The protocols that use ICMPv6, such as Neighbor Discovery Protocol (NDP), do implement retransmissions when packet delivery is important. However, in some cases, such as with ICMPv6 Packet Too Big or Time Exceeded, there are no retransmissions. Therefore, for example, firewalls should not be as limiting as in IPv4 and should not filter ICMPv6. Filtering of ICMPv6 can cause unexpected issues for IPv6. For example, for IPv6 Path MTU Discovery procedure that depends on the ICMPv6, and will not work if ICMPv6 packets are dropped along the way (see Section 3.2.3).

Table 3.2 summarizes ICMPv6 messages that have been standardized this far. The table contains section references for the majority of the messages that are described in more detail in dedicated sections within this book.

After the design of the ICMPv6, need arose to have ways to extend the already defined messages. RFC 4884 introduces a concept of multi-part operation as a way to extend ICMPv6 messages with additional information [57]. The enhancement brought by RFC 4884 is essentially introduction of a length-field to several ICMPv6 messages and an extension header appended at the end of an ICMPv6 message. This extension header can then contain extension objects, such as those defined in 'ICMP Extensions for Multiprotocol Label Switching' [58] and 'Extending ICMP for Interface and Next-Hop Identification' [59] RFCs.

Roughly the ICMPv6 messages can be divided into two categories – *ICMPv6 error*, and *ICMPv6 information* messages. We will look at these two categories in more detail in the following.

3.3.1 Error Messages

A set of ICMPv6 error messages has been defined that can be used by network routers and destination nodes for informing sending nodes about different error conditions. The four widely used messages are described below.

Destination Unreachable

A router on a network or a destination node may generate an ICMPv6 Destination Unreachable message. A router generates the Destination Unreachable when it has no route to a destination, communication with a destination is administratively prohibited, destination is beyond the scope of the source address, the destination address is unreachable, the source address failed ingress or egress filtering policy, or the router has a reject route to a destination. A destination node typically generates a Destination Unreachable message only when the destination port is not in use [56].

Packet Too Big

A router that is unable to forward a packet because the packet is too big to fit in the next hop link sends an ICMPv6 Packet Too Big to the packet originator and with the same

Table 3.2 ICMPv6 messages types

ICMPv6 message	Code	Content in brief
Error: Destination Unreachable	1	Various reasons for unreachability [56]
Error: Packet Too Big	2	MTU of the link packet did not fit [56]
Error: Time Exceeded	3	Timeout or the hop count exceeded [56]
Error: Parameter Problem	4	Incomprehensible IPv6 parameter [56]
Error: Private experimentation	100	Depends on experiment [56]
Error: Private experimentation	101	Depends on experiment [56]
Error: Reserved for expansion	127	Reserved for the future [56]
Echo Request	128	Identifier for the Echo request [56]
Echo Reply	129	Echo request's identifier [56]
Multicast Listener Query	130	See Section 3.2.4 [44]
Multicast Listener Report	131	See Section 3.2.4 [44]
Multicast Listener Done	132	See Section 3.2.4 [44]
Router Solicitation	133	See Section 3.4 [41]
Router Advertisement	134	See Section 3.4 [41]
Neighbor Solicitation	135	See Section 3.4.4 [41]
Neighbor Advertisement	136	See Section 3.4.4 [41]
Redirect Message	137	See Section 3.4 [41]
Router Renumbering	138	Commands for router renumbering [60]
Node Information Query	139	Information that is being queried [53]
Node Information Response	140	Response data to the query [53]
Inverse Neighbor Discovery Solicitation	141	See Section 3.4.4 [61]
Inverse Neighbor Discovery Advertisement	142	See Section 3.4.4 [61]
Version 2 Multicast Listener Report	143	See Section 3.2.4 [45]
Home Agent Address Discovery Request	144	See Section 3.7.2 [35]
Home Agent Address Discovery Reply	145	See Section 3.7.2 [35]
Mobile Prefix Solicitation	146	See Section 3.7.2 [35]
Mobile Prefix Advertisement	147	See Section 3.7.2 [35]
Certification Path Solicitation	148	See Section 3.4.9 [12]
Certification Path Advertisement	149	See Section 3.4.9 [12]
Experimental mobility protocols	150	For experimental mobility protocols [62]
Multicast Router Advertisement	151	See Section 3.2.4 [55]
Multicast Router Solicitation	152	See Section 3.2.4 [55]
Multicast Router Termination	153	See Section 3.2.4 [55]
FMIPv6 Messages	154	See Section 3.7.2 [63]
RPL Control Message	155	See Section 3.10.4 [64]
Private experimentation	200	Depends on experiment [56]
Private experimentation	201	Depends on experiment [56]
Reserved for expansion	255	Reserved for the future [56]

message indicates what is the MTU of the link that the packet did not fit into. This will help the originator to decrease its Path MTU estimation and hence emit smaller packets (see Section 3.2.3).

Time Exceeded

A router sends an ICMPv6 Time Exceeded error message when the Hop Limit of the packet reaches zero and hence the packet has to be discarded. The Hop Limit may reach zero, for example, due to misconfiguration errors such as routing loops.

The ICMPv6 Time Exceeded message may also be sent when a fragmented packet's reassembly process has timed out. Typically this would happen when one or more fragments have been lost in the transit, making the reassembly impossible, but it could also happen when a network is very congested or otherwise too slow to transmit all of the fragments. The reassembly timeout typically happens at the receiving node, but could occur in an intermediate router that performs reassembly, for example, for policy control reasons.

Parameter Problem

In case a router or a destination node is unable to fully parse IPv6 packet or extension headers, the packet will be dropped and an ICMPv6 Parameter Problem message generated.

3.3.2 Informational Messages

The ICMPv6 defines two important informational messages for network debugging purposes. These are ICMPv6 Echo Request and Reply for 'pinging' a destination, and ICMPv6 Node Information Query for asking for information about a node.

The ICMPv6 Echo Request is used to ask if a destination node is alive and to measure the RTT it takes for a message to traverse to the destination and back. The ICMPv6 Echo Request can also be used for trace routing purposes by purposefully sending several ICMPv6 Echo Requests with increasing Hop Limit values, starting from one. This causes each visible router on the way to the destination to drop the packet that reaches zero Hop Limit at the particular router and generate an ICMPv6 Time Exceeded message. By collecting the received ICMPv6 error packets and mapping them to the sent packets and used Hop Limit values, the originating host can find out what routers are visible on the path to the destination. The path may include invisible hops, such as bridges, Neighbor Discovery Proxies or firewalls, which perform their tricks without decrementing the Hop Limit field – the originating host will not learn of the presence of such entities. In addition, some routers can be partially invisible. This can happen if a router does decrement the Hop Limit field, but is configured not to send any ICMPv6 errors, or ICMPv6 rate limiting is restricting sending of ICMPv6 messages, or the ICMPv6 message is just lost in either direction. The originating node can learn about the presence of partially invisible routers even without getting ICMPv6 errors, if the originator gets ICMPv6 errors from subsequent routers for messages sent with a higher Hop Limit value.

The ICMPv6 Node Information query can be used to poll a node for information regarding its hostname or fully qualified domain name. This functionality is partially overlapping with reverse Domain Name System (DNS) queries (see Section 3.10.2), but has the benefit of not requiring any servers for the resolution. Additionally, the Node Information query can be used to ask for information about the destination's IPv6 and IPv4 addresses.

3.4 Neighbor Discovery Protocol

The most important user of the ICMPv6 is the IPv6 Neighbor Discovery Protocol (NDP) [41]. The IPv6 Neighbor Discovery is used to find other nodes and routers that reside on the same link as an IPv6 node. The key functions, which are covered in following sections, make use of five different ICMPv6 packet types:

Router Solicitation: A node may multicast a Router Solicitation message to the link in order to trigger routers to send Router Advertisement messages faster.

Router Advertisement: A router sends Router Advertisement messages either periodically, or solicited after receiving a Router Solicitation message. This message contains information that a node needs to configure various IPv6 parameters, such as default router and the node's IPv6 address itself.

Neighbor Solicitation: A node uses Neighbor Solicitation to find a link-layer address of a neighbor, to check the presence of a neighbor, and for DAD.

Neighbor Advertisement: A node sends Neighbor Advertisement as a response to the Neighbor Solicitation sent for the node, or to announce changes in the node's link-layer addresses.

Redirect: Routers may send a Redirect message when a more suitable next hop device is available than the receiving router itself.

The used link type has implications for Neighbor Discovery procedures, and these are addressed in Section 3.6.

One of the key benefits of IPv6, in addition to increased address space, is its improved capability for host address autoconfiguration. The SLAAC system is build around Neighbor Discovery Protocol and in particular around the ICMPv6 Router Advertisement message. We have dedicated the whole Section 3.5 to addressing this topic.

3.4.1 Router Discovery

Each node is required to learn information about at least one router in order to be able to communicate beyond the link to which the node is attached. The node finds routers by either listening to the unsolicited Router Advertisement messages sent periodically by routers on the link, or by explicitly soliciting for Router Advertisements by sending a Router Solication messages.

Attacks with Router Advertisements

As the Router Advertisement is a key ICMPv6 message for configuring hosts, it is also an attractive tool to use for attacking hosts – or confusing hosts due to accidental configuration errors on hosts or routers [65]. An unwanted Router Advertisement on a link is referred to as a *Rogue Router Advertisement*. For example, a misconfigured host performing 'tethering' could be sending Rogue Router Advertisements advertising automatically generated Connection of IPv6 domains via IPv4 clouds (6to4) prefixes (6to4 is an IPv6 transition mechanism used for automated tunneling, see RFC 3056 [66]) to a network link that may or may not have any 'legitimate' IPv6 routers. In either case, hosts will receive the Rogue Router Advertisement from this misconfigured host, and statelessly configure IPv6 addresses based on the received prefix and other information. The misconfigured IPv6 address can cause hosts to initiate data sessions and transport data through the misconfigured host, thus causing various issues such as lack of connectivity altogether, low bandwidth, and longer than necessary latencies. An attacker sending Rogue Router Advertisements can initiate man-in-the-middle or denial-of-service attacks. The IETF has designed Secure Neighbor Discovery (SEND) to protect against such threats (see Section 3.4.9). However, unfortunately SEND is currently not deployed.

For unmanaged nodes receiving Rogue Router Advertisements, there is very little they can do until SEND is perhaps supported one day. The nodes need to be able to hastily switch to use alternative routers and addresses (including IPv4 addresses) if addresses or routers first configured do not yield properly working connectivity (see also the Happy Eyeballs feature described in Section 3.9.4). Managed nodes could be provisioned with, for example, firewall rules allowing Router Advertisements only from whitelisted sources. For routers, IETF has defined tools in RFC 6105 that can be used to protect hosts. Effectively, managed layer 2 switches could drop Rogue Router Advertisements (e.g. based on the direction from where a Router Advertisement arrived at a switch – this is called *Router Advertisement Guard (RA-Guard* in RFC 6105), or routers could be running 'prefix depreciation tools' that would detect when Rogue Router Advertisement messages were sent to hosts and then would immediately send legitimate Router Advertisements to counter the effects of the Rogue Router Advertisements, such as providing hosts with more preferred prefixes. In addition to protocol-level tools, administrators can monitor the network for the presence of Rogue Router Advertisements and then initiate actions such as blacklisting hosts sending Rogue Router Advertisements. For a full toolbox please see RFC 6104 [65] and RFC 6105 [67].

3.4.2 Parameter Discovery

The Router Advertisement message that nodes receive contains the set of information that nodes need for global communications (link-local communications do not require Router Advertisement, or the presence of routers).

The mandatory set of parameter information that may be received via Router Advertisements is the following:

Hop Limit: The default value that nodes should use in the Hop Limit field of packets they send [41].

Address Autoconfiguration Protocol bit (M-bit): Indicates whether Stateful Address Autoconfiguration (Section 3.5.2) services are available on the link (and that the router knows about it) [41].

Other Configuration Available bit (O-bit): Indicates whether other configuration information is available with Stateless DHCPv6 (and that the router knows it) (Section 3.5.2) [41].

Home Agent bit: Indicates whether the router sending the Router Advertisement is acting as a Mobile IPv6 home agent [35].

Default Router Preference bits: Indicate the relative preference of the router over other default routers on the same link – see more in Section 3.5.6 and [68].

Router Lifetime: Indicates how long the router can be used for as a default router, and if at all (lifetime value of zero indicates that the router should not be used as a default router) [41].

Reachable Time: The time other nodes can be assumed to stay reachable after positive reachable confirmation. A value of zero means unspecified by the router sending the Router Advertisement [41].

Retransmit Timer: The retransmit time between consecutive Neighbor Solicitation messages. A value of zero means unspecified, by the router sending the Router Advertisement [41].

Additionally, a number of options have also been standardized for Router Advertisements. Some of these are used only in specific circumstances, such as those defined for use with Mobile IPv6 (see Section 3.7.2) or Secure Neighbor Discovery (see Section 3.4.9).

Source Link-Layer address: The link-layer address of the router that sent the Router Advertisement [41].

Link MTU: The MTU of the attached link (see Section 3.2.3) [41].

Prefix information: List of prefixes that nodes may possibly use for on-link determination (if 'L'-flag is set), possibly for SLAAC (if 'A'-flag is set) [41], and in the case of Mobile IPv6, may additionally communicate the complete IPv6 address of the sending router [35].

DNS configuration: List of recursive DNS server addresses and DNS search lists that a node can utilize for DNS queries [69].

Route Information option: Information about more specific routes and their preferences on the router listed within the option [68]. This information is used for address and router selection purposes.

IPv6 Router Advertisement flags option: This option extends the number of flag bits that can be carried in Router Advertisement [70].

Advertisement interval: Mobile IPv6 uses this option to communicate to nodes the send interval of unsolicited Router Advertisements [35]. This information can be used for movement detection.

Home Agent information: Mobile IPv6 uses this option to communicate preferences between Home Agents and the time that a Home Agent may be used.

CGA: SEND uses the CGA option to make it possible for a node to verify a router's cryptographically generated address [12].

RSA signature: SEND uses the RSA signature option to allow attachment of a public key-based signature into the Router Advertisement [12].

Proxy signature: SEND uses the Proxy signature option for secure Neighbor Discovery Proxy function [71].

Timestamp: SEND uses the Timestamp option to protect against replay attacks.

Nonce: SEND uses the Nonce option to ensure that the message received by a node is recent.

3.4.3   On-link Determination

When sending an IPv6 packet, a node needs to know whether the packet is destined to another node on the same link or not. If the destination is on the same link, the node needs to send the packet directly to the destination, otherwise the node needs to send the packet to a router for forwarding. The determination of whether a destination is on the same link is based on knowledge about prefixes being used on the link that the node is attached to, or via explicit information about the destination node. The node learns, via Prefix Information Options (PIOs) in Router Advertisements, the prefixes and whether they can be used for on-link determination (from 'on-link' flag bit 'L'). Additionally, a node can learn that a specific IPv6 address is on-link by receiving an ICMPv6 Redirect message from a router showing that the destination is actually on-link.

It is important to note that even if a destination's IPv6 address matches a prefix used on the link, or with the IPv6 address that a sending node has been configured with, it does not necessarily mean that the destination is actually on-link [8]. This kind of scenario can happen, for example, in network links having star topology and thus where each packet has to traverse through the central node. The on-link determination can only be made based on information received via Router Advertisements, ICMPv6 Redirects, or by some other explicit means such as by manual configuration. By default, nodes must assume that destinations are off-link.

3.4.4   Link-layer Address Resolution

Before a node can send IPv6 packets over an underlying data link layer, a link layer address of the next hop entity needs to be known – assuming that the link has link-layer addresses in the first place. The node can use the Neighbor Solicitation message destined to a solicited node multicast address to request the owner of the destination IPv6 address to report its corresponding link-layer address in a response Neighbor Advertisement message. The resolving node originates a Neighbor Solicitation message with a source address set to any of the configured unicast addresses on the used outbound interface and a destination address set to the Solicited-Node multicast address of the to-be-resolved IPv6 address (see Section 3.2.4). Additionally, neighbors' addresses can be learned by receiving unsolicited Neighbor Advertisement or Router Advertisement messages.

It is also possible to perform the Neighbor Discovery procedure in reverse: resolve a node's IPv6 addresses on the link when the link-layer address is already known. This procedure is called Inverse Neighbor Discovery and is specified in RFC 3122 [61].

Each IPv6 node maintains a *Neighbor Cache* that contains the list of its known neighbors, and the reachability state for each neighbor. The different reachability states are described in the following.

INCOMPLETE: Neighbor Solicitation has been sent, but no response received yet. After a response is received, the state is changed to REACHABLE.

REACHABLE: Neighbor is reachable. The state will change to STALE if no communication takes place for a specified time, which is configured via Router Advertisement (see Section 3.4.3) and is link specific.

STALE: More than a configured time has passed since the last positive confirmation of the neighbor's reachability, and hence reachability is unknown. The neighbor's reachability will be checked when the neighbor is to be contacted next time, in which the case state changes to DELAY.

DELAY: Upper layer protocol has sent a packet, but confirmation of reachability is not yet received. This state lasts five seconds, and if no confirmation is received, the state changes to PROBE.

PROBE: Reachability of the neighbor is being actively checked with periodic Neighbor Solicitation messages. Once the neighbor is reachable, the state is changed to REACH-ABLE.

Attacks Against Neighbor Discovery

It has been discovered that the Neighbor Discovery procedure can be used by attackers to perform denial-of-service attacks [72]. Effectively, attackers would try to get nodes to reserve memory for Neighbor Cache for addresses whose resolution will not complete (state remains INCOMPLETE), causing fixed sized Neighbor Caches to either fill up, or just to cause difficulties for setting up new legitimate IPv6 communication. This attack is based on remote nodes sending packets destined to addresses having the 64-bit prefix used on a link, for which the router will attempt to find a link-layer address before forwarding the packet. As the destinations are not there, the Neighbor Caches will fill up. The documented means for mitigation involve the following (and more): rate limiting of Neighbor Discovery messages, smart IPv6 packet filtering, prioritization of different Neighbor Discovery messages, and by smarter Neighbor Cache management (cleaning up of older entries in INCOMPLETE state and so forth).

3.4.5 Neighbor Unreachability Detection

IPv6 nodes keep track of reachability of their neighbors, including nodes and routers. The reachability status of neighbors can be indirectly updated based on successful IPv6 communications, but in the case of no communications or failing communications, a node may use Neighbor Unreachability Detection (NUD) procedures to check whether a given neighbor is still reachable. The explicit reachability check is performed by sending a Neighbor Solicitation message with a source address set to any of the configured unicast addresses on the used outbound interface and the destination set to the unicast address whose reachability is being checked. The neighbor is considered unreachable if it does not respond with a Neighbor Advertisement message and have the solicited 'S' bit set.

The Neighbor Cache is updated when a node receives ICMPv6 Neighbor Solicitations, Neighbor Advertisements, Router Solicitations, or Router Advertisements with Source Link-Layer Address option indicating the link-layer address of the sender. The Neighbor Cache is not updated just based on MAC-addresses seen in use at the Ethernet layer.

Additionally, reachability updates are made when upper layer protocols, such as TCP, have the ability to provide positive reachability information.

3.4.6   Next-hop Determination

Whenever a node is sending an IPv6 packet, it must determine what is the IPv6 address of the next hop to which the packet is to be sent [41]. This process involves the determination of whether the destination is actually on the same link as the sending host, or further away. If the destination is on the same link, the packet can be sent directly to the destination. Otherwise the packet must be sent to a next-hop router for forwarding. The selection of the a router is based on selecting a router that is known to be reachable, or probably reachable (see description of states in Section 3.4.5). Alternatively, selection in a round-robin fashion from the default routers list. RFC 4191 further improves the router selection by introduction of preferences for default, and more specific, router selection as we discuss in Section 3.5.6 [68].

The on-link determination is done by matching the destination IPv6 address to the prefixes that node has received via Router Advertisements, assuming the prefixes are allowed to be used for on-link determination purposes (see Section 3.4.2). The link-local prefix `fe80::/10` and multicast prefixes are always considered to be on-link.

Once the IPv6 address of the next hop is known, the node needs to find out what is the corresponding link-layer address (for the case where the link layer has link layer addresses, see Section 3.6). This information is fetched from the Neighbor Cache (see Section 3.4.5), or if not available then with help of Neighbor Discovery procedures.

After the link-layer address of the next-hop destination is available, it is stored into a *Destination Cache*. The Destination Cache exists to assist sending of the follow-up IPv6 packets to the same address – if an entry is found in the Destination Cache the Neighbor Cache checking and possible Neighbor Discovery procedures are avoided.

The Destination Cache may be updated if ICMPv6 Redirect messages are received (see Section 3.4.8), and may also be updated if the NUD detects changes in link-layer addresses related to specific destinations (e.g. due to default router changes). A mere disconnection to the neighbor that has been used as next-hop does not cause cache entry flushing, as the entry may have information that is useful once the connectivity is regained. This information can be related, for example, to Path MTU, depending on the implementation.

3.4.7   Duplicate Address Detection

Before any address can be assigned to an interface, a node must ensure the address is unique on the attached link [73]. This requirement applies equally to all unicast addresses including the link-local addresses.

Before a node can check the uniqueness of an address, the node has to use MLD to join the 'Solicited-node multicast address' (see Section 3.2.4) of the tentative address node has (see more about address states in Section 3.5.1). Uniqueness of a tentative address is tested by sending an ICMPv6 Neighbor Solicitation [41] message, sourced from an undefined address `::` if the node does not have address yet, to the Solicited-node multicast address [4] of the tentative address. The number of times the Neighbor Solicitation is sent may be configurable via nodes' `DupAddrDetectTransmits` configuration variable [73].

The default value is one indicating no retransmissions (configuration variable value zero would skip DAD). If there is no reply after the last (and by default the only) Neighbor Solicitation message, the node can start using the new address. If there is a Neighbor Advertisement reply indicating that the tentative address is already used by another node, the node behavior depends on how the IID was configured. If the IID was based on a hardware address, the failed DAD means there is a conflict of hardware addresses and hence IPv6 operations on a link should be stopped. If the IID was randomly selected, the node can just randomize a new identifier and repeat the DAD procedure with it.

The DAD procedure can be optimized by using a technology called Optimistic DAD. Optimistic DAD makes the address configuration faster in the successful case. It essentially assumes no conflicts, and in case of one, it will back off fast [74]. This is a very effective improvement as address conflicts are very unlikely in real life scenarios where unique hardware-based or pseudorandom IIDs are used. Thus, the Optimistic DAD allows the node to proceed faster to actual data transmission phase.

3.4.8 Redirect

A router can inform a sending node with ICMPv6 Redirect message about a better next-hop destination than itself. This is a useful property if a router knows another router that is more preferred to route packets destined to certain addresses. Furthermore, a router can use ICMPv6 Redirect to inform a sending node that the destination is actually on-link, and hence the node can send packets directly to its peer. With the ICMPv6 Redirect the router can help the sending node by directly including the link-layer address of the better next-hop destination in the Redirect message.

A node that receives an ICMPv6 Redirect message will update its Destination Cache accordingly. The following packets to the same IPv6 address will be sent to the redirected link-layer and destination address (see more about next-hop determination in Section 3.4.6).

3.4.9 Secure Neighbor Discovery

The NDP is an attractive target for on-link attacks, as it is essential for IPv6 node setup and function. The IETF has attempted to mitigate attacks against NDP by defining 'SEcure Neighbor Discovery (SEND)' [12]. The SEND uses public-private key cryptography and certification paths to provide integrity protection and to certify the authority of message senders, essentially the authority of routers. However, the requirement to have global or local trust anchors, routers provisioned with certificates from trust anchors, and hosts being configured with pointers to these trust anchors, has caused significant deployment challenges. Due to these deployments challenges, the SEND has not been able to really take off. Another reason for the lack of deployment is that enough security may also be provided by a link layer (such as is the case in 3GPP networks) and/or by higher layers, such as by Transport Layer Security (TLS), thus diminishing the need to secure NDP messages at layer 3.

The SEND defines four Neighbor Discovery options that are used to protect NDP messages: Cryptographically Generated Address (CGA), RSA Signature, Timestamp, and Nonce options. Furthermore, the SEND uses CGAs as nodes' addresses [75]. The CGAs are used to provide proof that the sender of the message is actually who it claims to be. Two ICMPv6 messages are also defined: Certification Path Solicitation (CPS) and Certification Path Advertisement (CPA).

The CGA option is used for verification of the sender's CGA while the RSA Signature option is then used to protect the messages related to Router and Neighbor discovery. Timestamp and Nonce options are used to prevent replay attacks.

The CPS and CPA messages are used to pass information required for discovery of the certification path without loading Router Solicitation and Router Advertisement messages with that information. A node sends CPS to request certification path a router has with one or more nodes' trust anchors included in a Trust Anchor option. The router replies with CPA and includes Trust Anchor options in addition to Certificate option containing a certificate node can use to establish a certification path to a trust anchor.

3.4.10 Neighbor Discovery Proxies

In certain deployments, IPv6 nodes that are on the same logical link may not be directly connected into the same physical medium. For example, multiple IPv6 nodes may be connected via point-to-point links to a central node, which hears everybody connected through point-to-point links, but none of the nodes on the point-to-point links can hear each other. Another example is a setup where a central node is between different link types, such as between a shared link and a point-to-point link. In such a case nodes in the shared link can hear each other, but not directly the node in the point-to-point link. If allowed by link-layer types, simple bridging would be enough to enable direct node-to-node communications. Unfortunately simple bridging does not work if the link types are different.

In order to facilitate communications between nodes that do not hear each other directly, the central node can implement an experimental functionality called 'Neighbor Discovery Proxy' illustrated in Figure 3.14 [76]. The Neighbor Discovery Proxy functionality works so that the Neighbor Discovery Proxy node listens for all multicast traffic on proxied interfaces (such as the shared and point-to-point links of Figure 3.14), and selectively forwards or proxies multicast packets to other links. The Neighbor Discovery Proxy node keeps a track of all nodes connected to all interfaces used for packet proxying. This allows optional selective multicast forwarding only to the link, or links, that the Neighbor Discovery Proxy node knows has the right receivers. The selective forwarding is useful when resources

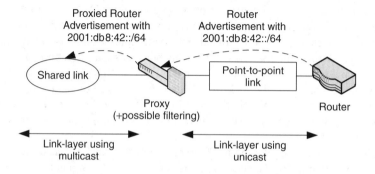

Figure 3.14 Illustration of Neighbor Discovery Proxy.

on the interfaces need to be conserved and when simple forwarding would only consume resources, such as energy and bandwidth, without bringing any benefits. For example, in Figure 3.14 the point-to-point link could be an energy-consuming cellular connection. The multicast packets of special interest for proxying are ICMPv6 Router Advertisements, Redirects, Neighbor Solicitations, and Neighbor Advertisements. These four messages transport link-layer addresses in their payloads and hence require special attention from the Neighbor Discovery Proxy. All other multicast traffic is always proxied as it is to all interfaces. In order to further optimize, for example to conserve power, the central node may implement additional selective multicast filtering and forwarding policies.

When proxying, with few exceptions, the Neighbor Discovery Proxy leaves the packet otherwise intact, but the source link-layer address is changed to the one owned by the proxy node. The destination's link layer address will be looked on from the Neighbor Discovery Proxy's Neighbor Cache and it will be the one corresponding to the destination IPv6 address. Furthermore, link-layer addresses included in the protocol payload of Router Solicitation, Redirects, Neighbor Solicitations, and Neighbor Advertisements will be replaced with Neighbor Discovery Proxy's link-layer address. It is worth noting that the Hop Limit field is never updated, hence the Neighbor Discovery Proxy does not show as an IPv6 router for other nodes. The only instance where the Neighbor Discovery Proxy appears as a node, a router, to others is the case where the Neighbor Discovery Proxy has to emit ICMPv6 Packet Too Big message to inform a sender about an MTU problem on the next link. In the case of proxying Router Advertisement messages, the Neighbor Discovery Proxy has to set a 'P' bit on in the flags field of Router Advertisement. The 'P' bit indicates to the receiver that the Router Advertisement is proxied, and cannot be proxied any further. This feature is needed to avoid the formation of network loops. Consequently, if a node that would like to act as a Neighbor Discovery Proxy receives a Router Advertisement with 'P'-bit on, it cannot engage in proxying functions.

The great benefit of the Neighbor Discovery Proxy technology is that it significantly decreases the need to implement IPv6 Network Address Translations (NATs), which otherwise might be needed to connect different link types, and also in scenarios where there are not enough IPv6 addresses locally available (e.g. if DHCPv6 Prefix Delegation (PD), described in Section 3.5.2, is not available for use).

3.5 Address Configuration and Selection Approaches

IPv6 comes with two major automatic address configuration protocols. The one developed first, and that has more widespread support, is 'IPv6 Stateless Address autoconfiguration' [73]. The fundamental idea of the SLAAC is to enable hosts to configure addresses without a centrally administered address allocation scheme, per-host states in the network, and address management network entities. The second tool, stateful address autoconfiguration, is 'Dynamic Host Configuration Protocol for IPv6 (DHCPv6)' [54]. The approach of DHCPv6 is familiar to people accustomed to Dynamic Host Configuration Protocol version 4 (DHCPv4), but there are differences as well – such as DHCPv6 relying on ICMPv6 Router Advertisements to deliver default router information. Originally the IPv6 was agnostic on which protocol would be used to implement the 'Managed address configuration' (see Section 3.4.3), but at least by now, and for the foreseeable future, the DHCPv6 is **the** managed protocol. The main pull for DHCPv6 technology comes from specific deployments, such as in enterprises, where managed address configuration schemes are commonly preferred by the network administrators.

3.5.1 Stateless Address Autoconfiguration

In this section, we introduce the procedure for IPv6 SLAAC, which is the main mechanism for hosts to number themselves [41, 73]. The stateless approach is always required for link-local address configuration, and it is a lighter procedure for global address autoconfiguration than the DHCPv6. The SLAAC essentially has to be supported by all general-purpose IPv6 hosts. Other kinds of nodes, such as routers, special build hosts, or hosts in specific deployment scenarios, may be managed with manual address configuration or with DHCPv6-only.

The overall operation of SLAAC is illustrated in Figure 3.15 and briefly described herein before we discuss some of the details in more depth. The figure illustrates how a router is sending unsolicited Router Advertisements before the example's host is attached to a link (first message). Once a host joins a link, it first selects an IID for itself, creates a link-local IPv6 address, and does DAD for it (not shown in the figure). The host sends a Router Solicitation as it proceeds to configure more information for itself (the second message shown), in order to receive Router Advertisement faster and from all routers. Router Solicitation also ensures that received Router Advertisements are complete, as unsolicited Router Advertisements do not have to contain all the information [41].

After the host receives solicited, or unsolicited, Router Advertisement back (the third message shown), it combines the IPv6 prefix received from the Router Advertisement with the IID that the host has chosen. The host needs to perform DAD for the newly formed address, but it is not shown in Figure 3.15. This will result in a /128 IPv6 address configured for a host, which can be used for communications.

In the following sections we will look in more depth at some of the details related to IPv6 address management. In Section 3.12.1 we show a detailed example of SLAAC operation based on real-life packet capture, and explain how the address configuration procedures happen in practice.

Address States

During the SLAAC procedure, IPv6 addresses traverse through different states, which indicate the usability of the address for communications [41]. We will be referring to

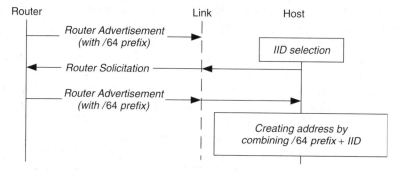

Figure 3.15 Overall operation of stateless address autoconfiguration.

these states later within this section when we describe the address autoconfiguration procedure, and in the rest of the book. The state descriptions are the following:

Tentative address: The uniqueness of the address on a link is being verified. The address can not yet be used for outbound traffic, and inbound traffic is discarded, except traffic related to DAD (see Section 3.4.7). The tentative address becomes preferred or deprecated once the DAD has succeeded.

Optimistic address: An address that is available for use but that has not yet completed DAD. This state was introduced by RFC 4429 [74] and is not yet supported by all IPv6 stack implementations. The stack should treat optimistic addresses as deprecated, and not to use them if better addresses are available. An optimistic address must not be used to send any packets that may influence neighbors' Neighbor Caches. The address becomes preferred or deprecated once the DAD succeeds. If the DAD fails, the use of the address is ceased immediately – which can possibly cause issues for upper layer protocols that have already started using the optimistic address. For full details of optimistic address usage, please refer to RFC 4429.

Preferred address: An address that can be fully utilized for all IPv6 communications. The preferred address becomes deprecated once the *preferred lifetime* expires.

Deprecated address: An address of which the *preferred lifetime* has expired. This address should not be used to initiate new outbound communications, if better alternatives are available. Existing sessions (e.g., TCP sessions) can continue to use the address, and incoming communications towards this address should be allowed. A deprecated address becomes invalid after the *valid lifetime* expires.

Invalid address: An address of which *valid lifetime* has expired, and which cannot be used for any communications.

Interface Identifiers

Before any address autoconfiguration mechanisms are used, a node generates an IID as was discussed in Section 3.1.6 that uniquely identifies the node on a link. The method for generating the IID depends on the access technology used. For example, in Ethernet-based interfaces the identifier is often built from a 48-bit IEEE 802 hardware address. The IID may also be chosen randomly, as is done when privacy extensions are used (see Section 3.5.5). Whatever the means for selecting an IID are, a node has to ensure that it does not conflict with any of the reserved IPv6 IIDs listed in RFC 5453 [77].

Creation of Link-local Addresses

A link-local address is created, initially to 'tentative' state, by combining an IID with a well-known prefix `fe80::/10` [4] so that the 10 left most bits are the well-known prefix, the 64 right most bits are the IID, and the bits between the prefix and the IID are set to zero. The resulting tentative address is checked for uniqueness using DAD (see Section 3.4.7) before allocating the address to an interface. However, immediately before starting to perform the DAD, the node must join to the Solicited node multicast address (see Section 3.2.4) group of the tentative link-local address using MLD.

Creation of Global Addresses

A global scope IPv6 address is created by combining a prefix obtained from Router Advertisement's PIO option, which has an autonomous 'A'-flag bit set, with an IID. A host may configure a multitude of IPv6 addresses from a single prefix, and in a case multiple prefixes are available, multiple addresses from different prefixes. Multitude of prefixes may be received via a single Router Advertisement or from multiple Router Advertisements received from multiple routers. The PIOs also contain *valid lifetime* and *preferred lifetime* information for each prefix, and these are used for advancing an address through the states listed above.

Once a 'tentative' state global scoped address is created, the DAD procedure is started just as for link-local address.

3.5.2   Dynamic Host Configuration Protocol Version 6

In this section we will briefly introduce Stateful DHCPv6, as it is a significant part of IPv6 suite. 3GPP-based hosts very commonly support Ethernet-based interfaces, such as IEEE 802.11 Wireless Local Area Network (WLAN) networks, where Stateful Address Autoconfiguration is often used. In 3GPP networks, stateful DHCPv6 is not supported for configuring addresses for hosts' cellular interfaces. For a long time, only Stateless DHCPv6 was supported for parameter configuration, but since 3GPP Release-10, prefix delegation has been part of 3GPP specifications (see more in Section 4.4.6). Hence we will explain the DHCPv6 PD in more depth.

While the DHCPv6 is an integral part of the IPv6 suite, it is running on top of UDP and hence could be considered to be an application layer protocol. The UDP ports used by the DHCPv6 are '547' for servers and relays, and '546' for clients. As with any UDP-based application, if the overall packet size exceeds the Path MTU, fragmentation and reassembly is required. To increase reliability and simplicity, DHCPv6 servers may try to keep message size under the minimum Path MTU of 1280 bytes.

Stateful DHCPv6

According to the specifications, a host that supports a DHCPv6 client, initiates Stateful Address Autoconfiguration procedures after detecting the M-bit on a Router Advertisement (see Section 3.4.3). Effectively, before this happens, a host already has autoconfigured a link-local address for itself with SLAAC.

The first DHCPv6 protocol step that the host performs is to send a DHCPv6 SOLICIT multicast message containing indications about the client interest to configure addresses for itself. The indication is given by the client including 'Identity association for non-temporary addresses (IA_NA)', or 'Identity association for temporary addresses (IA_TA)', or both, options in the SOLICIT message. If the client prefers to use a single message exchange, it may include a Rapid Commit option in the SOLICIT message [54].

DHCPv6 servers that receive the SOLICIT message sent by a client may respond with a DHCPv6 ADVERTISE message. From the ADVERTISE messages the client learns what DHCPv6 servers are available and what addresses they could provide for the client.

Once the client has chosen which DHCPv6 server to use, the client sends a DHCPv6 REQUEST unicast message to the DHCPv6 server. The client may populate the REQUEST with information it has earlier received in a DHCPv6 ADVERTISE message. After receiving the DHCPv6 REQUEST the DHCPv6 server will respond with a DHCPv6 REPLY message that contains addresses allocated to the client and possible other configuration information. The client needs to perform DAD to offered addresses similarly as for any other IPv6 address candidate that a host ends up with. If the DAD fails, the client has to send a DECLINE message to the server.

The DHCPv6 framework allows a DHCPv6 server to be located at a different link than DHCPv6 clients. In such a case, the link that the client is connected to has a DHCPv6 relay entity, which forwards messages between the DHCPv6 client and the server.

In addition to the messages used in the four-way handshake for address configuration (SOLICIT, ADVERTISE, REQUEST, and REPLY), a number of additional DHCPv6 messages have been defined as listed below. The DHCPv6 REPLY is a general purpose message used by the server in response to clients' SOLICIT, REQUEST, RENEW, REBIND, INFORMATION-REQUEST, CONFIRM, RELEASE, and DECLINE messages.

CONFIRM: A host that moves to a new link, for example due to changing physical connection or due to a reboot, can use the CONFIRM message to verify from DHCPv6 servers whether the addresses and configuration information it has is still valid [54].

RENEW: A host uses the RENEW message to extend the remaining lifetimes of its leases. The RENEW is unicasted to the DHCPv6 server from where the host got its address allocation [54].

REBIND: If a host has failed to renew its existing binding with the DHCPv6 server it has been communicating with, the host multicasts a REBIND message in an attempt to get some other DHCPv6 server to extend the lifetime of current lease [54].

DECLINE: A host may reject the address is offered by a DHCPv6 server, if the host detects the address is already being used on the same link [54].

RELEASE: A host releases addresses that it has been allocated with a RELEASE message [54].

RECONFIGURE: A DHCPv6 server can send an unicast RECONFIGURE message to a host in order to trigger the host to renew its address(es) and configuration data [54].

INFORMATION-REQUEST: A host may use INFORMATION-REQUEST to ask for various configuration parameters, such as recursive DNS server addresses [54].

RELAY-FORWARD: A DHCPv6 relay uses a RELAY-FORWARD message to encapsulate the original DHCPv6 message sent by a host [54].

RELAY-REPLY: A DHCPv6 server replying to a DHCPv6 message received within a RELAY-FORWARD message uses a RELAY-REPLY to encapsulate the response message [54].

LEASEQUERY: This message can be used to poll a DHCPv6 server for lease information regarding a specific host, based on an address. The information includes the contents of DHCPv6 options provided for the host of interest. This query does not affect states of the DHCPv6 server [78].

LEASEQUERY-REPLY: A reply message to the LEASEQUERY request. This option contains the information requested about a host, which may be for example, the host's identifier, the host's IPv6 address, and a delegated prefix [78].

LEASEQUERY-DATA: A bulk leasequery mechanism has been defined to allow efficient queries that result in more than one response [79]. The LEASEQUERY-DATA message is used alongside LEASEQUERY-REPLY to transport additional responses.

LEASEQUERY-DONE: The queried DHCPv6 server sends LEASEQUERY-DONE, when it has finished sending the bulk leasequery results [79].

DHCP Unique Identifier

Each node implementing DHCPv6 must have a unique DHCP Unique IDentifier (DUID). The DUID is the identifier used by DHCPv6 clients and servers to uniquely identify each other over time [54]. The following four DUID types have been defined so far:

DUID Link-Layer address (DUID-LL): DUID based on a link-layer address only [54].

DUID Link-Layer address plus Time (DUID-LLT): DUID based on a link-layer address and time [54].

DUID vendor-assigned unique identifier based on Enterprise Number (DUID-EN): Vendor-assigned DUID to a device based on vendor's enterprise number [54].

DUID Universally Unique IDentifier (DUID-UUID): DUID based on Universally Unique IDentifier [80].

Implementations may use the DUID type that suits them best – the peers they communicate with will only compare the values passed in these options. What is important is that nodes keep the DUID the same over time – so that the DHCPv6 server sees a host as one, and not an ever-increasing number of hosts.

DHCPv6 Prefix Delegation

A node that forwards packets between its uplink and downlink network interfaces needs a way to number the downlink interfaces. The official tool for that in IPv6 is DHCPv6 PD [81].

The DHCPv6 client that requests prefixes with DHCPv6 PD is called a Requesting Router (RR) and the DHCPv6 server delegating the prefixes is called a Delegating Router (DR). What is important to notice is that the use of DHCPv6 PD is orthogonal to the method RR uses to configure addresses for its uplink interface. This means that RR may have its uplink configured with link-local addresses only, addresses configured with SLAAC, or addresses configured with DHCPv6.

The prefix delegation works by the RR sending a DHCPv6 SOLICIT message with an Identity Association for Prefix Delegation (IA_PD) option. DRs that could delegate prefixes to the requester then respond with a DHCPv6 ADVERTISE message. The RR

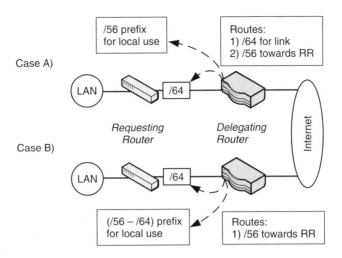

Figure 3.16   Example of DHCPv6 PD (case A) and DHCPv6 PD Exclude Option (case B).

chooses the DR it wishes to use, and sends the DHCPv6 REQUEST message, which DR replies with prefixes and possibly other information. After the successful message exchange, the network will start routing packets destined to the delegated prefix to the RR. Case A of Figure 3.16 illustrates the basic DHCPv6 PD in action: the DR has allocated a /64 prefix for the link between it and the RR. The DR has also delegated a /56 prefix for the RR, which the RR may utilize for numbering downlink Local Area Networks (LANs). Case A illustrates also how the DR has two routes towards one RR. This example of case A is based on fixed-network architecture where the link, such as a Virtual Local Area Network (VLAN), between RR and DR is numbered with global addresses. In some deployments, the link between RR and DR could be numbered with just link-local addresses, in which case the example's DR would only have one route.

In some networks, a special extension for PD is needed due to addressing limitations on said networks. An option called DHCPv6 PD Exclude Option may be used to exclude a prefix out from the delegated prefix, as defined in RFC 6603 [82]. The prefix that is excluded is commonly the prefix gateway (which may or may not be the DR) uses in the Router Advertisements sent to the link towards the RR and that may be used by RR for statelessly configuring its uplink's IPv6 addresses. Case B of Figure 3.16 illustrates the use of the DHCPv6 PD Exclude Option. The DR has excluded the /64 prefix used on the link between DR and RR from the prefix delegated to the RR. Therefore, the RR has one /64 prefix less to use on the downlink LANs when compared to Case A, but on the other hand, the gateway has to maintain only a single route towards the RR. The saving of this one route can be significant in some deployments (e.g. in 3GPP, see Section 4.4.6).

An important detail related to PD is that the RR assigns an Identity Association IDentifier (IAID) for each IA_PD option – this IAID is used consistently when the RR communicates with DR. The IAID is needed when the delegated prefix is being renewed, or if rebinding is needed due to, for example, RR reboot or RR's uplink connection disconnect/re-establishment cycle. Without IAID the delegated prefix could change too easily and too often, causing renumbering in LANs supported by the RR, and possibly causing unnecessary resource (address space) consumption in the DR.

Stateless DHCPv6

In many IPv6 deployment scenarios stateful address autoconfiguration is not required, or wanted. However, often the hosts need configuration for various pieces of information, for which Router Advertisement (Section 3.4.3) is not a suitable tool. The solution for such cases is to use a stateless variant of DHCPv6, the 'Stateless Dynamic Host Configuration Protocol (DHCP) Service for IPv6' [83].

The Stateless DHCPv6 is a subset of DHCPv6 RFC 3315. In essence, Stateless DHCPv6 defines that hosts must support DHCPv6 INFORMATION-REQUEST and corresponding reply messages. With the INFORMATION-REQUEST message hosts can ask for configuration information – the same information that the hosts could ask for with full stateful DHCPv6 implementation, but without having to implement all of the DHCPv6 protocol.

The lack of state in the stateless approach may cause issues if the configuration information changes on the network. In order to get hosts with stateless clients to refresh the information, a DHCPv6 server may include the Information Refresh Time option in the INFORMATION-REPLY message, if the client had indicated support for the option in the request message [84]. The Information Refresh Time option provides a maximum time that a client should wait before refreshing the configuration information, and hence setting the time with suitable value makes it possible to update configuration information on clients.

DHCPv6 Options

At the time of writing, more than 70 DHCPv6 options have been defined. In this section we have chosen to describe only the options that are discussed in this book or well-known to be useful in 3GPP access networks. Obviously, as only a subset of options is listed here, deployments, even some 3GPP networks, will make use of other options as well. A full and up-to-date option code list is available on the IANA web page: *http://www.iana.org/assignments/dhcpv6-parameters/dhcpv6-parameters.xml*.

DNS Recursive Name Server option: 'DNS Configuration options for Dynamic Host Configuration Protocol for IPv6 (DHCPv6)' describes options that hosts can use to obtain Recursive DNS Server (RDNSS) addresses and domain search lists [85].

RDNSS Selection option: A new DHCPv6 option for RDNSS selection is specified in RFC 6731 [86]. The RDNSS Selection option configures nodes with information which domains and networks a specific RDNSS has special knowledge of. The client can then use the information when selecting to which RDNSS a DNS query should be sent.

Rapid Commit option: The Rapid Commit option can be used to optimize DHCPv6 signaling into two messages: DHCPv6 SOLICIT and reply a for it. The optimizations save an extra RTT that would otherwise be spent on signaling. The client sending a Rapid Commit option commits to accept the information sent by a DHCPv6 server that also accepts the commitment with a single message exchange [54].

Session Initiation Protocol (SIP) Server Domain Name List option: A host implementing an IP Multimedia Subsystem (IMS) client can use this DHCPv6 option to find out a list of Proxy Call Session Control Function (P-CSCF)(s) domain names [87].

SIP Servers IPv6 Address List option: A host implementing an IMS client can use this DHCPv6 option to find out a list of P-CSCF(s) IPv6 addresses [87].

Option Request option: A host indicates to a server DHCPv6 options it wishes to receive by listing the option codes in the Option Request option [54].

Identity Association for PD option (IA_PD): A host performing RR functionality includes the IA_PD option to indicate its identity. This option can include other options, such as an IA_PD Prefix option [81].

IA_PD Prefix option: This option is transported inside an IA_PD option to communicate a prefix delegated to RR [81].

Prefix Exclude option: This option communicates a portion of a prefix on an IA_PD Prefix option that is excluded, and that must not be used by an RR for numbering the RR's downlink network [82]. The excluded prefix may be used on the RR's uplink, as is a case with 3GPP.

Information Refresh Time option: This option specifies an upper bound for the time after a client should refresh the information it has received from DHCPv6 [84].

Server Unicast option: DHCPv6 server uses this option to indicate to the DHCPv6 client that it is allowed to send unicast messages to this DHCPv6 server [54].

Port Control Protocol (PCP) Server Name option: This option communicates a PCP server's domain name (see Section 6.4.3). At the time of writing, the option is still under standardization by the IETF [88].

Address Selection option: This option communicates site-specific address selection policies. The option is under standardization by the IETF [89].

3.5.3 IKEv2

Even though Internet Key Exchange version 2 (IKEv2) is, as the name suggests, designed for key management (see Section 3.8.3), it also supports single /128-bit IPv6 address configuration [90]. Unfortunately, in the standard track IKEv2 RFC 5996 the IPv6 support is excessively simple, and for example does not support DHCPv6 or MLDv2 at all. An experimental RFC 5739 has been created to address the shortcomings [91]. Here we discuss only the experimental RFC's features, as that is the way IPv6 should probably be supported in the IKEv2 in the future.

In the case of IKEv2, the IPsec gateway can assign IPv6 prefixes for clients that establish Virtual Private Network (VPN) tunnels [91]. This enables the use of IPv6 on top of a VPN link and presentation of IPv6 as a virtual interface. As the configuration information related to client's address configuration comes in IKEv2, there is no need to use SLAAC or DHCPv6 for address configuration purposes. However, a client may ask for additional configuration information with DHCPv6.

The IKEv2 always assigns a unique /64 prefix for a client, and as the IIDs are negotiated, there is no need for a client to perform DAD procedures. Furthermore, as the IKEv2 configured link has only two nodes and no link-layer addressing, there is no need for

ICMPv6 Redirects or link-layer address resolution either. In this regard, the connection negotiated with IKEv2 is very similar to Point to Point Protocol (PPP) and 3GPP Packet Data Network (PDN) Connection.

3.5.4 Address Selection

IPv6 is designed to have fluent support for multiple simultaneously configured unicast addresses. Even in the simplest scenario, IPv6 hosts usually have at least two addresses configured: a link-local scoped address and a global scoped address. It is also very natural for a host to have multiple global addresses configured at the same time. For example, a network that a host is attached to may have multiple uplink Internet connection providers, each providing their own IPv6 prefixes for a host to configure addresses from. Another common example is a host with multiple network interfaces active at the same time: set of addresses will be configured from each interface.

The problem that arises from the multitude of available addresses is called *address selection*. Not only can the host itself have many addresses, but so can the destination for which a new connection needs to be created. The host initiating connection can obtain a list of addresses for a destination, for example, via DNS.

IETF has defined default source and destination address selection algorithms that hosts should follow when deciding which pairs of source and destination addresses to use and in which order [92]. In practice, this prioritization is commonly done by the host's operating system at the time when an application requests DNS resolution or calls a connection creation method. The operating system either automatically creates a connection using, at first, addresses of the highest preference, or returns a list of destination addresses in priority order to the application calling the DNS resolution API function.

Default Policy Table

To help prioritizing IPv6 addresses – both source and destination – a default policy table has been defined in RFC 6724 [92] and illustrated in Table 3.3. The table lists the IPv6 prefix, a related precedence of that prefix over other prefixes (higher value means higher precedence), a label used to find pairs of source and destination addresses that have a matching label, and a short description.

The default policy table is, as the name states, a default. Hosts may override specific rules if they "know better", and it will also be possible to administratively alter the policy table on a host: the IETF is defining a new DHCPv6 option for provisioning hosts with site-specific address selection policies [89].

Source Address Selection

The algorithm for source address selection works by taking a single destination IPv6 address at a time, and listing all source addresses that could be used to communicate with the destination address. The rules for selecting the preferred source address for the destination are briefly the following, in the priority order:

1. Prefer a source address that is the same as the destination address. In this case, communications are internal to a host.

Table 3.3 IPv6 address selection default policy table

Prefix	Precedence	Label	Description
::1/128	50	0	IPv6 loopback address [4]
::/0	40	1	Global IPv6 addresses [4]
::ffff:0:0:/96	35	4	IPv4-mapped IPv6 addresses [4]
2002::/16	30	2	6to4 addresses [66]
2001::/32	5	5	Teredo addresses [14]
fc00::/7	3	13	ULA addresses [7]
::/96	1	3	IPv4-compatible IPv6 addresses (deprecated) [4]
fec0::/10	1	11	Site-local addresses (deprecated) [4]
3ffe::/16	1	12	6bone addresses (phased out) [93]

2. Prefer a source address with the same or higher scope as the destination address. For example, prefer a global scope source address when a destination is also of global scope.
3. Avoid deprecated addresses, if preferred addresses are available (see Section 3.5.1).
4. Prefer a Mobile IP Home Address (HoA) over a Care-of Address (CoA) (see Section 3.7.2).
5. Prefer addresses from the interface that will be used to send a packet to a given destination.
6. Prefer a matching label. This means, for example, that 6to4 source addresses are preferred when the destination address is also 6to4 address.
7. Prefer temporary addresses over public addresses, by default (see Section 3.5.5).
8. Prefer a source address that has the longest matching prefix with the destination address.

Destination Address Selection

The algorithm for prioritizing a list of destination addresses uses the source address selection algorithm, described above, to find out what source address the host would use to communicate to a given destination address. The rules for prioritizing destination addresses are briefly the following, in priority order:

1. Avoid unusable destination addresses: these destination addresses may be such for which a host does not have suitable source addresses available, or destination addresses that are known to be unreachable.
2. Prefer addresses with matching scope: a destination address is preferred if a host has a source address available with the same scope.
3. Avoid deprecated source addresses: a destination address is preferred for which a host has a non-deprecated source address available (see Section 3.5.1).
4. Prefer a Mobile IP HoA: a destination addresses is preferred for which a host would use an HoA as a source address.
5. Prefer matching label: a destination address is preferred for which a host has a source address with a matching label.
6. Prefer addresses with higher precedence: a destination address that has higher preference based on the default address selection policy table is preferred.

7. Prefer native transport: a destination address that does not require encapsulation or protocol translation is preferred.
8. Prefer smaller scope: a destination address with the lowest possible scope is preferred, for example, a link-local scoped address pair is preferred to a global scoped address pair.
9. Use longest matching prefix: prefer a destination address that has the longest matching prefix with available source addresses in order to prefer addresses closest to each other in the network topology.
10. Leave the original order – for example the order as received from the DNS – of the destination address list unchanged.

On Sending Packets via Correct Network Interface

The source address selection rule 5, about preferring source addresses from the network interface that will be used for sending a packet to a destination, has special importance in certain deployments. Namely, if a packet is sent over a different network interface than where the source address is configured, packet drops may occur. Firstly, a reply for the sent packet will arrive on the interface where the address is actually configured. This is called asymmetric routing. In the case of asymmetric routing, the network on the path of the reply packet has not seen the outgoing packets, and hence may, based on firewall ingress filtering policies, drop the packet as being 'unsolicited'. Secondly, a router on the network used for sending a packet may employ ingress filtering for the packets coming from hosts in the same network, as described in RFC 2827 (BCP 38) [94] and in Source Address Validation Improvements (SAVI) [95, 96]. Essentially, the ingress filter on a router will ensure that a host cannot send packets with a source address that has not been given by that network. The main intent is to prevent misbehaving or malicious hosts from sending packets with somebody else's IP address. By doing this the filtering routers protects the Internet against attacks utilizing source address spoofing. Somewhat unfortunately, in the case of multi-interfaced hosts, the router performing filtering on one network might be unaware of the other legitimate addresses the host has configured from other networks to which the host is also connected at the same time. In some scenarios, essentially those that do not involve dynamic network selection, ingress filtering rules can be adapted to support multi-interfacing behavior – see RFC 3704 (BCP 84) [97]. In any case, if a host is dynamically selecting networks it attaches to, and hence networks have no way to adapt their ingress filters based on the hosts changing state, the host has to be careful to send packets only with source addresses legitimate on the interface used to send the packets.

3.5.5 *Privacy and Cryptographically Generated Addresses*

The source IP address used by a node in the packets it sends, by design, points back to the node. In some cases, such as when Network Address Translators (NAT) are in the network – as is increasingly often the case with IPv4 – the IP address seen on the Internet points to the NAT entity thus hiding the actual source of the packet. A static and public IP address comes with benefits, but also with some cons. The main benefits are stability and better connectivity options. The downside is decreased privacy: a static address can be used to identify a node in the Internet, thus enabling things such as user tracking and user identification.

Tracking based on IP address is a more serious issue in the case of IPv6. In IPv4, NATs are becoming increasingly common, and the IPv4 address is always bound to a specific network (unless Mobile IPv4 is used [162]). However, with IPv6 and SLAAC, a node can create an IPv6 address by using an IID derived from network interface's unique identifier (see Section 3.1.6), such as the link layer's MAC address [73]. The MAC addresses are usually burned into hardware and hence very static in nature. The problem with hardware-derived IIDs is that even if the node's point of network attachment changes, and an IPv6 prefix changes, the IID remains static! Hence with IPv6, it is possible to track a node in the Internet even when node a changes its point of network attachment (the tracking would ignore IPv6 prefixes and just track IIDs). The solution for this problem, in the case of SLAAC, is presented in RFC 4941 [11]. This RFC introduces the concept of Temporary Addresses, which are used only for a short while – from hours to days – after which a new Temporary Address is generated.

In some networks, IPv6 address changing nodes can become problematic for the network administrators, for example, if access control lists require IPv6 address information [98]. The problems related to combining needs for privacy and management calls for additional solutions. IETF is currently working on stable but privacy enhanced addresses for SLAAC. This solution proposes an algorithm to generate IIDs in a manner that results in constant IIDs within each subnet. The algorithm takes a set of input parameters, including a network prefix, a secret key, a network identifier, such as Service Set Identifier (SSID), the interface index, and the DAD counter. The algorithm produces a random looking result, which is very hard to guess, but is always the same given the same input parameters. This solution of 'stable privacy enhanced addresses' is still at the working group stage, and may significantly change prior to publication as standard.

The tracking enabled by static addressing is not only a problem with SLAAC, but can also appear to a lesser extent with other address allocation schemes. For example, if DHCPv6 would always provide the same IPv6 address for a node, tracking is made easier as long as the node stays within a network. To avoid problems with static addresses, DHCPv6 supports allocation of both Temporary Addresses (DHCPv6 IA_TA option) and Non-temporary Addresses (DHCPv6 IA_NA) option. A node wishing to increase privacy would prefer using temporary addresses over non-temporary addresses.

In addition to the abovementioned solutions, implementations may sometimes use other approaches. These include randomizing IID when connecting to a network, but not changing the IID during the lifetime of a network connection. This fully solves the problem of tracking host movement through different access points, and also within a network as long as network connection lifetime is not excessively long. In some other scenarios, like 3GPP, the IID may also be received via layer 2, in which case the responsibility of avoiding static IIDs is left for the layer 2 implementation (for the network entity providing the IID to a host).

3.5.6 Router Selection

Internet hosts are sometimes connected to networks that have more than one first-hop router, and sometimes the hosts are simultaneously connected to multiple networks.

Multiple routers can be present on one link, for example, in order to provide redundancy or to provide access to different networks – such as the Internet and a private network. Hosts will configure IPv6 addresses from all prefixes they receive advertisements for, independent whether they originated from one or more routers. From the multitude of configured addresses, hosts choose the address to use based on source address selection algorithms. However, when a host is sending packets, it needs to determine which router to send the packets. Note that a host is not mandated to keep track which router advertised each prefix, which may cause issues (see Section 3.5.4). A tool for improved router selection is defined in RFC 4191, 'Default Router Preferences and More-Specific Routes' [68].

RFC 4191 defines preferences for both default routers and for more specific routes. Specifically, ICMPv6 Router Advertisement is enhanced with bits that define whether a router is to be consider as a high, medium (default), or low priority default router. With these bits a network administrator can define which routers the hosts should prefer when communicating to the Internet.

For the scenarios where more fine-grained router selection is needed, RFC 4191 allows definition of specific routes with the help of the Route Information Option (RIO). This option allows configuration of variable length prefixes and preferences for using the router when sending packets to matching destinations. The preference values for specific routes are the same as for default router configuration [68].

The combination of default and more specific rules allow configurations, for example, where a router is configured to be a low priority default router, but a high priority router for traffic destined to specific networks. This property can be very useful, for example, in cellular networks for implementing a traffic offloading configuration, if a mobile handset is simultaneously connected to a cellular and a local area network. The low priority default router in the cellular network effectively guides the mobile host to prefer local area networks for Internet traffic (provided that the router in the local area network is indicating either medium or high priority default routing services).

The router selection influences source address selection procedures, because for egress and ingress filtering reasons use of an IPv6 address configured from one access network on another access network usually does not yield a working connection (see Section 3.5.4). Even if the network would allow asymmetric traffic, the purpose of router prioritization would be partially lost as downlink traffic would likely come via a less preferred router.

3.6 IPv6 Link Types and Models

The IPv6 is a layer 3 protocol in the Open System Interconnect (OSI) model. As such, IPv6 can be transported over great variety of layer 2 protocols. From the IPv6 point of view, the layer 2 protocols and the physical link types they provide can be divided into two main categories: point-to-point and shared links. In this section, we will focus only on the physical link types, listed in Table 3.4, that are commonly used by the 3GPP networks and often supported by 3GPP capable mobile handsets.

There are also specifications for transmitting IPv6 over various other kinds of link types, which we are not addressing in this book. These specifications might be of interest for readers and so are listed in Table 3.5.

In addition to transporting IPv6 over physical mediums, various solutions for transporting IPv6 over IPv4 or other protocols have been defined. In this book, we will further discuss tunneling based approaches in Chapter 5.

Table 3.4 Link types commonly supported by 3GPP handsets

RFC	Description	Reference
RFC 2464	Transmission of IPv6 Packets over Ethernet Networks	[99]
RFC 5072	IP Version 6 over PPP	[100]
RFC 6459	IPv6 in 3GPP EPS	[101]

Table 3.5 Link types not covered by this book

RFC	Description	Reference
RFC 2467	IPv6 over FDDI Networks	[102]
RFC 2470	IPv6 over Token Ring Networks	[103]
RFC 2491	IPv6 over Non-Broadcast Multiple Access (NBMA) networks	[10]
RFC 2492	IPv6 over ATM Networks	[104]
RFC 2590	IPv6 over Frame Relay Networks	[105]
RFC 3146	IPv6 over IEEE 1394 Networks	[106]
RFC 4338	IPv6, IPv4, and ARP over Fibre Channel	[107]
RFC 4944	IPv6 over IEEE 802.15.4 Networks	[108]
RFC 5121	IPv6 over IPv6 Convergence Sublayer over IEEE 802.16 Networks	[109]
RFC 5692	IP over Ethernet over IEEE 802.16 Networks	[110]

Perhaps to make some self-parody of IETF community's interest in defining *IPv6 over foo*, where the foo is practically anything that could theoretically transport IPv6 packets, IETF has published April 1st RFCs about how to transport IPv6 over avian carriers [111] and social networks [112].

3.6.1 IPv6 over Point-to-point Links

In IPv6 protocol-wise the most significant implications of point-to-point links are the ones caused to IPv6 Neighbor Discovery procedures. Specifically, as by definition a point-to-point link has only two peers, link-layer addresses are not necessary and hence do not need to be discovered either. Furthermore, multicasting loses some meaning in these types of links – the sent packets always end up with the other peer. Of course, nodes keep on listening to multicast addresses, and packets can be sent to 'all-nodes', 'all-routers', and similar multicast addresses. Support for multicast is required to allow IPv6 stacks to behave in link type agnostic manner. The lack of need for link-layer address discovery on these type of links has led, in some cases, to implementation optimizations as described in RFC 3316 [3]. Namely, implementations have been able to cut down on address resolution and next-hop determination logic. This has caused some issues that we discuss in Section 4.9.1.

The point-to-point link setup protocols often come with features for negotiating some higher layer parameters. For PPP the possible parameters are listed below, and for the 3GPP network's links in Section 4.4.7. The IKEv2 can also be used to negotiate a point-to-point link, with client and RDNSS addresses, as we discussed in Section 3.5.3.

Point-to-Point Protocol

The 3GPP bearer model is point-to-point. This link type significantly resembles IPv6 over PPP [100] and is described in depth by 3GPP 23.060 [113] and 23.401 [114]. In fact, the 3GPP specifications did actually support PPP type of Packet Data Protocol (PDP) context until Release-8, but it saw very little to no real use. However, IPv6 over PPP [100] could be used on circuit switched data connections.

While the Internet Protocol Control Protocol (IPCP) [115] of PPP allows negotiation of an IPv4 address itself, the IPv6 Control Protocol (IPv6CP) [100] is used only to negotiate IPv6 IIDs. The IID negotiations allow skipping of DAD procedures for IPv6 addresses constructed using negotiated IID, as that IID is known to be unique on the link.

The current IPv6 over 3GPP system is described in depth in 'IPv6 in 3GPP Evolved Packet System (EPS)' RFC 6459 [101]. In 3GPP access to the IID is negotiated during the bearer establishment phase very similarly to PPP, but also other information such as RDNSS addresses and P-CSCF addresses can be negotiated.

As the PPP type of PDP context has been dropped from the 3GPP specifications, we will not look in any more depth in this book at how IPv6 is transported over PPP.

3.6.2 IPv6 over Shared Media

The cellular link used in 3GPP networks is always a non-shared point-to-point link. Of course, it is very common for mobile handsets to have a shared link type available, when the handsets are attached to non-3GPP accesses, such as home or hotspot WLAN networks. It is also fairly common to have a shared local area link, for example in tethering scenerios, or when a shared medium is used for peer-to-peer communications.

The majority of the shared types of links are based on Ethernet technology, for which the IPv6 usage model is defined in RFC 2464 'Transmission of IPv6 Packets over Ethernet Networks' [99]. The most common forms of Ethernet technology are wired and wireless LANs, defined by IEEE 802.3 and 802.11, respectively. The same Ethernet technology, with slight adaptations, is also used by other link layers, such as IEEE 802.15.4 [108] and possibly in Bluetooth Low-Energy as well, for which standardization work is currently ongoing in IETF [116]. The link-layer addresses in Ethernet are 48-bit IEEE 802 addresses.

For transmitting IPv6 packets over Ethernet networks, nodes need to find out a link-layer address corresponding to a destination IPv6 address. Please see Section 3.4 for details about how a link-layer address for an IPv6 address is resolved. In the case of multicast addresses, a special Ethernet multicast address of $33:33:xx:xx:xx:xx$ is reserved. A function that maps IPv6 multicast addresses to Ethernet multicast addresses works by appending the last 32 bits of an IPv6 multicast address to fixed $33:33$, hence creating a 48-bit multicast MAC address. An additional feature to this technique is defined in RFC 6085, which allows mapping of IPv6 multicast addresses to unicast Ethernet link-layer addresses in cases where the sending node knows that it is enough to send the message as a unicast to only one receiving device [117]. Obviously, for this to work, the receiving node must not drop multicast IPv6 packets that it has received on its unicast MAC address.

3.6.3   Link Numbering

As per the IPv6 addressing architecture, described in Section 3.1, each IPv6 interface typically has to have a link-local address (with few exceptions, e.g. with some tunneling solutions [160]). Then nodes on a link will always have at least a link-local addresses. This means that IPv6 does not have a concept of *unnumbered* links. Whether a link has addresses with a wider scope than link-local in use depends on the time and deployment scenario. When a link is without routers advertising global or ULA prefixes, the link-local addresses are the only means of communication. In some deployments inter-router links may not have other than link-local addresses by design, as for IPv6 packet forwarding link-local addresses are enough.

Whenever a link has a prefix, the prefix is typically 64 bits long. In many link types, such as Ethernet, the prefix length of 64 bits is fixed [99]. Every now and then, mostly from the old habit of extreme IPv4 address conservation needs, people tend to propose longer than 64-bit prefixes for scenarios where only few nodes are using the prefix. In IPv6, addresses are so plentiful that such extreme conservation is not necessary, especially when it comes with increased complexity. However, there are also other reasons why the prefix length is best kept at the current state. Having a 64-bit prefix on all link types allows simplification of IPv6 stack implementations, and also ensures that features such as privacy addresses (see Section 3.5.5) and Neighbor Discovery Proxying (see Section 3.4.10) are available for hosts.

A special case of the above, and a requirement for network link numbering, has appeared for inter-router point-to-point links and is described in RFC 6164 [118]. If a 64-bit global prefix is used in a point-to-point link not using link-layer addresses, a ping-pong attack could be launched by sending a packet that would keep bouncing between routers. For link-layers using links-layer addresses resource consumption attack is possible by forcing routers to create Neighbor Cache entries to INCOMPLETE (see Section 3.4.4) state. An additional yet small benefit of /127-prefix usage is an IPv6 address space conservation. The Neighbor Cache resource consumption attack vector can impact hosts as well, especially those that are performing tethering functions. These hosts need to implement some means for mitigating these resource-consuming attacks, such as by rate limiting ICMPv6 messages and cleaning caches from INCOMPLETE entries [72].

Numbering of Star Topology

On some occasions, designers build networks that have a single link layer implemented on top of a star topology consisting of point-to-point links from a central point towards end nodes. As the network is considered – from an IPv6 perspective – to be a single link, it also has a shared prefix for all nodes on the branches. In such a deployment all Neighbor Discovery messages, such as those for DAD, have to be bridged via a central point and multiplied to all branches. Simple bridging of all multicast traffic to all branches may consume resource unnecessarily, and hence solutions such as switching may be used. An efficient switching requires the central point to keep perfect awareness of addresses in use in the various branches, and that increases the required state and risks of failures. Therefore, the best known practice to number a star topology is to allocate a unique prefix for each branch of the topology, as is done in 3GPP networks.

3.6.4   Bridging of Link Types

Bridging of Ethernet-type of links together is a fairly common procedure and usually works quite easily. The challenges are mostly related to differences in MTU and possible address conflicts occurring at the moment of bridge establishment. However, bridging shared links with point-to-point links, or links with link-layer addresses to links that do not have link-layer addressing, is a much more troublesome task. The difficulties mostly arise from the nodes' assumptions about link characteristics, which cease to be correct after a bridge has been established. For example, a node that only sees a point-to-point link without link-layer addresses and is unaware of that link being bridged to a shared link that has link-layer addresses, may cause problems for other nodes' DAD procedures due to its assumption that DAD is never needed on a link it is attached to. The challenges related to bridging of different types of links continues to be topical in IETF.

3.7   Mobile IP

An IPv6 address is anchored to a specific point in the Internet's network topology – the Internet's routing system delivers packets to the location in the network topology indicated by the packets' destination addresses. Generally speaking, as a node moves from the first network point of attachment to the second, the node ceases to be able to receive packets sent to the address that the node had at the first attachment point. Furthermore, if the node attempts to use the address from the first network in the second network, the packets are probably going to be dropped by the second network's ingress filters, and even if not, the return packets will be delivered to the first network, not to the second network where the node actually is. For applications and transport protocols, the change of IPv6 address during a node's movement results in disconnected transport layer sessions. These transport layer sessions, such as UDP and TCP, are always bound to specific IPv6 addresses. The disconnecting transport layer sessions and challenges in receiving incoming connections are the key problems for moving nodes, that is, the MNs.

A lot of research effort has been spent on finding good network layer solutions for node mobility problems. In the early days of IPv6, the Mobile IPv6 was planned to be an integral part of IPv6. This goal has proven to be too challenging. Lacking proper network layer solutions, moving nodes have generally relied on application layer solutions, such as fast session re-establishments when addresses change and connections are lost. In this section we will provide brief summary of the network layer approaches IPv6 protocol suite currently has to offer.

3.7.1   Detecting Network Attachment

The first step in solving problems caused by changes in the point of network attachment is to detect that the change has in fact occurred. The simplest approach, from an IPv6 point of view, is to receive Router Advertisements that contain information not previously known. A new Router Advertisement indicates that something has changed – at least a new router has become visible. In some scenarios, early indications of possible changes can be received from layer 2 indications, such as change from a cellular network to a WLAN network. However, due to developments in network based mobility approaches, it is not always that straightforward to rely on layer 2 events. Furthermore, sometimes the layer 2 can change without impacting layer 3.

IETF has defined a layer 3 based tool for faster detection of changes on a network point of attachment. The solution is called Detecting Network Attachment (DNA) [119]. The DNA procedures are triggered from layer 2 indications about *possible* change events taking place, such as layer 2 connection being (re)established. The first step that DNA takes is to assume that all network link specific information may have become obsolete and may need to be verified. The verification happens by performing IPv6 Neighbor Discovery procedures, such as sending unicast Neighbor Solicitations for previously learned routers in parallel to multicast Router Solicitation message to quickly learn about new routers. A DHCPv6-using node would need to verify that the information learned via DHCPv6 is valid by reaching the DHCPv6 servers.

If the responses to solicitations indicate that the attachment point to a network has not changed, or has changed but without implications to the networking layer, the node can continue using the information it previously had. However, if the point of attachment has changed, fully or partially, the node must flush the information that is no longer valid, such as IPv6 addresses that the node has autoconfigured from routers that are no longer reachable. It is worth noting that host movement may not warrant flushing of all information, but only pieces of information received from routers and DHCPv6 servers that are no longer reachable.

The benefit for applications of (fast) detection of network attachment change is that applications will quickly be informed about disconnected transport layer sessions, and hence can immediately attempt reconnection. Obviously, the DNA can be used to speed up any mobility solutions that the node may be supporting.

3.7.2 Host-based Mobile IP

One significant solution approach to the IP mobility problem is to make MNs establish and maintain tunnels to the network that their addresses topologically belong. The mainstream solution for this approach is called Mobile IP, which was originally designed for IPv4 [162]. The first MIPv6 [120] specification was published by the IETF in 2004. Since then the base specification has been revisited in 2011 [35] and a significant dual-stack feature introduced in RFC 5555 in 2009 [121]. Besides these major documents, a set of additional and supportive standards has been created, for example, related to network mobility, fast handovers, route optimization, prefix delegation, management information base, TLS-based security, and flow based mobility. MIPv6 with its extensions is a large protocol worthy of a dedicated book. In this section we will only describe the functionality on a very high level.

In the world of MIPv6, MNs are configured with IPv6 addresses from the home network, called HoAs, and from visited networks, called CoAs. The HoAs are the ones that provide mobility and reachability services for transport layer sessions and applications. In order to support MIPv6, a host implements functions specified for an MIPv6 MN, and a router in its home network implements a function called MIPv6 Home Agent (HA). The HA provides an anchor point for HoAs and tunneling services. The MN establishes IPv6 over IPv6 (or IPv6 over IPv4) tunnels from the visited networks to the home network by using the CoA. All the packets sent with the HoA are encapsulated and sent to the HA.

When HA receives these encapsulated packets, it performs decapsulation and forwards the decapsulated packet to the Internet. Similarly, the HA tunnels all downlink packets destined to the MN's HoA to the MN's current CoA.

Figure 3.17 illustrates MIPv6 in action. In the step 1, an MN is connected directly to the home network, where the MN is using IPv6 HoA without using MIPv6 – there is no need to tunnel as the MN is at home. Step 2 illustrates MN movement to a visited network and establishment of a tunnel to the HA. In step 2 the MN can continue using HA for reachability and mobility services, but all packets sent using HoAs have to be routed via the HA. Any sessions that the MN had ongoing during step 1 will continue working after step 2, as the traffic using HoA is switched to using the tunnel.

To allow creation of advanced MIPv6-aware applications and user-space implementations, MIPv6 specifies extensions to the socket API in RFC 4584 [122]. We will not look at the details of this API here, but instead recommend interested readers to look at the reference.

3.7.3 Network-based Mobile IP

One reason why the host-based Mobile IP has faced deployment challenges is the requirement for explicit support on hosts. In order to alleviate that concern, IETF standardized a network-based IPv6 mobility solution called Proxy Mobile IPv6

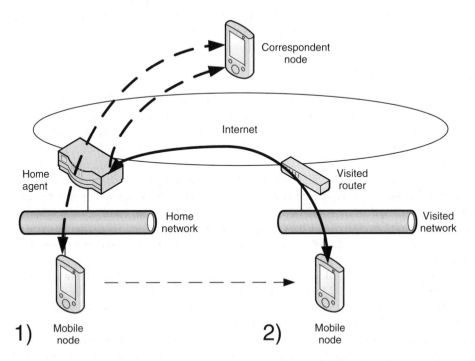

Figure 3.17 MIPv6 illustration. Case 1 shows data flow for the mobile node at home, and case 2 for the mobile node on a visited network.

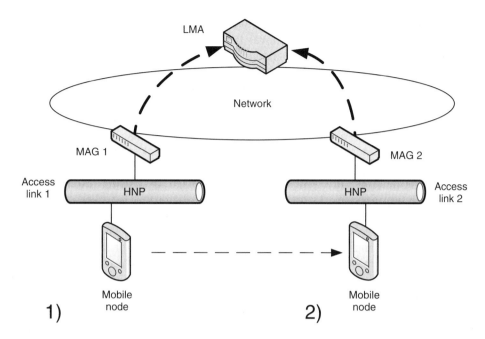

Figure 3.18   Illustration of PMIPv6 and an MN moving while retaining HNP.

(PMIPv6) [123], and extended that to also support mobility for IPv4 [124]. The essence of PMIPv6 is the idea of placing routers with Mobile Access Gateway (MAG) roles into access networks for serving moving hosts with the same Home Network Prefix (HNP) at each point of network attachment. MAGs are controlled and coordinated by a Local Mobility Anchor (LMA), which acts as an anchor and a tunnel endpoint very much like the Home Agent (HA) in MIPv6. Figure 3.18 illustrates the PMIPv6 setup by showing one LMA and two access networks, both having an MAG. Initially, MAG '1' obtains an HNP from the LMA and advertises that for the MN. When the MN moves from the first access network to the second, MAG '2' of the second access network learns from the LMA what is the HNP that should be advertised to the MN.

The mobility between accesses can be further divided into two categories: inter-access mobility and intra-access mobility. Inter-access mobility refers to scenarios where the access networks, between which the MN changes, are using different technologies, or are otherwise seen by the MN as different points of network attachment (for example, WLAN networks with different SSIDs). The intra-access mobility refers to scenarios where, from the MN's point of view, the point of network attachment does not change. An example of intra-access mobility is MN changing between WLAN access points while being connected to the same SSID.

While the original PMIPv6's idea of providing mobility without any host impacts sounded tempting, the reality proved to be more complex. The PMIPv6 can indeed ensure that the same HNP is available in the new access link that the MN moves to, but while that works for intra-access mobility scenarios, it is unfortunately not enough to provide mobility also for inter-access scenarios. The inter-access scenarios are problematic, because several operating systems are not prepared for such address stability provided by PMIPv6. For example, if the handover between two access networks is done in a break-before-make manner, some operating systems release configured IPv6 addresses at the instant of the first access network being disconnected, and even if the same IPv6 address is configured again a moment later from the second access network, all transport layer sessions have already been terminated and hence seamless mobility is not achieved. In some operating systems, sockets are strictly bound to network interfaces, and hence even if address stability would be maintained, the sockets are unable to switch using the second interface. Furthermore, in the case of make-before-break, many operating systems refuse to configure the same IPv6 address for more than one network interface at a time.

The inter-access mobility problem could be solved by adding explicit support capabilities to host operating systems. Several proposals for that have been proposed in the IETF, but none has yet to become a standard. The proposals are essentially proposing means to hide inter-access mobility from an IPv6 stack by introducing an additional layer between IPv6 and access network interfaces. This layer is often called a 'logical interface', whose purpose is to hide changes and details of real physical interfaces, thus hiding possible breaks during access switching and thereby avoiding issues with the current stacks' assumptions. The work in IETF has not been completed due to various issues related to such schemes: how are MTU differences and changes handled, how will Neighbor Discovery be managed, how can point-to-point links and shared links be combined seamlessly, how are multicast transmissions handled, and so on? It remains to be seen if PMIPv6-based inter-access mobility happens.

3.8 IP Security

In the early days of IPv6, support for IPsec was designed to be mandatory for all IPv6 implementations. The mandatory security support requirement was one justification for why IPv6 ought to be superior to IPv4. However, as the years passed, IPsec and especially deployment of shared Public Key Infrastructure (PKI) system, which would have been needed to get the envisioned security properties, proved to be too complex to deploy. With IPv6 implementations emerging, it soon became clear that the market decided that IPsec should be an optional feature for the stack and usually supported only when needed for Virtual Private Network (VPN) uses. Furthermore, since the 1990s, other options, such as TLS and secure shell, had become popular and diminished the need for IPsec. These facts are nowadays acknowledged even by the latest revision of the 'IPv6 Node Requirements' RFC 6434 [1], which states that nodes should – not anymore 'must' – support IPsec [125].

IPsec is described only superficially in this book, as there are no mandatory host requirements for IPsec in the 3GPP. Generally in 3GPP hosts, as with any hosts, IPsec is only used for VPNs. Readers who are interested in IPsec should look at specific IPsec literature.

While perhaps obvious, it is worth noting that while IPsec is optional, secure and solid design and implementation of any component of the IPv6 suite is a hard requirement in today's world.

The IPsec architecture RFC 4301 [125] builds on the following key components: security protocols, security associations, key management, and cryptographic algorithms. These are briefly introduced in the following sections.

3.8.1   Security Protocols

The IPsec protocol suite offers services for authentication, integrity protection, anti-replay, access control, and encryption. All the other services except encryption can be achieved by use of AH [38]. With ESP [39] all the benefits of AH can be achieved, but encryption is also provided. This is one reason why RFC 6434 requires IPsec capable nodes to implement ESP, while AH is optional [1].

Both AH and ESP can be used in two modes: transport and tunnel. In the transport mode the security is provided for the next layer protocols, such as for TCP and UDP, while in the tunnel mode the full IP packet is secured. The selection of the usage mode depends on the deployment scenario. For example, tunnel mode can be used to hide a tunneled packet's sender and recipient identities from eavesdroppers, who would only see the tunnel endpoint addresses. Tunnel mode can also be used to transport an IP address family that is not otherwise transportable; for example, IPv6 packets can be transported inside IPsec tunnels over IPv4 networks. One practical and typical use case for tunnel mode is an IPsec tunnel that securely tunnels and connects all IP-traffic between an enterprise's sites.

The header structure of AH is illustrated in Figure 3.19 and ESP in Figure 3.20. An interesting property of the ESP header is that the payload field may actually contain a full IP packet in the tunnel mode. In the transport mode the ESP header's payload contains destination options, transport layer protocol and the payload of the transport layer protocol. The AH header is always inserted between the external header and the encapsulated packet in the case of tunnel mode, and before the destinations options header in the case of transport mode. Figure 3.21 illustrates how the AH and ESP headers are located in both transport and tunnel modes.

3.8.2   Security Associations

A decision as to which traffic IPsec processing should be performed is based on Security Associations (SAs), which may have been set up with IKEv2. An SA, identified by a Security Parameters Index (SPI), is set up for one direction and it describes to which destination address (or destination and source address) the SPI applies, whether AH or ESP is to be used, what is the next layer protocol that the SA relates to, which keys are used, and to which Traffic Class the SPI applies. The SA is stored in a Security Association Database (SAD). It is important to note that two security associations, one for each direction, are needed for protecting bi-directional traffic.

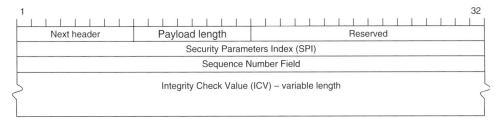

Figure 3.19   Authentication header.

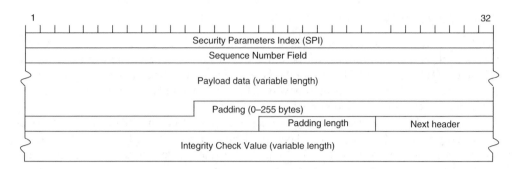

Figure 3.20   Encapsulating security payload header.

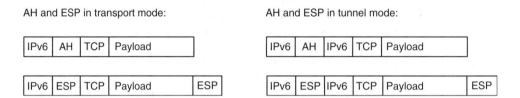

Figure 3.21   AH and ESP header placement illustrative example.

3.8.3   Key Management

All security services provided by the IPsec require distribution of keys. The distribution can be manual, which does not scale very well, or performed by a key management protocol. The scalability issues related to manual key distribution and lack of globally available automated key distribution mechanisms were the main reason why markets decided to not widely implement IPsec – except for VPN uses. The IKEv2 [90] is the main protocol for key management. IKEv2 can also be used to set up SAs between nodes.

3.8.4   Cryptographic Algorithms

To enable interoperability between different nodes, a set of mandatory cryptographic algorithms has been listed in RFC 4835 [126]. The mandatory encryption algorithms are: NULL, AES-CBC with 128-bit keys, and TripleDES-CBC. The mandatory authentication algorithm is HMAC-SHA1-96. In this book we will not look at the details of these algorithms.

3.8.5   MOBIKE

As described in Section 3.7, nodes can move and in the case of movement, IPv6 addresses change. The host-based mobility solutions utilize tunneling to enable use of static HoA. The IPsec in tunnel mode is architecturally very close to host-based mobile IP, and therefore 'IKEv2 Mobility and Multihoming (MOBIKE)' [127] extensions have been defined to allow changing of the outermost, 'CoA', IPv6 address used to encapsulate the tunneled IPv6 traffic. As MOBIKE allows the updating of SAs after movement, or if new IP addresses become available on additional network interfaces, both break-before-make and make-before-break are supported.

MOBIKE is an attractive technology to provide mobility in the VPN scenarios, as it does not require new network entities and as the security gateway already provides an anchor point in the network. Furthermore, while the host changes are a problem for host-based mobile IP, the changes are not as big as a problem for MOBIKE as the host has to be already modified to support IPsec tunneling.

3.9 Application Programming Interfaces

The introduction of IPv6 brings changes also to Application Programming Interfaces (API), including users of APIs and to the implementations of APIs. Depending on the style of the interface the changes vary from minor to significant. The APIs that provide IP address family agnostic usage approaches do not necessarily need any visible API changes other than those required for allowing explicit use of IPv6 by applications that do care. On the other hand, APIs that always require explicit IP-version handling by applications are the ones with most impacts.

In all cases, actual implementations of APIs have to become IPv6-aware, except special cases where an API implementation uses an additional layer of API and thus can remain IP-version unaware. The abstract API implementations, in the end, have to use an IP-version-aware API to actually transmit data. In that API layer possible issues with IP-versions arise, and APIs may need to implement functionality such as Happy Eyeballs – see Section 3.9.4.

3.9.1 Socket APIs

One of the APIs that are impacted a lot by IPv6 is the Portable Operating System Interface for uniX (POSIX) API managed by the Austin Group. The POSIX API is a textbook example of an API that requires extreme awareness of IP-version from the API users. IETF has documented implications to 'socket' style interfaces in several RFCs, including 'Basic Socket Interface Extensions for IPv6' [128], 'Advanced Sockets API for IPv6' [129], 'Socket Interface Extensions for Multicast Source Filters' [130], and 'IPv6 Socket API for Source Address Selection' [131]. The API implications are not limited to the listed documents and the work is still ongoing, for example, to add API functions for improved multihoming support [132].

3.9.2 Address Family Agnostic APIs

Many, if not most, of the modern operating systems provide high level APIs for Internet communications. These APIs completely hide IP address family related matters from application programmers. Often, a connectivity API is provided that only requires the name and port number of the destination.

An example of an API that allows 'connection by name' is provided by Qt's QAbstractSocket class, which provides APIs for address family agnostic program-ming, while allowing socket API-style programming as well (see specification in web `http://qt-project.org/doc/qt-4.8/functions.html`). An application using `QAbstactSocket`'s `connectoToHost` function can remain address family agnostic by providing just the domain name and port number of the remote host. The Qt framework then performs the necessary DNS queries and chooses whether to use IPv4 or IPv6 for the communications, and performs possibly required fallbacks in the case when the preferred address does not yield a working connection. Once a connection has been set up, the application can start using the connection.

3.9.3 IP Address Literals and Unique Resource Identifiers

Applications may encounter textual IP address literals, such as IPv4 addresses like `192.0.2.1` or IPv6 addresses such as `2001:db8:0000:42::30`, entered via user interfaces (not an API as such, but nevertheless an interface) or passed as URIs (see below). Applications have to be prepared to handle both types of addresses, including reserving space for displaying long IPv6 addresses (if the application displays addresses). In the face of literal IP addresses, applications may need to be able to use address family aware APIs correctly, unless the underlying API allows simply passing of IP literals without the caller being itself aware of the addresses.

URIs are used by several application level protocols, such as HyperText Transfer Protocol (HTTP). The IPv6 addressing format used in URIs is defined in 'Uniform Resource Identifier (URI): Generic Syntax' [133], which basically states that IPv6 address literals are to be presented inside brackets as discussed in Section 3.1.9. The brackets help to distiguish colons used inside of the IPv6 addresses from the colon used to mark the used port. For example, an IPv6 address of `2001:db8::2` is presented as `[2001:db8::2]` and when combined with a port number, for example '80', the address is presented as `[2001:db8::2]:80`. In the case of IPv6, a full HTTP URL could look like: `http://[2001:db8::2]:80/index.html`. When it comes to applications, an IPv6 formatted URI does not always require IP address family awareness, as an IPv6 URI may be passed to an HTTP loader API as simple strings. It will then be the implementation behind HTTP loader API that has to be able to handle the parsing of an address literal given through its API. However, if an application performs any parsing of the URI by itself, it must be aware that, in the case of IPv6, the URI may contain brackets and more than one colon, and that the presence of a colon does not by itself indicate the presence of a port number.

3.9.4 Happy Eyeballs

In the Internet, IPv4 and IPv6 may traverse significantly different paths between two dual-stack enabled peers. This means that the path properties and functionalities may be different for IPv4 and IPv6. Sometimes either IPv4 or IPv6 may be fully broken, have significantly different latencies, or may be not working between some specific peers. IETF has defined the concept of 'Happy Eyeballs' around this issue [134].

The Happy Eyeballs concept means that the end user is kept as unaware as possible about possible failures in one address family. In practice a host, and often the implementation of an API, must perform quick fallback to another address family if the first one tried does not yield successful results. For example, if an application first attempts to create a connection to a destination with IPv6, but it does not succeed reasonably

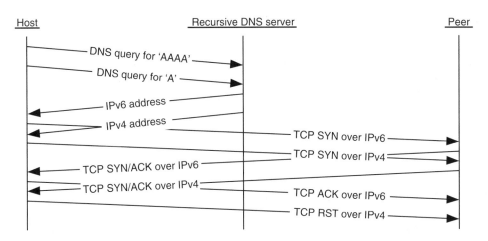

Figure 3.22   An example of the Happy Eyeballs procedure.

fast, the application has to fallback to using IPv4. Figure 3.22 illustrates Happy Eyeballs in a normally functioning case. In the figure, a host is connecting to a peer by using TCP. The host starts by sending out both DNS IPv6 address record (AAAA) query for the peer's IPv6 address and the DNS IPv4 address record (A) query for the peer's IPv4 address. In the example, a recursive DNS resolver returns the addresses in the same order as queried, while it could happen the other way as well. The host in this example initiates TCP connections as soon as it gets responses in order to get a working connection up as soon as possible – in reality some implementations may want to wait for all addresses before initiating the first transport session connection. In this example, the host sends TCP SYN-messages over both IPv6 and IPv4 before it receives any response back from the peer. It could happen that the network reorders packets and TCP SYN/ACK-messages may arrive in a different order from the SYN-messages sent, but in this example the ordering is not changed. Instead, the IPv6 TCP connection completes first, and the host can start to use it. When the IPv4 connection completes, the host, in this example, discards it immediately by sending a TCP RST-message. However, the host might just as easily have taken the IPv4 connection into use as well, or made a decision later whether IPv4 or IPv6 seem to work better for the application. There are several different scenarios that can happen: for example the IPv4 connection could complete faster than IPv6, even if TCP SYN was first sent over IPv6.

It is quite obvious that always trying both address families consumes twice the resources. This resource 'waste' can be optimized with some intelligence. For example, the host might allow some time for the first connection to succeed, and only after that try the second address family. This time can be as short as expected time for a TCP handshake to complete. RFC 6555 recommends a wait time of around 150–250 milliseconds [134]. Implementations can also go beyond this, for example, by keeping a record of which address families have been successful generally or for particular destinations, and starting connections with the address family that has worked better.

The authors expect that it will take a couple of years before the best implementation strategies for Happy Eyeballs are found – and it is even possible that time proves that Happy Eyeballs-strategies are unnecessary.

3.10   Implications of IPv6 for Other Protocols

In this section we will describe implications of IPv6 for some of the key protocols used in the Internet, as well as describing generally what kinds of implications IPv6 brings to higher level protocols. This section will by no means give an exhaustive list of affected protocols.

3.10.1   Transport Layer Protocols

All protocols directly above IP face changes due to the introduction of IPv6. The three most widely used transport layer protocol are UDP [135], TCP [136], and ICMPv6 (discussed in depth in Section 3.3). There are also other transport layer protocols, such as Stream Control Transmission Protocol (SCTP) [137].

A major implication for all upper layer protocols is caused by the design decision of not having a checksum field as part of the IPv6 header, as illustrated in Figure 3.8. The upper layer protocols, if they wish to verify the correctness of the IPv6 header, have to include a so-called IPv6 pseudo header, illustrated in Figure 3.23, in their checksum calculation. Inclusion of the pseudo header is important, for example, in the case of ICMPv6 which has no other means to verify that the packet was received at the intended address and that the sender address was received without error.

In the pseudo header of Figure 3.23, the destination's IPv6 address is the final destination (not any intermediary address, e.g. if routing headers are used), the next header value points to an upper layer protocol such as TCP (but not to IPv6 extension headers), and the packet length field content depends on the upper layer. In the case of UDP, a protocol that carries its own length information, the length used in pseudo header is directly the UDP's length. In the case of TCP, as TCP headers do not have a length field, the payload length from the IPv6 header minus any extension headers is used.

RFC 2460 states that UDP must include the checksum field that covers both the UDP itself and the pseudo header. However, certain classes of applications can benefit of receiving packets that contain some errors: audio and video codecs can manage with some errors. For tunneling protocols it is usually enough to have checksum calculation only

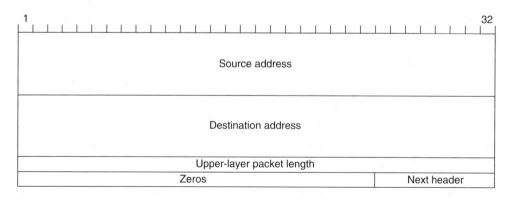

Figure 3.23   IPv6 pseudo header [21].

for the tunneled (inner) packets and there are benefits for not calculating a checksum for outer packets. For those applications, UDP-Lite has been defined, which allows checksum protection to cover only sensitive parts of a packet [139]. As UDP-Lite is a separate transport layer protocol, even if very similar to UDP, it faces deployment challenges as not all devices such as firewalls support it. As a consequence, IETF is standardizing ways to allow UDP packets without checksum calculation [140].

Besides the checksum calculation, the ability to work with longer IPv6 addresses, and the capability required to parse error codes provided by ICMPv6, the transport layer protocols do not have to change. That said, transport layer protocols may want to take the existence of IPv6 addresses into account for other reasons. For example, SCTP allows communication of nodes' IPv4 and IPv6 addresses to the peer within an 'Initiation'-message. Hence, the SCTP protocol by design can utilize both IPv4 and IPv6 addresses at the same time. The transport layer protocols may also assist in finding the Path MTU, as described in Section 3.2.3, and by providing peer reachability information for the Neighbor Cache as mentioned in Section 3.4.5.

3.10.2 Domain Name System

IPv6 has significant implications for the DNS [141], which must be known by everyone working with IPv6. The main point of reference is the 'DNS Extensions to Support IP Version 6' RFC 3596 [142]. The most important implication of IPv6 for DNS is the introduction of the new AAAA DNS resource record type. The quadruple 'A' indicates that the 128-bit IPv6 address contained in this resource record type is four times the size of the 32-bit IPv4 address stored in DNS A resource record type. Figure 3.24 illustrates how both AAAA and IPv6 type of Pointer Record (PTR) records can be configured on a DNS server by using Internet Systems Consortiums' BIND-style format. The name BIND comes from Berkeley Internet Name Domain – see for more information at http://www.isc.org.

The increased length of IPv6 addresses when compared to IPv4 results in larger DNS response messages. The maximum length of DNS protocol payload is 512 bytes over UDP transport [141], which can become a limiting factor and hence result in truncated messages. Therefore, it is highly recommended that implementations support 'Extension Mechanism for DNS (EDNS0)' [143], which enables requestors to inform recursive DNS servers about the capability to receive larger responses. The maximum theoretical size for DNS messages is increased to 64KB by the ENDS0, but due to fragmentation and Path MTU issues it is recommended that requestors advertise capability to receive packets with

AAAA record:

www.example.com   IN AAAA   2001:db8:0:0:0:0:1234:abcd

PTR record:

```
$ORIGIN 0.0.0.8.b.d.0.1.0.0.2.IP6.ARPA.
1.0.0.0.0.0.0.0.0.0.0.0.0.0.0.0.1.0.0.0      IN     PTR     ns1.example.com.
2.0.0.0.0.0.0.0.0.0.0.0.0.0.0.0.1.0.0.0      IN     PTR     mail.example.com.
```

Figure 3.24 BIND-style examples of AAAA and PTR type of DNS resource records.

size of 1280 bytes, or more if Path MTU discovery between DNS requestor and server indicates the network capability to transport larger packets.

A6 Record Type

While today AAAA records are dominant, there was a competing record type called A6. The A6 was originally defined in RFC 2874 [144], but was eventually declared historic by RFC 6563 [145], which discusses in depth what were the reasons for A6 record type's failure. These included, for example, the general problem of having two tools for the same thing – either AAAA or A6 had to go – slower latency to resolve A6 queries, and increased security risks when compared to AAAA.

Forward DNS Queries

As stated above, a new AAAA resource record type is defined for storing a single IPv6 address. A node making a AAAA DNS query may receive a list of addresses, if the queried fully qualified domain name maps to multiple DNS records. This list of addresses is then utilized by the host's address selection algorithms (see Section 3.5.4). An example of AAAA record is given in Figure 3.24.

Reverse DNS Queries

For the scenarios where a fully qualified domain name needs to be resolved based on an IPv6 address literal, a domain for IPv6 reverse DNS lookups has been defined. The domain for reverse DNS lookups is hosted at IP6.ARPA. The IPv6 address is represented in the IP6.ARPA domain by listing each nibble separated by dots, starting from least significant nibble, and ending the domain with suffix `ip6.arpa`, as illustrated in Figure 3.24.

Recursive DNS Server Discovery

The history related to finding recursive DNS servers' addresses is a very interesting one. Even though the core of IPv6 was defined in the 1990s, it took a long time to agree how nodes ought to discover IPv6 addresses of RDNSSs. This was not initially a very pressing topic, as AAAA and PTR resource records can be queried by talking to DNS servers over IPv4. However, for IPv6-only modes and just for allowing the use of IPv6 for DNS communications, a method was needed to find IPv6 addresses for RDNSSs.

Standardization of the DNS server discovery mechanism was a long battle between Router Advertisement and DHCPv6-based approaches. The biggest point of principle was whether IPv6 Router Advertisement should be used for configuring nodes with any other information than what is needed for router discovery and the nodes' stateless and stateful IPv6 address autoconfiguration. In particular, the fear was that if DNS configuration were added to Router Advertisement, soon there would be plethora of proposals for adding other pieces of configuration information to Router Advertisements (and in fact, there has been and are proposals for adding other pieces of information to Router Advertisements). The desire, especially from DHCPv6 advocates, is to centrally provide all configuration from DHCPv6 servers. On the other hand, many were against mandating support of the DHCPv6 itself, which sometimes is considered as significant additional implementation effort. Especially mandating full DHCPv6 support just to provide bits of configuration information was seen as undesired. Partially due these discussions, a stateless variant of DHCPv6 emerged in the form of RFC 3736 [83], which allows implementation of only a minimal subset of DHCPv6 – essentially just INFORMATION-REQUEST and corresponding reply messages (Section 3.5.2). The DNS configuration options for DHCPv6 were standardized as early as 2003 [85]. However, the advocates for Router Advertisement based approach were very persistent, and managed to get Experimental RFC out in 2007, and evolved that into Standards Track RFC in 2010 [69]. The seven years between standards track RFCs has had an effect: hosts' and routers' support for a DHCPv6-based approach is more widespread than for the Router Advertisement-based approach. However, as time passes, implementations will likely, and may have to, support both of these approaches.

The DHCPv6 approach for RDNSS discovery was updated at 2012 with support for learning about RDNSSs that have special knowledge of some particular domain [86]. The new 'RDNSS Selection Option' communicates the IPv6 address of an RDNSS with a list of domains and network prefixes that the server has particular information about. With this information, a node can more wisely choose to which RDNSS it should send queries. This solution was developed for multi-interfaced use cases, where a node needs to decide which network interface to use for DNS queries.

In addition to the abovementioned approaches, there was a proposal at November 2001 from Dave Thaler and Jun-ichiro itojun Hagino, which proposed a stateless approach and use of well-known site-local anycast addresses for DNS servers. These addresses were `fec0:0:0:ffff::1`, `fec0:0:0:ffff::2`, and `fec0:0:0:ffff::3`. Due to lack of any better mechanisms, some vendors chose to support these well-known addresses. Even today some operating systems use these addresses as the last resort means for communicating with DNS servers. Hence, if one sees these addresses in action, it is a hint that the node has failed to obtain DNS server addresses by other means. Despite these implementations, IETF chose not to standardize this approach and the draft specification expired. Furthermore, the whole concept of site-local addressing was made obsolete in RFC 3879 [6].

DNS Queries for Both IPv4 and IPv6 Resource Records

As the transition to IPv6 involves a long period of coexistence with IPv4, hosts need to be able to use both IPv4 and IPv6 for communications. A significant implication for DNS of the coexistence is that dual-stack hosts need to be able to resolve destination's IPv4 and IPv6 addresses. An important thing to note is that the type of DNS query, be that A or AAAA, is orthogonal to the IP address family used to transmit the actual DNS queries.

Due to the limitations in the DNS protocol, it is not feasible to ask for different types of records in a single DNS query. The main reason for that is the DNS protocol's lack of capability to indicate different result codes for different queries in a single response; for example, if an A resource record would be available but AAAA would not, the DNS could not differentiate which record resulted in success and which in failure. Therefore, the client implementations will always send separate requests for A and AAAA resource records. This approach increases the number of DNS messages that nodes emit and hence signaling load on a network, but with help of DNS caches (often residing already on nodes themselves) the issues are not that significant – and nevertheless unavoidable.

A variety of approaches for resolving both IPv4 and IPv6 addresses are deployed in different operating systems. There are four main approaches for sending both A and AAAA queries:

A first and, after response, AAAA: This approach, used by a popular desktop operating system, is the safest in the sense that the A query is finished before the AAAA is started, and issues described in RFC 4074 are avoided [146]. In the past, networks used to have old and buggy recursive DNS resolvers that could misbehave if they received an AAAA query from a client while resolution for an A query was not complete. Effectively the A query failed to resolve correctly. However, with this approach, if the A resolution takes time, the AAAA query will be delayed.

AAAA first and, after response, A: A popular open source operating system utilizes this approach. While this order is not very useful as of today, in the future when the world is IPv6 dominant this may become the main approach – but at that time the A query may not be sent at all unless AAAA has failed to resolve a usable address. This approach has issues if the AAAA query does not resolve as expected [146] – the sending of an A query would be delayed.

A first and AAAA immediately after: Essentially the parallel DNS queries are needed for the fastest resolution of both, or either, of IPv4 and IPv6 addresses. This approach is used by another popular desktop operating system in an attempt to resolve both address families faster – roughly in half of the time – than in the two approaches described above. On some occasions issues described by RFC 4074 may also cause problems for this approach [146], but luckily the issues with AAAA records and name servers have been significantly diminishing during recent years, as IPv6 deployments have become more common and server software has been updated.

AAAA first and A immediately after: This approach does not have significant difference from the one above, other than that the small preference is given to IPv6 over IPv4. On some occasions issues described by RFC 4074 may also cause problems for this approach [146].

Once the DNS results are available, hosts utilize source and destination address selection algorithms to choose which addresses to try first – see Section 3.5.4. Implementations that attempt to provide the fastest response time sometimes start using the addresses as soon as they arrive, and if those do not yield a working connectivity, fall back to the addresses that were slower to resolve.

3.10.3 Applications

IPv6 has a significant impact on many applications, and in particular to those that are, for whatever reason, IP version aware. File Transfer Protocol (FTP) is a classic and famous protocol used on the Internet for a very long time. As the FTP [147] is IP-version aware and passes IP address literals in its protocol payloads, it is a handy protocol for us to highlight typical implications that IPv6 can have for upper layer protocols – both those right above layer 3, and those that are in the application layer. RFC 2428, FTP Extensions for IPv6 and NATs, describes in depth the implications for FTP [148], and here we provide a short summary.

The original specification for FTP, quite fairly for that time, assumed that the IP addresses were always of 32 bits in length. As IPv6 address are four times the length, 128 bits, issues were certain to arise. Any piece of code that is participating somehow at the IP address management must change, and similarly any part of protocol specifications that touches IP addresses must be updated. The problems are made even harder as endpoints may be addressed with both IPv4 and IPv6 addresses. In the case of FTP, the abovementioned issues have the following implications:

1. An FTP program has to be able to set up a TCP connection to a remote peer with IPv6, if the DNS resolution for the peer's name results in an IPv6 address or a program is given an IPv6 address literal as input. In some cases the program has to be able to fallback from a failure to establish a connection with the first address family (e.g., IPv6) to a second address family (e.g. IPv4). The preferences and fallback logic related to address family selections, depending on network conditions, is a typical IP-version-aware application protocol requirement.
2. FTP needs means of communicating IPv6 related endpoint information. In practice this means replacement of the FTP command PASV with Enhanced PASV – EPSV and PORT with Enhanced PORT – EPRT commands. Generally speaking, this means that whatever signaling a protocol is using in relation to IP addresses, the protocol must support the existence of both address families.
3. The EPRT command enhances the PORT command with additional fields for communicating which address family is used (IPv4 or IPv6) and an IP address in address family specific format. The command can also fail in new ways: the remote peer can fail with an error indicating unknown or unsupported address family. Such errors usually trigger fallback to another address family, if available. This kind of approach for passing protocol family and address in specific format is typical for protocols involving IP addresses, although often lists of both IPv4 and IPv6 addresses are provided.
4. The EPSV command enhances the PASV command only with placeholders for address family and IP address. Interestingly enough, the FTP passive mode listens only to one address family – the one that the control channel is using. However, in many protocols listening for incoming connections needs to be conducted on both IPv4 and IPv6, as the remote peers may have only either available.

3.10.4 Internet Routing

Introduction of IPv6 into the Internet required changes to the existing IPv4 routing infrastructure. All routers had to, and have to, be modified to support IPv6 related forwarding functions. Additionally, routing protocols had to be enhanced with IPv6 capabilities. It is also worth noting that the mere existence of both IPv4 and IPv6 routes increases quite significantly resource (mostly memory) requirements for routers' routing table management, but that is not a problem we are focusing on in this book. Besides the details related to the new address family, there are no significant implications to routing protocol logics. There are a set of more subtle and minor changes, but those are not covered here. We recommend readers interested in routing to study literature specifically focusing on routing.

The first IPv6 related change to routing standards in the IETF was to introduce IPv6 into the Routing Information Protocol (RIP), resulting in a new revision called Routing Information Protocol next generation (RIPng) in 1997 [149]. The RIPng is a *distance-vector*-based routing protocol designed for Autonomous Systems (ASs)' interior routing purposes. More powerful routing protocols have been developed since RIPng for exterior routing purposes and for more efficient interior routing purposes. These include Multi-Protocol Border Gateway Protocol (MP-BGP) (first in 2000, updated in 2007 [150]), Open Shortest Path First version 3 (OSPFv3) (in 2008 [138]) and Intermediate System to Intermediate System (IS-IS) (in 2008 [151]), of which the last two are the routing protocols commonly used for the internal IPv6 route configuration in the 3GPP networks. The MP-BGP is the routing protocol for external (between ASs) communications. The OSPFv3 and the IS-IS differ from the RIPng, and to some extent from the MP-BGP, by not using a distance-vector approach but instead they are *link-state* routing protocols. The MP-BGP is not quite a distance-vector protocol, as in addition to distance vectors it uses other pieces of information for routing decisions.

The distance-vector based routing protocols work by routers sharing their routing tables with neighbors. The distance-vector information gets propagated through a network and eventually all routers have route entries with metric values towards each IPv6 prefix and next hop pointers to their neighboring routers. In the link-state routing protocols, routers share – or flood – information about their reachable neighbors throughout the network to all other routers, and hence each individual router is able to independently build a topology map of the network. A router using a link-state protocol advertising changes in its neighbors to the network, will quickly cause recalculation of the network topology graphs in the other routers.

From the routing protocols we describe here, the RIPng is effectively not used in 3GPP networks, the Border Gateway Protocol (BGP) is usually used only for routing between ASs, even if it can be configured to work also as an interior routing protocol, and for internal routing purposes OSPFv3 and IS-IS are used due to their superior route convergence times. The OSPFv3 and IS-IS cannot be used for inter-Autonomous System routing purposes.

Changes to RIPng

Changes were required for RIPng routers to store the following IPv6-related information in their routing table: an IPv6 prefix for a destination, a metric value representing the cost of sending packets to the destination, the next hop address towards the destination, a flag indicating if the route has changed recently, and a set of timers. For the neighbors a RIPng router tells its each IPv6 route table entry consisting of a prefix and its length, a metric, and a route tag entry.

Changes to BGP for IPv6

IETF defined IPv6 changes for the BGP in RFC 4760, which defines a protocol called MP-BGP [150]. The core of the BGP protocol did not require any changes, but what was required for IPv6 was the addition of IP address family independent ways to bind a network layer protocol to the next hop and the network layer reachability information. The network layer reachability information includes an IPv6 prefix and its length. Additionally, due to the scoped addressing architecture of IPv6, the MP-BGP needs to be aware not to advertise the AS's internal prefixes (such as those using ULA and link-local prefixes) to other networks [152].

Changes to OSPF for IPv6

The OSPF version supporting IPv6 is version three – often referred as OSPFv3 [138]. The main changes that IPv6 caused were related to addressing semantics for supporting longer addresses, and for the protocol to work on a per-link instead of a per-subnet basis. In addition to the changes caused by longer addresses, other changes were also made, such as changing to use ESP instead of OSPF's own authentication methods. However, the fundamental mechanisms did not need to change.

Changes to IS-IS

An implication of IPv6 to IS-IS is the need for two new options, as defined in RFC 5308 [151]. The first option is called *IPv6 Reachability*, and it is used to pass information regarding the IPv6 prefix and its length, a metric, and a couple of control bits including one describing whether the information has been 'externally' learned from another routing protocol. The second option is called the *IPv6 Interface Address* that is effectively otherwise similar to the corresponding IPv4 option, but allows addresses of 128 bits. IS-IS works on top of layer 2, and hence IPv6 does not impact the transportation of this protocol.

IPv6 Routing for Low-power and Ad-hoc Networks

Special focus for IPv6 routing has been given to low-power wireless networks and ad-hoc networks, which often do not even support IPv4. In these networks mesh topologies are commonly used, and routing has to consider the very constrained nature of the routing nodes. Protocols that have been developed in this area include RPL [64]. As this book is not about low-power networks, we do not discuss how these protocols work. We suggest the readers to study the literature related to sensor networks and routing.

3.10.5 *Management Information Base*

The Simple Network Management Protocol (SNMP) is the standard for managing network nodes in the Internet [153]. The information that SNMP uses for management is defined in Management Information Bases (MIBs). There are hundreds of MIBs and hence also some for IPv6. The most relevant MIBs for IPv6 are RFC 4292 'IP Forwarding Table MIB' [154] and RFC 4293 'Management Information Base for the Internet Protocol (IP)' [155]. The MIBs are supported by nodes that implement SNMP agents, which can effectively mean all kinds of nodes present in the internet: hosts, routers, switches, and so forth.

The information that IPv6 MIBs provide about a node include IPv6 routes, default router table, setting for enabling and disabling forwarding, setting values to Hop Limits, providing interface and system statistics, what IPv6 prefixes the node has, what IPv6 addresses node has configured, IPv6 address to physical address mapping table, ICMPv6 statistic tables, and so on. The information delivered with the help of IPv6 MIBs is essential for managing IPv6 networks.

The IETF has also defined a set of IPv6 related MIBs in addition to the core IPv6 MIBs, and more are in the pipeline. The already existing additional MIBs include, for example, 'RADIUS Authentication Server MIB for IPv6' [156], 'Proxy Mobile IPv6 Management Information Base' [157], and 'IP Multicast MIB' [158].

In this book we will not look at the SNMP or MIBs in any more depth than what has already been said. Interested readers should look at the referenced RFCs or at textbooks dedicated to SNMP MIB.

3.11 Validation and Certification

In traditional IETF style, the IPv6 protocol suite does not have any facilities in place for official certification or type approval. However, protocol test suites do exist to help validate implementations, conformance against IPv6 standards and to ensure interoperability between implementations. In this section we will give a brief view of what is available in the marketplace.

3.11.1 Test Suites

A very well known free IPv6 conformance test suite was created and has been maintained by the TAHI project, see *http://www.tahi.org*. The TAHI project was established in Japan at 1998 by three organizations: The University of Tokyo, Yokogawa Electric Corporation, and YDC Corporation, the last of which is no longer participating in the project. The suite covers all the core IPv6 specifications and also additional features such as IPsec and MIPv6. In September 2012 the project announced that it would conclude its activities in December 2012, which it did.

Commercial test suites are available from multiple vendors. These include, but are not limited to, 'Codenomicom Defensics Core Internet package', see more at *http://www.codenomicom.com*, and 'Ixia IxN2X IPv4 and IPv6 Protocol Conformance Test Suites', see more at *http://www.ixiacom.com*.

3.11.2 IPv6 Ready Logo

The IPv6 Forum, *http://www.ipv6forum.com* has established an 'IPv6 Ready Logo Program' to help validate implementation conformance with self-test tools and to indicate the level of implementation's conformance to people and organization. The program uses TAHI's conformance test suites for validation, and passing of the tests is a requirement for obtaining logos. The golden 'Phase-2 IPv6 Ready' logo is shown in Figure 3.25,

Figure 3.25   Golden IPv6 Ready logo from IPv6 Forum. Reproduced by permission of The IPv6 Forum.

Table 3.6   IPv6 Ready Logo Phase-2 Test Specification and Test Tools

Category	Targets	Notes
IPv6 Core Protocols	Router, Host	
DHCPv6	Client, Server, Relay agent	
IPSec	End-node, Security Gateway	
IKEv2	End-node, Security Gateway	
MLDv2	Router, Listener	Experimental
SNMP	Agent, Manager	
CE Router	CE Router	Under public review
IMS UE	UE	Experimental
MIPv6	Correspondent Node, Home Agent, Mobile Node	Experimental
NEMO	Home Agent, Mobile Router	Experimental
SIP	UA, Endpoint, B2BUA, Proxy, Registrar	Experimental

although in a black and white version here. A silver 'Phase-1 IPv6 Ready' logo was used in the past, but is now being phased out.

For an implementation to achieve the golden 'IPv6 Ready' logo it has to pass relevant test suites, such as IPv6 Core Protocols' suite, and apply for the logo. The categories at the time of writing are listed in Table 3.6. The table is updated as specifications evolve, so newest version should be checked from the IPv6 Forum pages. After passing reviews by the IPv6 Ready Logo Technical Group and IPv6 Forum Logo Committee, permission to use the logo is given. At the time of writing, the phase-2 golden logo has been given for 794 different implementations, and the older phase-1 silver logo for 480 different implementations.

3.12   Example IPv6 Packet Flows

In this section we are going to provide real-life examples of IPv6 packet flows. The examples include SLAAC, DNS, and transport layer sessions. We have used the *Wireshark* packet capture and analysis tool that is freely available at *http://www.wireshark.org*.

The purpose of this section is to familiarize readers with looking at IPv6 headers and, while doing so, to look at some of the basic and key IPv6 concepts at very detailed level.

3.12.1 IPv6 on Ethernet

In this example, we will look in detail at how a popular operating system starts performing SLAAC after attaching to an Ethernet link. We will also use this opportunity to describe IPv6 packets in detail with reference to the sections above.

Subscription to Solicited-node Multicast Group

Figure 3.26 is our first step, and we look at the trace number five (leftmost column). From the log entries we have excluded the traffic that is not relevant to this example, such as all the IPv4 traffic that the operating system emits after joining an Ethernet network.

Before the highlighted step, the IP stack of a host has chosen an IID to use (see Section 3.5.1) for at least IPv6 link-local address creation. At this point, the link-local address will be in 'tentative' state (see Section 3.5.1). In this example, the IID is based on the interface's MAC address b8:27:eb:f1:8d:0d, as hinted at by the Solicited-node

```
No.. Time      Source                    Destination        Protocol  Info
   5 23.1693  ::                         ff02::16           ICMPv6 Multicast Listener Report Message v2
   6 23.7791  ::                         ff02::1:fff1:8d0d  ICMPv6 Neighbor solicitation
   7 24.5386  fe80::230:5ff:fe7f:fd5f    ff02::1            ICMPv6 Router advertisement
   8 24.6088  ::                         ff02::1:fff1:8d0d  ICMPv6 Neighbor solicitation
   9 24.7789  fc80::ba27:cbff:fcf1:8d0d  ff02::2            ICMPv6 Router solicitation
  11 25.1889  fe80::ba27:ebff:fef1:8d0d  ff02::16           ICMPv6 Multicast Listener Report Message v2
  12 27.7387  fe80::230:5ff:fe7f:fd5f    ff02::1            ICMPv6 Router advertisement
```

```
⊞ Frame 5 (90 bytes on wire, 90 bytes captured)
⊟ Ethernet II, Src: b8:27:eb:f1:8d:0d (b8:27:eb:f1:8d:0d), Dst: IPv6-Neighbor-Discovery_00:00:00:16
   ⊞ Destination: IPv6-Neighbor-Discovery_00:00:00:16 (33:33:00:00:00:16)
   ⊞ Source: b8:27:eb:f1:8d:0d (b8:27:eb:f1:8d:0d)
     Type: IPv6 (0x86dd)
⊟ Internet Protocol Version 6
     Version: 6
     Traffic class: 0x00
     Flowlabel: 0x00000
     Payload length: 36
     Next header: IPv6 hop-by-hop option (0x00)
     Hop limit: 1
     Source address: ::
     Destination address: ff02::16
⊟ Hop-by-hop Option Header
     Next header: ICMPv6 (0x3a)
     Length: 0 (8 bytes)
     Router alert: MLD (4 bytes)
     PadN: 2 bytes
⊟ Internet Control Message Protocol v6
     Type: 143 (Multicast Listener Report Message v2)
     Code: 0 (Should always be zero)
     Checksum: 0xe18b [correct]
   ⊟ Changed to exclude: ff02::1:fff1:8d0d
       Mode: Changed to exclude
       Aux data len: 0
       Multicast Address: ff02::1:fff1:8d0d
```

Figure 3.26   Example: joining a Solicited-node multicast address group from an unspecified IPv6 address.

multicast address seen inside the MLDv2 protocol, `ff02::1:fff1:8d0d`, where the three last bytes are the same as in the MAC address. From the following messages, which we will discuss shortly, we can see that the actual link-local address a host wants to configure into use is `fe80::ba27:ebff:fef1:8d0d`.

In packet five, the IP stack subscribes to the Solicited-node multicast group by sending an MLDv2 'Multicast Listener Report v2' message to 'All MLDv2-capable routers' multicast address (Section 3.5.1). It is important to note that the IPv6 packet sent to the IPv6 multicast address is also sent to Ethernet layer's multicast address `33:33:00:00:00:16`, as per Section 7 of the RFC 2464 [99]. The subsequent multicast messages of this example are also sent to the Ethernet multicast address starting with `33:33:`, with the bottom 48 bits changing according to a destination IPv6 multicast address. The Ethernet header indicates the protocol type of the transported packet, which in this case is `0x88dd`, i.e. IPv6. While the Ethernet header contains the source MAC address, the IPv6 source address is unspecified, as at this point, the node does not have any IPv6 address available for use [46].

The reason why this message is sent, even before any IPv6 address is available, is to ensure successful operation of DAD (Section 3.4.7). Specifically, this message is needed to inform MLD aware switches, possibly existing in the access network, that a node exists that listens for particular addresses. In some networks the MLD-snooping switches may not forward multicast packets at all, unless they have seen an MLD report being sent.

Next we will look at the details of the whole packet. For the subsequent packets we will not repeat the description of the IPv6 header, but only look at what is new in each message.

Starting with the IPv6 header itself, at the 'Internet Protocol Version 6' line we can see contents for the fields shown in Figure 3.8. IP-version field is set to '6' to indicate that this packet is IPv6. For this packet, Traffic Class and Flow Label are set to default, to zero. Payload Length indicates that there are 36 bytes of data after the main IPv6 header. The Next Header field indicates the presence of a Hop-by-hop option, and more on that later. The MLD packets' Hop Limit is always one, as shown also in this example. The MLD packets need not travel any further than the first-hop routers. As said before, the IPv6 source address has been set to unspecified because at this point the node does not have any source address available.

As required by the MLDv2 RFC 3810 Section 5, the Hop-by-hop option header (see Section 3.2.2) includes a Router Alert option [45]. The length field of the Hop-by-hop option contains zero, which in this case is specified to mean eight bytes (see RFC 2460 Section 4.3 [21]). The Router Alert then indicates to the router that this IPv6 packet contains an MLD message Section ([31] Section 2.1). One PadN padding option (Section 3.2.2), of two bytes in length, is included to make the whole Hop-by-hop option eight bytes long. The Next Header field on the Hop-by-hop option indicates ICMPv6, which is the protocol that actually contains the MLD message. The Router Alert option is required to notify routers that otherwise would not be interested in MLD.

The actual beef, so the speak, of this message is the ICMPv6 packet of type 'Multicast Listener Report Message v2'. With this packet the node informs routers that there are listeners for the Solicited-node multicast address `ff02::1:fff1:8d0d` matching to the link-local address `fe80::ba27:ebff:fef1:8d0d` that the host is going to configure for itself. The detailed contents of the ICMPv6 header are the 'message type' we mentioned, 'code', that in this case is always zero, and the checksum field. It is good to remember that this checksum covers the whole ICMPv6 packet and then the pseudo header, as described in Section 3.10.1, which includes key fields of the previous IPv6 header. The MLDv2 header indicates that this listener report message is of type 'exclude', but because the list of source address to exclude is empty ('Aux data length' field is zero), it effectively means that nothing is excluded and multicast packets from all possible sources destined to the ff02::1:fff1:8d0d group should be forwarded to this link.

Initiating DAD for IPv6 Link-local Address

After the MLD message is sent related to the link-local address that a node has chosen to the use, the node must perform the DAD procedure for the address as described in Section 3.4.7. This ensures that the address is not already in use in the link. As seen on the trace in Figure 3.27, the DAD is performed by sending ICMPv6 Neighbor Solicitation message as multicast to the Solicited-node multicast address ff02::1:fff1:8d0d of the link-local address that host wants to configure. The packet is not sent to 'all nodes' address of ff02::1, as then it would be unnecessarily delivered to all nodes on the link – and not just to those that might be in conflict. Please note that the destination address is the same node used previously when registering a multicast address for listening. This address is sent from an undefined IPv6 address in order not to pollute Neighbor Caches of other nodes on the link, in case there were conflicts.

In Figure 3.27 we have opened the IPv6 packet to show the next header field value of ICMPv6, and a more typical hop limit of '255'. The ICMPv6 message type is of course 'Neighbor Solicitation'. The 'code' field is zero, as always for this message type. The 'target' address field is the key here and it shows what is the exact IPv6 address that the host is soliciting for. It may be possible that a link has multiple nodes that have the

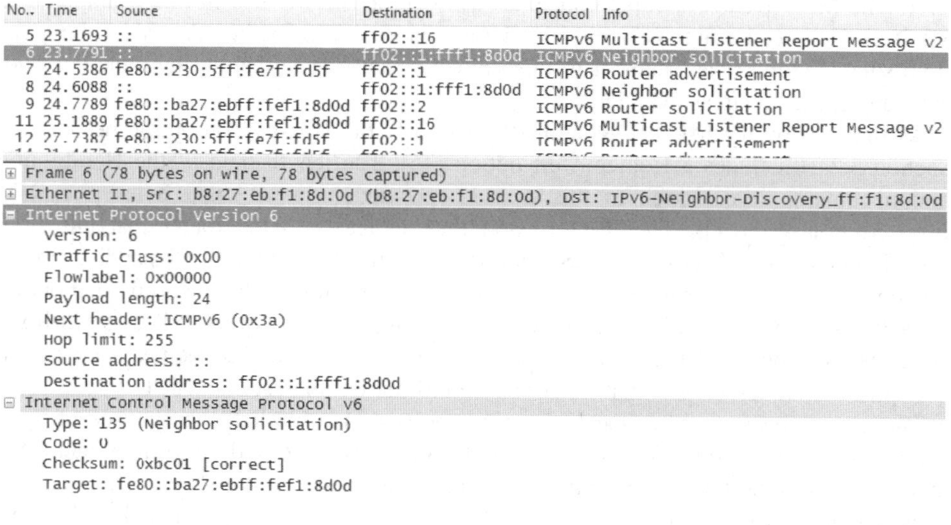

```
No.. Time          Source                  Destination        Protocol  Info
   5 23.1693 ::                            ff02::16           ICMPv6    Multicast Listener Report Message v2
   6 23.7791 ::                            ff02::1:fff1:8d0d  ICMPv6    Neighbor solicitation
   7 24.5386 fe80::230:5ff:fe7f:fd5f       ff02::1            ICMPv6    Router advertisement
   8 24.6088 ::                            ff02::1:fff1:8d0d  ICMPv6    Neighbor solicitation
   9 24.7789 fe80::ba27:ebff:fef1:8d0d     ff02::2            ICMPv6    Router solicitation
  11 25.1889 fe80::ba27:ebff:fef1:8d0d     ff02::16           ICMPv6    Multicast Listener Report Message v2
  12 27.7387 fe80::230:5ff:fe7f:fd5f       ff02::1            ICMPv6    Router advertisement
⊞ Frame 6 (78 bytes on wire, 78 bytes captured)
⊞ Ethernet II, Src: b8:27:eb:f1:8d:0d (b8:27:eb:f1:8d:0d), Dst: IPv6-Neighbor-Discovery_ff:f1:8d:0d
⊟ Internet Protocol Version 6
    Version: 6
    Traffic class: 0x00
    Flowlabel: 0x00000
    Payload length: 24
    Next header: ICMPv6 (0x3a)
    Hop limit: 255
    Source address: ::
    Destination address: ff02::1:fff1:8d0d
⊟ Internet Control Message Protocol v6
    Type: 135 (Neighbor solicitation)
    Code: 0
    Checksum: 0xbc01 [correct]
    Target: fe80::ba27:ebff:fef1:8d0d
```

Figure 3.27 Example: initiating DAD for an IPv6 link-local address.

same lowest 24 bits of the address – in this case `f1:8d:0d` – and hence multiple nodes might receive the solicitation. However, only one host could have the full solicited IPv6 address. Also from here we can for the first time see what the address the host is planning to configure for itself. In the previous MLDv2 message we saw only the Solicited-node multicast address and hence only the lowest 24 bits.

After sending the Neighbor Solicitation, the host must wait for a while before it can be fairly certain that there are no other nodes using this address. By default only one Neighbor Solicitation is sent ([73] Section 5.1), for which a response is waited for – for 1000 milliseconds [41]. The number of retransmissions and the time waited for a response can be different in various systems, and also may be configurable. We can see from the packet flow of Figure 3.27 that a Router Solicitation – which we will look in depth later – is sent from the link local address `fe80::ba27:ebff:fef1:8d0d` almost exactly one second after the Neighbor Solication (compare the times of trace numbers 6 and 9). This shows that the host waited for the default second, for possible ICMPv6 Neighbor Advertisements to arrive before taking the address into full use.

Reception of Unsolicited Router Advertisement

Trace entry 7 is interesting, as shown in Figure 3.28–the result of a bit of luck during packet capture. It is an unsolicited ICMPv6 Router Advertisement sent by a router from

```
No.. Time      Source                Destination        Protocol  Info
   5 23.1693  ::                     ff02::16           ICMPv6 Multicast Listener Report Message v2
   6 23.7791  ::                     ff02::1:fff1:8d0d  ICMPv6 Neighbor solicitation
   7 24.5386  fe80::230:5ff:fe7f:fd5f ff02::1           ICMPv6 Router advertisement
   8 24.6088  ::                     ff02::1:fff1:8d0d  ICMPv6 Neighbor solicitation
   9 24.7789  fe80::ba27:ebff:fef1:8d0d ff02::2         ICMPv6 Router solicitation
  11 25.1889  fe80::ba27:ebff:fef1:8d0d ff02::16        ICMPv6 Multicast Listener Report Message v2
  12 27.7387  fe80::230:5ff:fe7f:fd5f ff02::1           ICMPv6 Router advertisement

⊞ Frame 7 (110 bytes on wire, 110 bytes captured)
⊞ Ethernet II, Src: FujitsuS_7f:fd:5f (00:30:05:7f:fd:5f), Dst: IPv6-Neighbor-Discovery_00:00:00:01
⊞ Internet Protocol Version 6
⊟ Internet Control Message Protocol v6
    Type: 134 (Router advertisement)
    Code: 0
    Checksum: 0xe1b3 [correct]
    Cur hop limit: 64
  ⊞ Flags: 0x00
    Router lifetime: 60
    Reachable time: 10000
    Retrans timer: 10000
  ⊟ ICMPv6 option (Prefix information)
      Type: Prefix information (3)
      Length: 32
      Prefix length: 64
    ⊟ Flags: 0xe0
        1... .... = Onlink
        .1.. .... = Auto
        ..1. .... = Router Address
        ...0 .... = Not site prefix
      valid lifetime: 86400
      Preferred lifetime: 14400
      Prefix: 2001:14b8:138:42::
  ⊟ ICMPv6 option (Source link-layer address)
      Type: Source link-layer address (1)
      Length: 8
      Link-layer address: 00:30:05:7f:fd:5f
```

Figure 3.28   Example: reception of unsolicited router advertisement.

its `fe80::230:5ff:fe7f:fd5f` link local address to `ff02::1` all-nodes on a link multicast address. Even though this is an unsolicited Router Advertisement, our host does welcome it, and uses it, as shown in the trace entry 8 that we will look at in detail shortly.

We skip the IPv6 header, as there is nothing new there. The ICMPv6 protocol on the other hand is a rich pool of information. The beginning contains message type, code, and checksum, as discussed earlier. The Router Advertisement specific information starts from the 'Cur hop limit' field, where the router communicates to nodes on the link what Hop Limit they should use (see Section 3.4.3). The 'flags' field is zero, indicating that there are no stateless or stateful DHCPv6 services on the link, the router priority is medium (Section 3.5.6), and it is not an home agent (see explanations in Section 3.4.3).

The 'router lifetime' field value is 60, indicating that the router that sent the Router Advertisement can be considered as the default router for the next 60 seconds (see Section 3.4.3). 'Reachable time' tells how long a node should assume that other nodes on the link stay reachable after the node has received reachability confirmation (see also Section 3.4.5), such as Neighbor Solicitation message from the other node. The 'Retrans Timer' tells us how long nodes should wait between consecutive Neighbor Solicitations and how long to wait for DAD to complete. In this example both reachable and retransmit timers are 10000 milliseconds, which is a longish value in the case of a fixed Ethernet connection.

The shown Router Advertisement includes two options: the ICMPv6 PIO and the Source Link-Layer Address Option. The PIO option indicates that the prefix length is 64 bits, which is always the case in Ethernet links ([99] Section 4). Furthermore, the PIO tells via the 'L'-flag bit that the information within can be used for on-link determination (see Section 3.4.2) and via the 'A'-flag bit that the information can be used for SLAAC (see Section 3.5.1). In this trace we can see the Router Address' bit set as well (see Section 3.4.3), which in this environment is a Router Advertisement configuration bug: the system is not using MIPv6 and hence there is no need to communicate the full IPv6 address of the sending router in the PIO. Furthermore, the prefix carried in the PIO does not contain the full IPv6 but just the 64-bit prefix. This is a nice example of the policy of 'be liberal in what you receive', as this misconfiguration in Router Advertisement did not cause any issues for nodes connecting to the link. The 'Not site prefix' flag is a ghost from the past – around the year 2000 there was a proposal by Erik Nordmark a now expired document named *draft-ietf-ipngwg-site-prefixes-05.txt*, to have a bit to indicate if the included prefix should be accepted and preferred, as was described in the expired document. No current standard actually uses this bit.

The most interesting content of the PIO is the 64-bit prefix `2001:14b8:138:42::` that hosts will use for SLAAC as described in Section 3.5.1. The Valid lifetime' field indicates that the addresses configured from the included prefix can be used for 86400 seconds (one full day). The 'Preferred lifetime' field indicates how long an IPv6 address configured from this prefix remains in 'preferred' state (see Section 3.5.1) – in this case for 14400 seconds (for four hours).

The Source Link-Layer Address option tells to nodes what is the link layer address of the router, and allows nodes to directly update their Neighbor Caches. If the Source Link-Layer Address Option were not included in the Router Advertisement, the nodes would have to perform Neighbor Discovery (and *not* to pick the link-layer address from the Ethernet header). The address might be omitted to allow router load balancing, and of course if a link type did not have link-layer addresses, this option would not be present.

Initiating DAD for a Global IPv6 Address

The unsolicited Router Advertisement received on packet number 7 had an interesting effect on the stack. With the information from the Router Advertisement, the stack was able to configure a global IPv6 address based on the 64-bit prefix received in the PIO of the Router Advertisement and the same IID that was used for the link-local address (see Section 3.5.1). The resulting global IPv6 address was `2001:14b8:138:42:ba27:ebff:fef1:8d0d`, and as usual, it was initially in 'tentative' state. RFC 4862 says that DAD must be performed for all autoconfigured addresses. We can see the host stack initiating DAD for the global IPv6 address on packet number 8 illustrated in Figure 3.29.

The Neighbor Solicitation message is essentially the same as the previous Neighbor Solicitation sent for the link-local address. Also this Neighbor Solicitation is sent from the unspecified IPv6 address of all zeros, because the DAD for the link-local address is still ongoing.

Soliciting for Routers on the Link

At packet 9, illustrated in Figure 3.30, the host stack sends an ICMPv6 Router Solicitation message. One might wonder why the Router Solicitation is sent even though a Router Advertisement had (luckily) already been received. The reason is that the host may want to learn whether there are other routers on the same link, and also to ensure that it gets a complete Router Advertisement. It might be the case that the unsolicited periodic Router Advertisement does not contain all possibly available information, because RFC 4861 Section 6.2.3 [41] allows routers to send unsolicited Router Advertisements and omit some of the options that need not be sent every time.

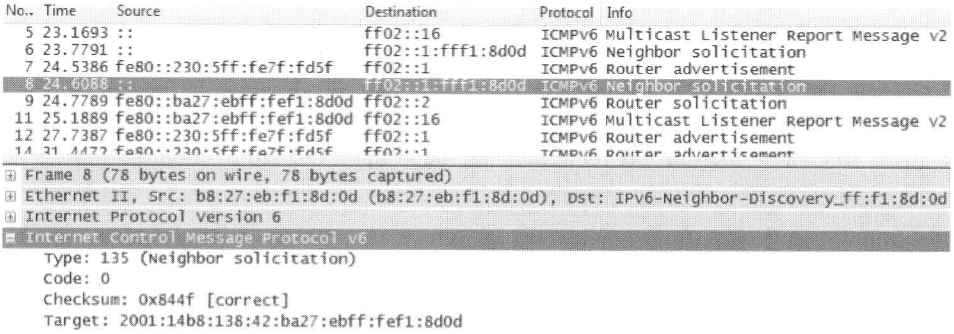

```
No.. Time      Source                      Destination          Protocol Info
   5 23.1693  ::                           ff02::16             ICMPv6 Multicast Listener Report Message v2
   6 23.7791  ::                           ff02::1:fff1:8d0d    ICMPv6 Neighbor solicitation
   7 24.5386  fe80::230:5ff:fe7f:fd5f      ff02::1              ICMPv6 Router advertisement
   8 24.6088  ::                           ff02::1:fff1:8d0d    ICMPv6 Neighbor solicitation
   9 24.7789  fe80::ba27:ebff:fef1:8d0d    ff02::2              ICMPv6 Router solicitation
  11 25.1889  fe80::ba27:ebff:fef1:8d0d    ff02::16             ICMPv6 Multicast Listener Report Message v2
  12 27.7387  fe80::230:5ff:fe7f:fd5f      ff02::1              ICMPv6 Router advertisement
  14 31.4472  fe80::230:5ff:fe7f:fd5f      ff02::1              ICMPv6 Router advertisement
⊞ Frame 8 (78 bytes on wire, 78 bytes captured)
⊞ Ethernet II, Src: b8:27:eb:f1:8d:0d (b8:27:eb:f1:8d:0d), Dst: IPv6-Neighbor-Discovery_ff:f1:8d:0d
⊞ Internet Protocol Version 6
⊟ Internet Control Message Protocol v6
    Type: 135 (Neighbor solicitation)
    Code: 0
    Checksum: 0x844f [correct]
    Target: 2001:14b8:138:42:ba27:ebff:fef1:8d0d
```

Figure 3.29 Example: initiating DAD for a global IPv6 address.

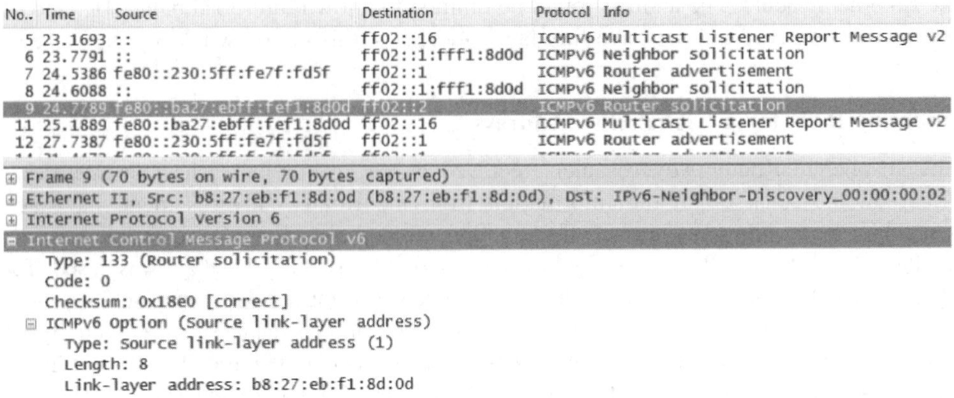

No..	Time	Source	Destination	Protocol	Info
5	23.1693	::	ff02::16	ICMPv6	Multicast Listener Report Message v2
6	23.7791	::	ff02::1:fff1:8d0d	ICMPv6	Neighbor solicitation
7	24.5386	fe80::230:5ff:fe7f:fd5f	ff02::1	ICMPv6	Router advertisement
8	24.6088	::	ff02::1:fff1:8d0d	ICMPv6	Neighbor solicitation
9	24.7789	fe80::ba27:ebff:fef1:8d0d	ff02::2	ICMPv6	Router solicitation
11	25.1889	fe80::ba27:ebff:fef1:8d0d	ff02::16	ICMPv6	Multicast Listener Report Message v2
12	27.7387	fe80::230:5ff:fe7f:fd5f	ff02::1	ICMPv6	Router advertisement

⊞ Frame 9 (70 bytes on wire, 70 bytes captured)
⊞ Ethernet II, Src: b8:27:eb:f1:8d:0d (b8:27:eb:f1:8d:0d), Dst: IPv6-Neighbor-Discovery_00:00:00:02
⊞ Internet Protocol Version 6
⊟ Internet Control Message Protocol v6
    Type: 133 (Router solicitation)
    Code: 0
    Checksum: 0x18e0 [correct]
  ⊟ ICMPv6 Option (Source link-layer address)
    Type: Source link-layer address (1)
    Length: 8
    Link-layer address: b8:27:eb:f1:8d:0d

Figure 3.30   Example: soliciting for routers on the link.

In a more typical Ethernet case there would not be an unsolicited Router Advertisement readily available, but a host would have to send a Router Solicitation to hasten reception of the Router Advertisement and hence to speed up address autoconfiguration procedures. On some other types of links, such as point-to-point, a router may send unsolicited Router Advertisement with full information immediately after the link layer has been established.

In the trace shown in Figure 3.30 we can see the source IPv6 address of the Router Solicitation packet to be the link-local IPv6 address stack starting to do DAD at trace step 6. We can note from the timestamps that a second has passed, which is the default time to wait for a Neighbor Advertisement to arrive. As there was no conflict detected, the link-local address was moved to 'preferred' state and hence became available for full use. The Router Solicitation also includes the Source Link-Layer Address option with the host's link-layer address, hence enabling message receivers to update their Neighbor Cache entries.

Joining Solicited-node Multicast Group Again

Another thing to perform after the DAD has provided a preferred address is to resend the 'Multicast Listener Report Message v2' for joining the Solicited-node multicast address group, but now from the configured link-local IPv6 address, as shown in Figure 3.31, and required by RFC 3810 Section 5.2.13. Another reason for retransmission is to ensure that all MLD routers on the link get the report, as stated in RFC 3810 Section 6.1 [45].

Reception of the Next Router Advertisement

In packet 12, the next unsolicited Router Advertisement is received. The new Router Advertisement is exactly the same as before and hence not expanded herein; it does not cause any more spectacular actions than refreshment of the various host's timers related to router and prefix lifetimes, and the router-related Neighbor Cache entry.

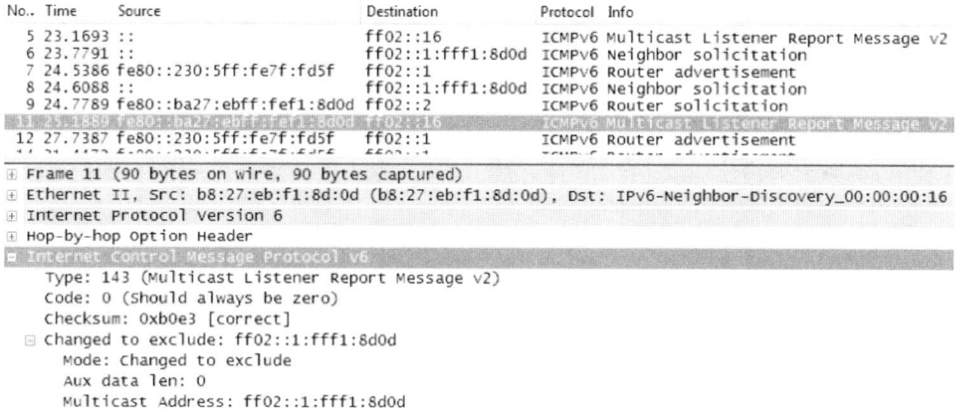

```
No.. Time     Source                    Destination           Protocol Info
   5 23.1693 ::                         ff02::16              ICMPv6  Multicast Listener Report Message v2
   6 23.7791 ::                         ff02::1:fff1:8d0d     ICMPv6  Neighbor solicitation
   7 24.5386 fe80::230:5ff:fe7f:fd5f    ff02::1               ICMPv6  Router advertisement
   8 24.6088 ::                         ff02::1:fff1:8d0d     ICMPv6  Neighbor solicitation
   9 24.7789 fe80::ba27:ebff:fef1:8d0d  ff02::2               ICMPv6  Router solicitation
  11 25.1889 fe80::ba27:ebff:fef1:8d0d  ff02::16              ICMPv6  Multicast Listener Report Message v2
  12 27.7387 fe80::230:5ff:fe7f:fd5f    ff02::1               ICMPv6  Router advertisement
```

⊞ Frame 11 (90 bytes on wire, 90 bytes captured)
⊞ Ethernet II, Src: b8:27:eb:f1:8d:0d (b8:27:eb:f1:8d:0d), Dst: IPv6-Neighbor-Discovery_00:00:00:16
⊞ Internet Protocol Version 6
⊞ Hop-by-hop Option Header
⊟ Internet Control Message Protocol v6
    Type: 143 (Multicast Listener Report Message v2)
    Code: 0 (Should always be zero)
    Checksum: 0xb0e3 [correct]
  ⊟ Changed to exclude: ff02::1:fff1:8d0d
     Mode: Changed to exclude
     Aux data len: 0
     Multicast Address: ff02::1:fff1:8d0d

Figure 3.31 Example: joining Solicited-node multicast address group from the link-local IPv6 address.

3.12.2 *IPv6 with DNS and TCP*

Another example that we are illustrating in this section is the one where a host performs DNS queries for the destination's Fully Qualified Domain Name (FQDN) and then initiates TCP connection to this site. In this scenario, a web browser in a popular operating system is requested by the user to go to the *http://www.ietf.org* website, which is the main page of the IETF. The website is also dual-stack enabled, that is it has both IPv4 and IPv6 addresses in the DNS. As shown in Figure 3.32 the DNS resolver on the operating system attempts to resolve both IPv4 and IPv6 addresses of the *www.ietf.org*. The resolver is utilizing the 'A resource record first and then the AAAA resource record' approach of those possible and listed in Section 3.10.2. It is worth noting that both DNS queries are

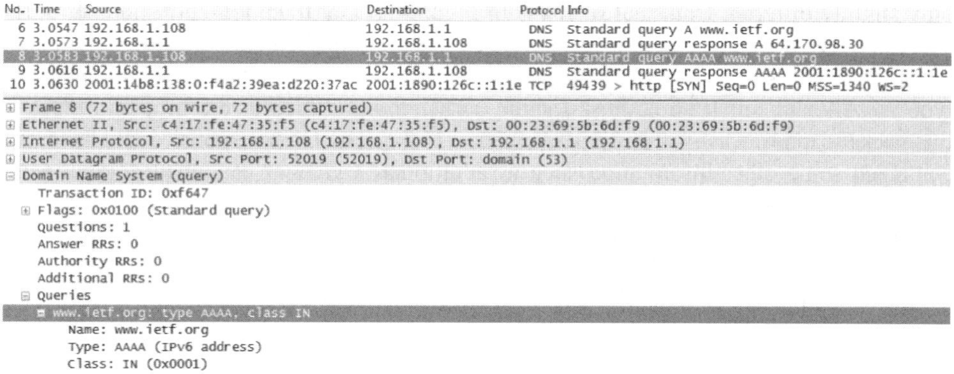

```
No. Time     Source                    Destination           Protocol Info
   6 3.0547 192.168.1.108              192.168.1.1           DNS  Standard query A www.ietf.org
   7 3.0573 192.168.1.1                192.168.1.108         DNS  Standard query response A 64.170.98.30
   8 3.0583 192.168.1.108              192.168.1.1           DNS  Standard query AAAA www.ietf.org
   9 3.0616 192.168.1.1                192.168.1.108         DNS  Standard query response AAAA 2001:1890:126c::1:1e
  10 3.0630 2001:14b8:138:0:f4a2:39ea:d220:37ac  2001:1890:126c::1:1e  TCP  49439 > http [SYN] Seq=0 Len=0 MSS=1340 WS=2
```

⊞ Frame 8 (72 bytes on wire, 72 bytes captured)
⊞ Ethernet II, Src: c4:17:fe:47:35:f5 (c4:17:fe:47:35:f5), Dst: 00:23:69:5b:6d:f9 (00:23:69:5b:6d:f9)
⊞ Internet Protocol, Src: 192.168.1.108 (192.168.1.108), Dst: 192.168.1.1 (192.168.1.1)
⊞ User Datagram Protocol, Src Port: 52019 (52019), Dst Port: domain (53)
⊟ Domain Name System (query)
    Transaction ID: 0xf647
  ⊞ Flags: 0x0100 (Standard query)
    Questions: 1
    Answer RRs: 0
    Authority RRs: 0
    Additional RRs: 0
  ⊟ Queries
    ⊟ www.ietf.org: type AAAA, class IN
      Name: www.ietf.org
      Type: AAAA (IPv6 address)
      Class: IN (0x0001)

Figure 3.32 Example: performing DNS queries for *www.ietf.org*.

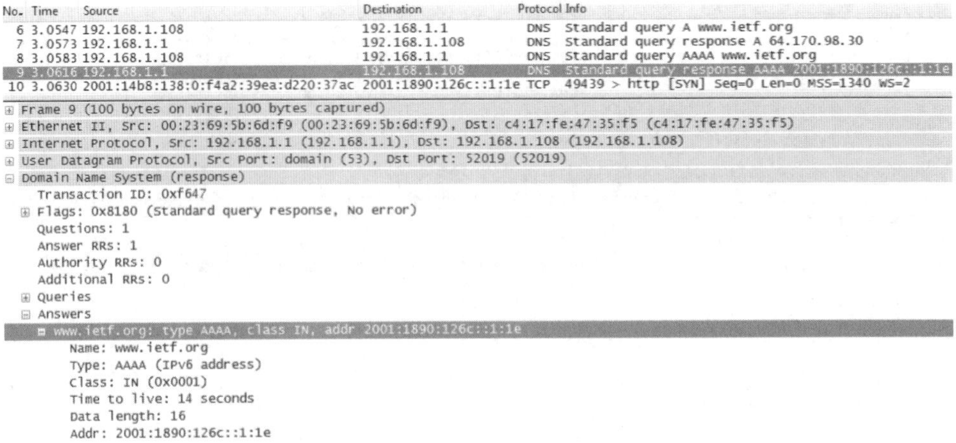

```
No. Time     Source              Destination       Protocol Info
   6 3.0547 192.168.1.108        192.168.1.1       DNS    Standard query A www.ietf.org
   7 3.0573 192.168.1.1          192.168.1.108     DNS    Standard query response A 64.170.98.30
   8 3.0583 192.168.1.108        192.168.1.1       DNS    Standard query AAAA www.ietf.org
   9 3.0616 192.168.1.1          192.168.1.108     DNS    Standard query response AAAA 2001:1890:126c::1:1e
  10 3.0630 2001:14b8:138:0:f4a2:39ea:d220:37ac 2001:1890:126c::1:1e TCP 49439 > http [SYN] Seq=0 Len=0 MSS=1340 WS=2

⊞ Frame 9 (100 bytes on wire, 100 bytes captured)
⊞ Ethernet II, Src: 00:23:69:5b:6d:f9 (00:23:69:5b:6d:f9), Dst: c4:17:fe:47:35:f5 (c4:17:fe:47:35:f5)
⊞ Internet Protocol, Src: 192.168.1.1 (192.168.1.1), Dst: 192.168.1.108 (192.168.1.108)
⊞ User Datagram Protocol, Src Port: domain (53), Dst Port: 52019 (52019)
⊟ Domain Name System (response)
     Transaction ID: 0xf647
  ⊞ Flags: 0x8180 (Standard query response, No error)
     Questions: 1
     Answer RRs: 1
     Authority RRs: 0
     Additional RRs: 0
  ⊞ Queries
  ⊟ Answers
     ⊟ www.ietf.org: type AAAA, class IN, addr 2001:1890:126c::1:1e
       Name: www.ietf.org
       Type: AAAA (IPv6 address)
       Class: IN (0x0001)
       Time to live: 14 seconds
       Data length: 16
       Addr: 2001:1890:126c::1:1e
```

Figure 3.33   Example: DNS response with AAAA record.

sent over the IPv4 independently of the address family being resolved – the query type has no relation to the address family used to transport the query. In this case the DNS queries are sent to an address of 192.168.1.1 that happens to be a local DNS proxy, as the nodes on the used dual-stack access network are not provisioned with any IPv6 address for DNS servers.

The first trace entry we have opened, trace log entry number 8, is the one showing the DNS query for an AAAA resource record. The queries for an A resource record are as performed in the usual IPv4 manner and hence we skipped those. In the highlighted entry, the implications of IPv6 to the DNS protocol are visible in the 'Queries' section, which is showing a type AAAA resource record being requested for the *www.ietf.org*. Of course one major implication of IPv6 to systems is the need to send a second DNS query in the first place.

A DNS response containing an IPv6 address in an AAAA type of resource record is only slightly different from that of IPv4. In Figure 3.33, at trace log entry number 9, the successful DNS reply containing an AAAA record is highlighted. The AAAA resource record is otherwise exactly as the A resource records, but it contains a identification of type AAAA and full IPv6 address for the *www.ietf.org*, which at the time and place of this packet capture was 2001:1890:126c::1:1e. The DNS response could have also contained some other information such as additional AAAA or canonical domain name records, but in this case the response contained only a single resource record.

After the host has successfully and quickly received both IPv4 and IPv6 addresses for the destination it is connecting to, the host performs address selection algorithms and, as usually happens for hosts having dual-stack connectivity, ends up preferring IPv6. In more complex real-life scenarios it can happen that either DNS query yields negative answers, or it is even lost in transit. In such cases the host should proceed after some time has passed with the possibly received positive answer, even if the resolution of the other address family would still be pending. The implementations vary on how they perform error recovery in these kinds of situations.

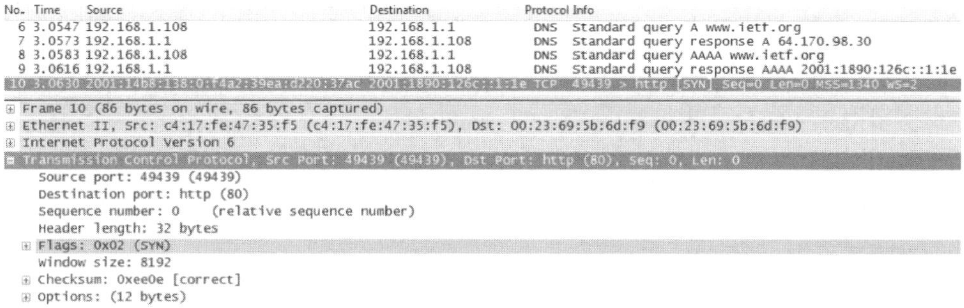

```
No.  Time    Source                          Destination              Protocol Info
 6 3.0547 192.168.1.108                    192.168.1.1              DNS   Standard query A www.ietf.org
 7 3.0573 192.168.1.1                      192.168.1.108            DNS   Standard query response A 64.170.98.30
 8 3.0583 192.168.1.108                    192.168.1.1              DNS   Standard query AAAA www.ietf.org
 9 3.0616 192.168.1.1                      192.168.1.108            DNS   Standard query response AAAA 2001:1890:126c::1:1e
10 3.0630 2001:14b8:138:0:f4a2:39ea:d220:37ac  2001:1890:126c::1:1e  TCP   49439 > http [SYN] Seq=0 Len=0 MSS=1340 WS=2

  Frame 10 (86 bytes on wire, 86 bytes captured)
  Ethernet II, Src: c4:17:fe:47:35:f5 (c4:17:fe:47:35:f5), Dst: 00:23:69:5b:6d:f9 (00:23:69:5b:6d:f9)
  Internet Protocol Version 6
  Transmission Control Protocol, Src Port: 49439 (49439), Dst Port: http (80), Seq: 0, Len: 0
      Source port: 49439 (49439)
      Destination port: http (80)
      Sequence number: 0    (relative sequence number)
      Header length: 32 bytes
  Flags: 0x02 (SYN)
      window size: 8192
  Checksum: 0xee0e [correct]
  options: (12 bytes)
```

Figure 3.34   Example: initiation of TCP handshake with the preferred address family.

In Figure 3.34 we can see the result of preferring of IPv6 over IPv4 at packet 10: the first message of the TCP handshake procedures is being sent to the IPv6 address of the *www.ietf.org* that was just learned over the DNS, and no IPv4 traffic whatsoever. Furthermore, this figure shows the main contents of the TCP header and we can see that there is no IPv6 specific content there. Of course, as described in Section 3.10.1, the checksum field also covers the IPv6 pseudo header in addition to the TCP packet headers.

If it happens that IPv6 connection establishment is excessively delayed, the host should try to use IPv4 instead. How fast this kind of error recovery is implemented is, as at the time of writing, an implementation-dependent decision.

The steps that follow the last entry of Figure 3.34 are the completion of the TCP handshake and initiation of data transfer from *www.ietf.org*. These are not shown. However, we have highlighted one packet from the actual website data transmission in Figure 3.35. In the link where this packet capture was taken from the MTU advertised in MTU option of Router Advertisement is 1400 bytes due to uplink limitations (Router Advertisement not shown). Because of this, the TCP data transmissions have been sized to fit the smallest MTU of the path, which in this case happens to be the MTU of the access link. From the 1400 bytes 40 bytes go to the IPv6 header itself, 20 bytes for the TCP header, and then 1340 bytes for the actual TCP segment data. We can determine the Path MTU used by the stack by adding 40 bytes (IPv6 header length) to the length shown inside the IPv6 header, and seeing from the trace that the amount of TCP data in transit significantly exceeds the 1340 bytes that happens to be in the payload of the highlighted packet.

3.13   Chapter Summary

In this chapter, we have been through the essential features of IPv6. We looked at how the IPv6 addressing architecture is set up, what kind of addresses there are, how they are statelessly and statefully allocated to hosts, and how prefix delegation works for the routing scenarios. The IPv6 header structure was analyzed from the main header to the extension headers and with a peek to transport protocol headers and how the transport layer checksum includes the IPv6 pseudo header. The ICMPv6 and the key protocols it provides, such as Neighbor Discovery, were studied. We had a look to IPsec, mobile IP, routing, and protocol verification, and gave attention to a key set of IPv6's companion protocols:

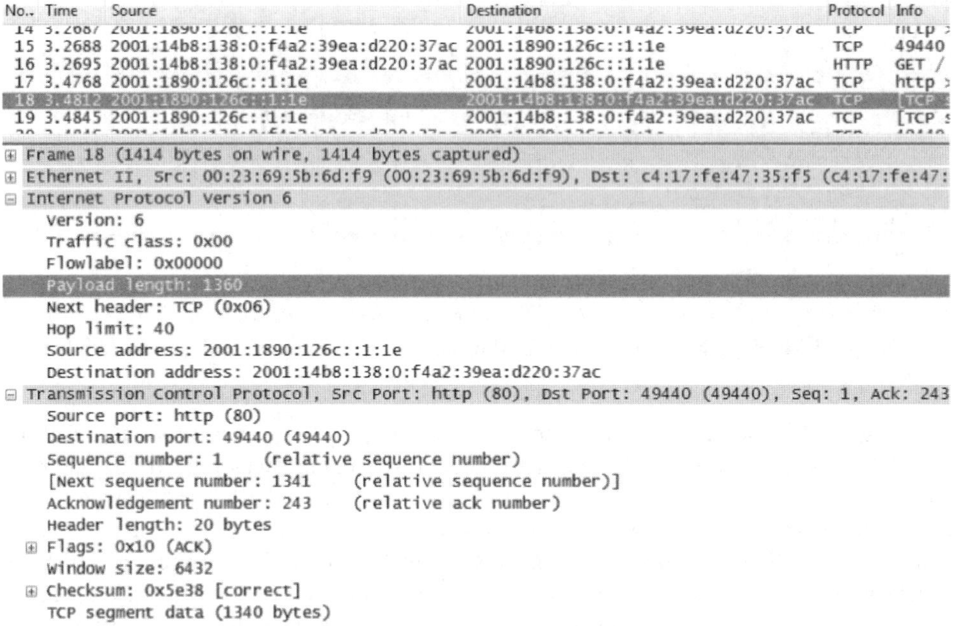

No..	Time	Source	Destination	Protocol	Info
14	3.2687	2001:1890:126c::1:1e	2001:14b8:138:0:f4a2:39ea:d220:37ac	TCP	http >
15	3.2688	2001:14b8:138:0:f4a2:39ea:d220:37ac	2001:1890:126c::1:1e	TCP	49440
16	3.2695	2001:14b8:138:0:f4a2:39ea:d220:37ac	2001:1890:126c::1:1e	HTTP	GET /
17	3.4768	2001:1890:126c::1:1e	2001:14b8:138:0:f4a2:39ea:d220:37ac	TCP	http >
18	3.4812	2001:1890:126c::1:1e	2001:14b8:138:0:f4a2:39ea:d220:37ac	TCP	[TCP ⟩
19	3.4845	2001:1890:126c::1:1e	2001:14b8:138:0:f4a2:39ea:d220:37ac	TCP	[TCP ⟩

⊞ Frame 18 (1414 bytes on wire, 1414 bytes captured)
⊞ Ethernet II, Src: 00:23:69:5b:6d:f9 (00:23:69:5b:6d:f9), Dst: c4:17:fe:47:35:f5 (c4:17:fe:47:
⊟ Internet Protocol Version 6
    Version: 6
    Traffic class: 0x00
    Flowlabel: 0x00000
    Payload length: 1360
    Next header: TCP (0x06)
    Hop limit: 40
    Source address: 2001:1890:126c::1:1e
    Destination address: 2001:14b8:138:0:f4a2:39ea:d220:37ac
⊟ Transmission Control Protocol, Src Port: http (80), Dst Port: 49440 (49440), Seq: 1, Ack: 243
    Source port: http (80)
    Destination port: 49440 (49440)
    Sequence number: 1   (relative sequence number)
    [Next sequence number: 1341   (relative sequence number)]
    Acknowledgement number: 243   (relative ack number)
    Header length: 20 bytes
  ⊞ Flags: 0x10 (ACK)
    Window size: 6432
  ⊞ Checksum: 0x5e38 [correct]
    TCP segment data (1340 bytes)

Figure 3.35   Example: highlighting TCP packet sizes and MTU.

the DHCPv6 and DNS. To really help readers jump into the 3GPP specific features in the next chapter, we also introduced different models of links that can transport IPv6.

The chapter ended with detailed real-life IPv6 packet captures and explanations of each message shown. The packet captures and their analysis should help to get up to speed with IPv6 traffic debugging and analysis – at least the packet traces will not look all that scary after studying these examples.

This chapter has provided general descriptions of IPv6 and will prove to be helpful when reading how IPv6 is used and applied on 3GPP accesses.

References

1. Jankiewicz, E., Loughney, J., and Narten, T. *IPv6 Node Requirements*. RFC 6434, Internet Engineering Task Force, December 2011.
2. Singh, H., Beebee, W., Donley, C., Stark, B., and Troan, O. *Basic Requirements for IPv6 Customer Edge Routers*. RFC 6204, Internet Engineering Task Force, April 2011.
3. Arkko, J., Kuijpers, G., Soliman, H., Loughney, J., and Wiljakka, J. *Internet Protocol Version 6 (IPv6) for Some Second and Third Generation Cellular Hosts*. RFC 3316, Internet Engineering Task Force, April 2003.
4. Hinden, R. and Deering, S. *IP Version 6 Addressing Architecture*. RFC 4291, Internet Engineering Task Force, February 2006.
5. Deering, S., Haberman, B., Jinmei, T., Nordmark, E., and Zill, B. *IPv6 Scoped Address Architecture*. RFC 4007, Internet Engineering Task Force, March 2005.
6. Huitema, C. and Carpenter, B. *Deprecating Site Local Addresses*. RFC 3879, Internet Engineering Task Force, September 2004.

7. Hinden, R. and Haberman, B. *Unique Local IPv6 Unicast Addresses*. RFC 4193, Internet Engineering Task Force, October 2005.

8. Singh, H., Beebee, W., and Nordmark, E. *IPv6 Subnet Model: The Relationship between Links and Subnet Prefixes*. RFC 5942, Internet Engineering Task Force, July 2010.

9. IEEE. *Guidelines for 64-bit Global Identifier (EUI-64) Registration Authority*. Technical report, IEEE Standards Association, 1997.

10. Armitage, G., Schulter, P., Jork, M., and Harter, G. *IPv6 over Non-Broadcast Multiple Access (NBMA) networks*. RFC 2491, Internet Engineering Task Force, January 1999.

11. Narten, T., Draves, R., and Krishnan, S. *Privacy Extensions for Stateless Address Autoconfiguration in IPv6*. RFC 4941, Internet Engineering Task Force, September 2007.

12. Arkko, J., Kempf, J., Zill, B., and Nikander, P. *SEcure Neighbor Discovery (SEND)*. RFC 3971, Internet Engineering Task Force, March 2005.

13. Bao, C., Huitema, C., Bagnulo, M., Boucadair, M., and Li, X. *IPv6 Addressing of IPv4/IPv6 Translators*. RFC 6052, Internet Engineering Task Force, October 2010.

14. Huitema, C. *Teredo: Tunneling IPv6 over UDP through Network Address Translations (NATs)*. RFC 4380, Internet Engineering Task Force, February 2006.

15. Huston, G., Lord, A., and Smith, P. *IPv6 Address Prefix Reserved for Documentation*. RFC 3849, Internet Engineering Task Force, July 2004.

16. Arkko, J., Cotton, M., and Vegoda, L. *IPv4 Address Blocks Reserved for Documentation*. RFC 5737, Internet Engineering Task Force, January 2010.

17. Venaas, S., Parekh, R., de Velde, G. Van, Chown, T., and Eubanks, M. *Multicast Addresses for Documentation*. RFC 6676, Internet Engineering Task Force, August 2012.

18. Kawamura, S. and Kawashima, M. *A Recommendation for IPv6 Address Text Representation*. RFC 5952, Internet Engineering Task Force, August 2010.

19. Fuller, V., Li, T., Yu, J., and Varadhan, K. *Classless Inter-Domain Routing (CIDR): an Address Assignment and Aggregation Strategy*. RFC 1519, Internet Engineering Task Force, September 1993.

20. Carpenter, B., Cheshire, S., and Hinden, R. *Representing IPv6 Zone Identifiers in Address Literals and Uniform Resource Identifiers*. Internet-Draft draft-ietf-6man-uri-zoneid-06, Internet Engineering Task Force, December 2012. Work in progress.

21. Deering, S. and Hinden, R. *Internet Protocol, Version 6 (IPv6) Specification*. RFC 2460, Internet Engineering Task Force, December 1998.

22. Nichols, K., Blake, S., Baker, F., and Black, D. *Definition of the Differentiated Services Field (DS Field) in the IPv4 and IPv6 Headers*. RFC 2474, Internet Engineering Task Force, December 1998.

23. Ramakrishnan, K., Floyd, S., and Black, D. *The Addition of Explicit Congestion Notification (ECN) to IP*. RFC 3168, Internet Engineering Task Force, September 2001.

24. Heinanen, J., Baker, F., Weiss, W., and Wroclawski, J. *Assured Forwarding PHB Group*. RFC 2597, Internet Engineering Task Force, June 1999.

25. Davie, B., Charny, A., Bennet, J. C. R., Benson, K., Boudec, J. Y. Le, Courtney, W., Davari, S., Firoiu, V., and Stiliadis, D. *An Expedited Forwarding PHB (Per-Hop Behavior)*. RFC 3246, Internet Engineering Task Force, March 2002.

26. Baker, F., Polk, J., and Dolly, M. *A Differentiated Services Code Point (DSCP) for Capacity-Admitted Traffic*. RFC 5865, Internet Engineering Task Force, May 2010.

27. Amante, S., Carpenter, B., and Jiang, S. *Rationale for Update to the IPv6 Flow Label Specification*. RFC 6436, Internet Engineering Task Force, November 2011.

28. Amante, S., Carpenter, B., Jiang, S., and Rajahalme, J. *IPv6 Flow Label Specification*. RFC 6437, Internet Engineering Task Force, November 2011.

29. Borman, D., Deering, S., and Hinden, R. *IPv6 Jumbograms*. RFC 2675, Internet Engineering Task Force, August 1999.

30. Conta, A. and Deering, S. *Generic Packet Tunneling in IPv6 Specification*. RFC 2473, Internet Engineering Task Force, December 1998.

31. Partridge, C. and Jackson, A. *IPv6 Router Alert Option*. RFC 2711, Internet Engineering Task Force, October 1999.

32. Floyd, S., Allman, M., Jain, A., and Sarolahti, P. *Quick-Start for TCP and IP*. RFC 4782, Internet Engineering Task Force, January 2007.

33. StJohns, M., Atkinson, R., and Thomas, G. *Common Architecture Label IPv6 Security Option (CALIPSO)*. RFC 5570, Internet Engineering Task Force, July 2009.

34. Macker, J. *Simplified Multicast Forwarding*. RFC 6621, Internet Engineering Task Force, May 2012.
35. Perkins, C., Johnson, D., and Arkko, J. *Mobility Support in IPv6*. RFC 6275, Internet Engineering Task Force, July 2011.
36. Fenner, B. *Experimental Values In IPv4, IPv6, ICMPv4, ICMPv6, UDP, and TCP Headers*. RFC 4727, Internet Engineering Task Force, November 2006.
37. Abley, J., Savola, P., and Neville-Neil, G. *Deprecation of Type 0 Routing Headers in IPv6*. RFC 5095, Internet Engineering Task Force, December 2007.
38. Kent, S. *IP Authentication Header*. RFC 4302, Internet Engineering Task Force, December 2005.
39. Kent, S. *IP Encapsulating Security Payload (ESP)*. RFC 4303, Internet Engineering Task Force, December 2005.
40. McCann, J., Deering, S., and Mogul, J. *Path MTU Discovery for IP version 6*. RFC 1981, Internet Engineering Task Force, August 1996.
41. Narten, T., Nordmark, E., Simpson, W., and Soliman, H. *Neighbor Discovery for IP version 6 (IPv6)*. RFC 4861, Internet Engineering Task Force, September 2007.
42. Mathis, M. and Heffner, J. *Packetization Layer Path MTU Discovery*. RFC 4821, Internet Engineering Task Force, March 2007.
43. Krishnan, S. *Handling of Overlapping IPv6 Fragments*. RFC 5722, Internet Engineering Task Force, December 2009.
44. Deering, S., Fenner, W., and Haberman, B. *Multicast Listener Discovery (MLD) for IPv6*. RFC 2710, Internet Engineering Task Force, October 1999.
45. Vida, R. and Costa, L. *Multicast Listener Discovery Version 2 (MLDv2) for IPv6*. RFC 3810, Internet Engineering Task Force, June 2004.
46. Haberman, B. *Source Address Selection for the Multicast Listener Discovery (MLD) Protocol*. RFC 3590, Internet Engineering Task Force, September 2003.
47. Holbrook, H., Cain, B., and Haberman, B. *Using Internet Group Management Protocol Version 3 (IGMPv3) and Multicast Listener Discovery Protocol Version 2 (MLDv2) for Source-Specific Multicast*. RFC 4604, Internet Engineering Task Force, August 2006.
48. Holbrook, H. and Cain, B. *Source-Specific Multicast for IP*. RFC 4607, Internet Engineering Task Force, August 2006.
49. Fenner, B., Handley, M., Holbrook, H., and Kouvelas, I. *Protocol Independent Multicast - Sparse Mode (PIM-SM): Protocol Specification (Revised)*. RFC 4601, Internet Engineering Task Force, August 2006.
50. Liu, H., Cao, W., and Asaeda, H. *Lightweight Internet Group Management Protocol Version 3 (IGMPv3) and Multicast Listener Discovery Version 2 (MLDv2) Protocols*. RFC 5790, Internet Engineering Task Force, February 2010.
51. Haberman, B. *Allocation Guidelines for IPv6 Multicast Addresses*. RFC 3307, Internet Engineering Task Force, August 2002.
52. Haberman, B. and Thaler, D. *Unicast-Prefix-based IPv6 Multicast Addresses*. RFC 3306, Internet Engineering Task Force, August 2002.
53. Crawford, M. and Haberman, B. *IPv6 Node Information Queries*. RFC 4620, Internet Engineering Task Force, August 2006.
54. Droms, R., Bound, J., Volz, B., Lemon, T., Perkins, C., and Carney, M. *Dynamic Host Configuration Protocol for IPv6 (DHCPv6)*. RFC 3315, Internet Engineering Task Force, July 2003.
55. Haberman, B. and Martin, J. *Multicast Router Discovery*. RFC 4286, Internet Engineering Task Force, December 2005.
56. Conta, A., Deering, S., and Gupta, M. *Internet Control Message Protocol (ICMPv6) for the Internet Protocol Version 6 (IPv6) Specification*. RFC 4443, Internet Engineering Task Force, March 2006.
57. Bonica, R., Gan, D., Tappan, D., and Pignataro, C. *Extended ICMP to Support Multi-Part Messages*. RFC 4884, Internet Engineering Task Force, April 2007.
58. Bonica, R., Gan, D., Tappan, D., and Pignataro, C. *ICMP Extensions for Multiprotocol Label Switching*. RFC 4950, Internet Engineering Task Force, August 2007.
59. Atlas, A., Bonica, R., Pignataro, C., Shen, N., and Rivers, JR. *Extending ICMP for Interface and Next-Hop Identification*. RFC 5837, Internet Engineering Task Force, April 2010.
60. Crawford, M. *Router Renumbering for IPv6*. RFC 2894, Internet Engineering Task Force, August 2000.
61. Conta, A. *Extensions to IPv6 Neighbor Discovery for Inverse Discovery Specification*. RFC 3122, Internet Engineering Task Force, June 2001.

62. Kempf, J. *Instructions for Seamoby and Experimental Mobility Protocol IANA Allocations*. RFC 4065, Internet Engineering Task Force, July 2005.

63. Koodli, R. *Mobile IPv6 Fast Handovers*. RFC 5568, Internet Engineering Task Force, July 2009.

64. Winter, T., Thubert, P., Brandt, A., Hui, J., Kelsey, R., Levis, P., Pister, K., Struik, R., Vasseur, JP., and Alexander, R. *RPL: IPv6 Routing Protocol for Low-Power and Lossy Networks*. RFC 6550, Internet Engineering Task Force, March 2012.

65. Chown, T. and Venaas, S. *Rogue IPv6 Router Advertisement Problem Statement*. RFC 6104, Internet Engineering Task Force, February 2011.

66. Carpenter, B. and Moore, K. *Connection of IPv6 Domains via IPv4 Clouds*. RFC 3056, Internet Engineering Task Force, February 2001.

67. Levy-Abegnoli, E., deVelde, G. Van, Popoviciu, C., and Mohacsi, J. *IPv6 Router Advertisement Guard*. RFC 6105, Internet Engineering Task Force, February 2011.

68. Draves, R. and Thaler, D. *Default Router Preferences and More-Specific Routes*. RFC 4191, Internet Engineering Task Force, November 2005.

69. Jeong, J., Park, S., Beloeil, L., and Madanapalli, S. *IPv6 Router Advertisement Options for DNS Configuration*. RFC 6106, Internet Engineering Task Force, November 2010.

70. Haberman, B. and Hinden, R. *IPv6 Router Advertisement Flags Option*. RFC 5175, Internet Engineering Task Force, March 2008.

71. Krishnan, S., Laganier, J., Bonola, M., and Garcia-Martinez, A. *Secure Proxy ND Support for SEcure Neighbor Discovery (SEND)*. RFC 6496, Internet Engineering Task Force, February 2012.

72. Gashinsky, I., Jaeggli, J., and Kumari, W. *Operational Neighbor Discovery Problems*. RFC 6583, Internet Engineering Task Force, March 2012.

73. Thomson, S., Narten, T., and Jinmei, T. *IPv6 Stateless Address Autoconfiguration*. RFC 4862, Internet Engineering Task Force, September 2007.

74. Moore, N. *Optimistic Duplicate Address Detection (DAD) for IPv6*. RFC 4429, Internet Engineering Task Force, April 2006.

75. Aura, T. *Cryptographically Generated Addresses (CGA)*. RFC 3972, Internet Engineering Task Force, March 2005.

76. Thaler, D., Talwar, M., and Patel, C. *Neighbor Discovery Proxies (ND Proxy)*. RFC 4389, Internet Engineering Task Force, April 2006.

77. Krishnan, S. *Reserved IPv6 Interface Identifiers*. RFC 5453, Internet Engineering Task Force, February 2009.

78. Brzozowski, J., Kinnear, K., Volz, B., and Zeng, S. *DHCPv6 Leasequery*. RFC 5007, Internet Engineering Task Force, September 2007.

79. Stapp, M. *DHCPv6 Bulk Leasequery*. RFC 5460, Internet Engineering Task Force, February 2009.

80. Narten, T. and Johnson, J. *Definition of the UUID-Based DHCPv6 Unique Identifier (DUID-UUID)*. RFC 6355, Internet Engineering Task Force, August 2011.

81. Troan, O. and Droms, R. *IPv6 Prefix Options for Dynamic Host Configuration Protocol (DHCP) version 6*. RFC 3633, Internet Engineering Task Force, December 2003.

82. Korhonen, J., Savolainen, T., Krishnan, S., and Troan, O. *Prefix Exclude Option for DHCPv6-based Prefix Delegation*. RFC 6603, Internet Engineering Task Force, May 2012.

83. Droms, R. *Stateless Dynamic Host Configuration Protocol (DHCP) Service for IPv6*. RFC 3736, Internet Engineering Task Force, April 2004.

84. Venaas, S., Chown, T., and Volz, B. *Information Refresh Time Option for Dynamic Host Configuration Protocol for IPv6 (DHCPv6)*. RFC 4242, Internet Engineering Task Force, November 2005.

85. Droms, R. *DNS Configuration options for Dynamic Host Configuration Protocol for IPv6 (DHCPv6)*. RFC 3646, Internet Engineering Task Force, December 2003.

86. Savolainen, T., Kato, J., and Lemon, T. *Improved Recursive DNS Server Selection for Multi-Interfaced Nodes*. RFC 6731, Internet Engineering Task Force, December 2012.

87. Schulzrinne, H. and Volz, B. *Dynamic Host Configuration Protocol (DHCPv6) Options for Session Initiation Protocol (SIP) Servers*. RFC 3319, Internet Engineering Task Force, July 2003.

88. Boucadair, M., Penno, R., and Wing, D. *DHCP Options for the Port Control Protocol (PCP)*. Internet-Draft draft-ietf-pcp-dhcp-05, Internet Engineering Task Force, September 2012. Work in progress.

89. Matsumoto, A., Fujisaki, T., and Chown, T. *Distributing Address Selection Policy using DHCPv6*. Internet-Draft draft-ietf-6man-addr-select-opt-08, Internet Engineering Task Force, January 2013. Work in progress.

90. Kaufman, C., Hoffman, P., Nir, Y., and Eronen, P. *Internet Key Exchange Protocol Version 2 (IKEv2)*. RFC 5996, Internet Engineering Task Force, September 2010.

91. Eronen, P., Laganier, J., and Madson, C. *IPv6 Configuration in Internet Key Exchange Protocol Version 2 (IKEv2)*. RFC 5739, Internet Engineering Task Force, February 2010.

92. Thaler, D., Draves, R., Matsumoto, A., and Chown, T. *Default Address Selection for Internet Protocol Version 6 (IPv6)*. RFC 6724, Internet Engineering Task Force, September 2012.

93. Fink, R. and Hinden, R. *6bone (IPv6 Testing Address Allocation) Phaseout*. RFC 3701, Internet Engineering Task Force, March 2004.

94. Ferguson, P. and Senie, D. *Network Ingress Filtering: Defeating Denial of Service Attacks which employ IP Source Address Spoofing*. RFC 2827, Internet Engineering Task Force, May 2000.

95. Nordmark, E., Bagnulo, M., and Levy-Abegnoli, E. *FCFS SAVI: First-Come, First-Served Source Address Validation Improvement for Locally Assigned IPv6 Addresses*. RFC 6620, Internet Engineering Task Force, May 2012.

96. Wu, J., Bi, J., Bagnulo, M., Baker, F., and Vogt, C. *Source Address Validation Improvement Framework*. Internet-Draft draft-ietf-savi-framework-06, Internet Engineering Task Force, January 2012. Work in progress.

97. Baker, F. and Savola, P. *Ingress Filtering for Multihomed Networks*. RFC 3704, Internet Engineering Task Force, March 2004.

98. Gont, F. *A method for Generating Stable Privacy-Enhanced Addresses with IPv6 Stateless Address Autoconfiguration (SLAAC)*. Internet-Draft draft-ietf-6man-stable-privacy-addresses-03, Internet Engineering Task Force, January 2013. Work in progress.

99. Crawford, M. *Transmission of IPv6 Packets over Ethernet Networks*. RFC 2464, Internet Engineering Task Force, December 1998.

100. S. Varada, Haskins, D., and Allen, E. *IP Version 6 over PPP*. RFC 5072, Internet Engineering Task Force, September 2007.

101. Korhonen, J., Soininen, J., Patil, B., Savolainen, T., Bajko, G., and Iisakkila, K. *IPv6 in 3rd Generation Partnership Project (3GPP) Evolved Packet System (EPS)*. RFC 6459, Internet Engineering Task Force, January 2012.

102. Crawford, M. *Transmission of IPv6 Packets over FDDI Networks*. RFC 2467, Internet Engineering Task Force, December 1998.

103. Crawford, M., Narten, T., and Thomas, S. *Transmission of IPv6 Packets over Token Ring Networks*. RFC 2470, Internet Engineering Task Force, December 1998.

104. Armitage, G., Schulter, P., and Jork, M. *IPv6 over ATM Networks*. RFC 2492, Internet Engineering Task Force, January 1999.

105. Conta, A., Malis, A., and Mueller, M. *Transmission of IPv6 Packets over Frame Relay Networks Specification*. RFC 2590, Internet Engineering Task Force, May 1999.

106. Fujisawa, K. and Onoe, A. *Transmission of IPv6 Packets over IEEE 1394 Networks*. RFC 3146, Internet Engineering Task Force, October 2001.

107. DeSanti, C., Carlson, C., and Nixon, R. *Transmission of IPv6, IPv4, and Address Resolution Protocol (ARP) Packets over Fibre Channel*. RFC 4338, Internet Engineering Task Force, January 2006.

108. Montenegro, G., Kushalnagar, N., Hui, J., and Culler, D. *Transmission of IPv6 Packets over IEEE 802.15.4 Networks*. RFC 4944, Internet Engineering Task Force, September 2007.

109. Patil, B., Xia, F., Sarikaya, B., Choi, JH., and Madanapalli, S. *Transmission of IPv6 via the IPv6 Convergence Sublayer over IEEE 802.16 Networks*. RFC 5121, Internet Engineering Task Force, February 2008.

110. Jeon, H., Jeong, S., and Riegel, M. *Transmission of IP over Ethernet over IEEE 802.16 Networks*. RFC 5692, Internet Engineering Task Force, October 2009.

111. Carpenter, B. and Hinden, R. *Adaptation of RFC 1149 for IPv6*. RFC 6214, Internet Engineering Task Force, April 2011.

112. Vyncke, E. *IPv6 over Social Networks*. RFC 5514, Internet Engineering Task Force, April 2009.

113. 3GPP, . General Packet Radio Service (GPRS); Service description; Stage 2. TS 23.060, 3rd Generation Partnership Project (3GPP), March 2012.

114. 3GPP, . General Packet Radio Service (GPRS) enhancements for Evolved Universal Terrestrial Radio Access Network (E-UTRAN) access. TS 23.401, 3rd Generation Partnership Project (3GPP), March 2012.

115. McGregor, G. *The PPP Internet Protocol Control Protocol (IPCP)*. RFC 1332, Internet Engineering Task Force, May 1992.

116. Nieminen, J., Patil, B., Savolainen, T., Isomaki, M., Shelby, Z., and Gomez, C. *Transmission of IPv6 Packets over BLUETOOTH Low Energy*. Internet-Draft draft-ietf-6lowpan-btle-11, Internet Engineering Task Force, October 2012. Work in progress.

117. Gundavelli, S., Townsley, M., Troan, O., and Dec, W. *Address Mapping of IPv6 Multicast Packets on Ethernet*. RFC 6085, Internet Engineering Task Force, January 2011.

118. Kohno, M., Nitzan, B., Bush, R., Matsuzaki, Y., Colitti, L., and Narten, T. *Using 127-Bit IPv6 Prefixes on Inter-Router Links*. RFC 6164, Internet Engineering Task Force, April 2011.

119. Krishnan, S. and Daley, G. *Simple Procedures for Detecting Network Attachment in IPv6*. RFC 6059, Internet Engineering Task Force, November 2010.

120. Johnson, D., Perkins, C., and Arkko, J. *Mobility Support in IPv6*. RFC 3775, Internet Engineering Task Force, June 2004.

121. Soliman, H. *Mobile IPv6 Support for Dual Stack Hosts and Routers*. RFC 5555, Internet Engineering Task Force, June 2009.

122. Chakrabarti, S. and Nordmark, E. *Extension to Sockets API for Mobile IPv6*. RFC 4584, Internet Engineering Task Force, July 2006.

123. Gundavelli, S., Leung, K., Devarapalli, V., Chowdhury, K., and Patil, B. *Proxy Mobile IPv6*. RFC 5213, Internet Engineering Task Force, August 2008.

124. Wakikawa, R. and Gundavelli, S. *IPv4 Support for Proxy Mobile IPv6*. RFC 5844, Internet Engineering Task Force, May 2010.

125. Kent, S. and Seo, K. *Security Architecture for the Internet Protocol*. RFC 4301, Internet Engineering Task Force, December 2005.

126. Manral, V. *Cryptographic Algorithm Implementation Requirements for Encapsulating Security Payload (ESP) and Authentication Header (AH)*. RFC 4835, Internet Engineering Task Force, April 2007.

127. Eronen, P. *IKEv2 Mobility and Multihoming Protocol (MOBIKE)*. RFC 4555, Internet Engineering Task Force, June 2006.

128. Gilligan, R., Thomson, S., Bound, J., McCann, J., and Stevens, W. *Basic Socket Interface Extensions for IPv6*. RFC 3493, Internet Engineering Task Force, February 2003.

129. Stevens, W., Thomas, M., Nordmark, E., and Jinmei, T. *Advanced Sockets Application Program Interface (API) for IPv6*. RFC 3542, Internet Engineering Task Force, May 2003.

130. Thaler, D., Fenner, B., and Quinn, B. *Socket Interface Extensions for Multicast Source Filters*. RFC 3678, Internet Engineering Task Force, January 2004.

131. Nordmark, E., Chakrabarti, S., and Laganier, J. *IPv6 Socket API for Source Address Selection*. RFC 5014, Internet Engineering Task Force, September 2007.

132. Liu, D., and Cao, Z. *MIF API consideration*. Internet-Draft draft-ietf-mif-api-extension-03, Internet Engineering Task Force, November 2012. Work in progress.

133. Berners-Lee, T., Fielding, R., and Masinter, L. *Uniform Resource Identifier (URI): Generic Syntax*. RFC 3986, Internet Engineering Task Force, January 2005.

134. Wing, D. and Yourtchenko, A. *Happy Eyeballs: Success with Dual-Stack Hosts*. RFC 6555, Internet Engineering Task Force, April 2012.

135. Postel, J. *User Datagram Protocol*. RFC 0768, Internet Engineering Task Force, August 1980.

136. Postel, J. *Transmission Control Protocol*. RFC 0793, Internet Engineering Task Force, September 1981.

137. Stewart, R., Xie, Q., Morneault, K., Sharp, C., Schwarzbauer, H., Taylor, T., Rytina, I., Kalla, M., Zhang, L., and Paxson, V. *Stream Control Transmission Protocol*. RFC 2960, Internet Engineering Task Force, October 2000.

138. Coltun, R., Ferguson, D., Moy, J., and Lindem, A. *OSPF for IPv6*. RFC 5340, Internet Engineering Task Force, July 2008.

139. Larzon, L-A., Degermark, M., Pink, S., Jonsson, L-E., and Fairhurst, G. *The Lightweight User Datagram Protocol (UDP-Lite)*. RFC 3828, Internet Engineering Task Force, July 2004.

140. Eubanks, M., Chimento, P., and Westerlund, M. *IPv6 and UDP Checksums for Tunneled Packets*. Internet-Draft draft-ietf-6man-udpchecksums-07, Internet Engineering Task Force, January 2013. Work in progress.

141. Mockapetris, P. V. *Domain names – implementation and specification*. RFC 1035, Internet Engineering Task Force, November 1987.

142. Thomson, S., Huitema, C., Ksinant, V., and Souissi, M. *DNS Extensions to Support IP Version 6*. RFC 3596, Internet Engineering Task Force, October 2003.

143. Vixie, P. *Extension Mechanisms for DNS (EDNS0)*. RFC 2671, Internet Engineering Task Force, August 1999.

144. Crawford, M. and Huitema, C. *DNS Extensions to Support IPv6 Address Aggregation and Renumbering*. RFC 2874, Internet Engineering Task Force, July 2000.

145. Jiang, S., Conrad, D., and Carpenter, B. *Moving A6 to Historic Status*. RFC 6563, Internet Engineering Task Force, March 2012.

146. Morishita, Y. and Jinmei, T. *Common Misbehavior Against DNS Queries for IPv6 Addresses*. RFC 4074, Internet Engineering Task Force, May 2005.

147. Postel, J. and Reynolds, J. *File Transfer Protocol*. RFC 0959, Internet Engineering Task Force, October 1985.

148. Allman, M., Ostermann, S., and Metz, C. *FTP Extensions for IPv6 and NATs*. RFC 2428, Internet Engineering Task Force, September 1998.

149. Malkin, G. and Minnear, R. *RIPng for IPv6*. RFC 2080, Internet Engineering Task Force, January 1997.

150. Bates, T., Chandra, R., Katz, D., and Rekhter, Y. *Multiprotocol Extensions for BGP-4*. RFC 4760, Internet Engineering Task Force, January 2007.

151. Hopps, C. *Routing IPv6 with IS-IS*. RFC 5308, Internet Engineering Task Force, October 2008.

152. Marques, P. and Dupont, F. *Use of BGP-4 Multiprotocol Extensions for IPv6 Inter-Domain Routing*. RFC 2545, Internet Engineering Task Force, March 1999.

153. Harrington, D., Presuhn, R., and Wijnen, B. *An Architecture for Describing Simple Network Management Protocol (SNMP) Management Frameworks*. RFC 3411, Internet Engineering Task Force, December 2002.

154. Haberman, B. *IP Forwarding Table MIB*. RFC 4292, Internet Engineering Task Force, April 2006.

155. Routhier, S. *Management Information Base for the Internet Protocol (IP)*. RFC 4293, Internet Engineering Task Force, April 2006.

156. Nelson, D. *RADIUS Authentication Server MIB for IPv6*. RFC 4669, Internet Engineering Task Force, August 2006.

157. Keeni, G., Koide, K., Gundavelli, S., and Wakikawa, R. *Proxy Mobile IPv6 Management Information Base*. RFC 6475, Internet Engineering Task Force, May 2012.

158. McWalter, D., Thaler, D., and Kessler, A. *IP Multicast MIB*. RFC 5132, Internet Engineering Task Force, December 2007.

159. Hui, J. and Vasseur, JP. *The Routing Protocol for Low-Power and Lossy Networks (RPL) Option for Carrying RPL Information in Data-Plane Datagrams*. RFC 6553, Internet Engineering Task Force, March 2012.

160. Townsley, W. and Troan, O. *IPv6 Rapid Deployment on IPv4 Infrastructures (6rd) – Protocol Specification*. RFC 5969, Internet Engineering Task Force, August 2010.

161. Savola, P. and Haberman, B. *Embedding the Rendezvous Point (RP) Address in an IPv6 Multicast Address*. RFC 3956, Internet Engineering Task Force, November 2004.

162. Perkins, C. *IP Mobility Support for IPv4, Revised*. RFC 5944, Internet Engineering Task Force, November 2010.

4

IPv6 in 3GPP Networks

This chapter introduces how Internet Protocol version 6 (IPv6) is implemented in 3rd Generation Partnership Project (3GPP) core networks and 3GPP compliant User Equipments (UEs). We will also take a look into network aspects that are specific to 3GPP. The network architectures of interest are the General Packet Radio Service (GPRS) and the Evolved Packet System (EPS), which both provide packet switched services. For readability reasons we will primarily use the terminology of the EPS, unless specifically noted when some technical aspect is particular to the GPRS.

4.1 PDN Connectivity Service

One of the basic concepts in 3GPP network architecture is the Packet Data Network (PDN) Connectivity Service, which offers Internet Protocol (IP) connectivity and services between the User Equipment (UE) and the Public Land Mobile Network (PLMN) external IP-based PDN network. The PDN Connectivity Service supports the transport of IP flow aggregates, consisting of one or more flows identified by various IP flow filters. The PDN can be an internal, walled-garden-like IP network within the mobile operator network, or any IP network outside operator administration, such as the Internet. PDN Connectivity is realized as an established PDN Connection. PDN Connection is the association between a UE and a PDN represented by an Access Point Name (APN). Each PDN Connection has an associated Internet Protocol version 4 (IPv4) address [1] and/or an IPv6 prefix [2]. This section describes the PDN Connectivity Services aspects that matter, from the IPv6 usage and deployment point of view. For simplicity we use the term *packet core* to mean both the GPRS and the Evolved Packet Core (EPC), unless otherwise stated.

The selection of a desired PDN is realized using an APN, which essentially refers to a gateway in the packet core that has connectivity to the PDN. The gateway in a 3GPP packet core is the Gateway GPRS Support Node (GGSN) or the Packet Data Network Gateway (PGW). For readability we will mostly use the term PGW as a synonym for a GGSN when there is no real functional difference between the two. Specific release, functional, or interface-related differences are separately noted.

Deploying IPv6 in 3GPP Networks: Evolving Mobile Broadband from 2G to LTE and Beyond, First Edition.
Jouni Korhonen, Teemu Savolainen and Jonne Soininen.
© 2013 John Wiley & Sons, Ltd. Published 2013 by John Wiley & Sons, Ltd.

Once a PDN Connection has been created, the required PDN Connection state is established in various packet core nodes, such as Serving Gateway Support Node (SGSN), Serving Gateway (SGW), Home Location Register (HLR), and so on. The PDN Connection state information contains addresses of peering gateway nodes, various GPRS Tunneling Protocol (GTP) related identifiers, Quality of Service (QoS) related parameter, but most importantly the IPv4 address and/or IPv6 prefix associated with the PDN Connection. Each PDN connection has its own IPv4 address and/or IPv6 prefix assigned to it by the PDN and topologically anchored in the corresponding gateway (GGSN or PGW) to the PDN. The PDN is responsible for the IPv4 address and/or IPv6 prefix allocation to the UE. On the UE, a PDN Connection is equivalent to a network interface. We will further discuss different types of PDN Connections in Section 4.1.2.

4.1.1 Bearer Concept

An important part of the PDN Connectivity service is the *bearer concept*. An EPS Bearer uniquely identifies traffic flows that receive a common QoS treatment between a UE and a PGW for GTP-based S5/S8-interfaces [3, 4], and between a UE and an SGW for PMIP-based S5/S8-interfaces [5–7]. A bearer can be further divided into radio bearers between the UE and the base station, S1-bearers between the base station and the SGW, and S5/S8-bearers (in case of GTP-based S5/S8-interfaces) between the SGW and the PGW. On the GPRS side, the equivalent concept to the EPS Bearer is the Packet Data Protocol (PDP) Context.

It is common to refer to PDN Connection and PDP Context as equivalent concepts, although that is not exactly correct. On the GPRS side, there are several architectural differences from the EPC, as discussed in Section 2.2, thus mapping a PDP Context to an EPS Bearer is not straightforward when it comes to bearer behavior details, functional details across various packet core elements, and the protocols used on various interfaces. For example, the GTP-based Gn/Gp-interfaces between the SGSN and the GGSN, and in the case of *direct tunneling* [8] between the Radio Network Controller (RNC) and the GGSN, use the older GPRS Tunneling Protocol version 1 (GTPv1), [9, 4]. The generic bearer concept for EPS is illustrated in Figure 4.1, when using GTP-based S5/S8-interfaces and for PMIP-based S5/S8-interfaces in Figure 4.2. For simplicity and overall clarity we will use the term *PDN Connection* to refer also to the *PDP Context*, unless there is a specific feature or functional difference that does not apply to both, and in this case we will use the exact terminology.

One EPS Bearer is immediately established when the UE powers on, attaches to a radio network, and connects to a PDN, and that remains established throughout the lifetime of the PDN Connection to provide the UE with an always-on IP connectivity to the PDN. This bearer is referred as the *default bearer* in EPS and historically as the *Primary PDP Context* in GPRS. Any additional PDP Context or EPS Bearer that is established for the same PDN Connection is referred to as a *dedicated bearer* in EPS and as a *Secondary PDP Context* in GPRS. The dedicated bearers share their fate with the default bearer (as a detail, the same does not hold for GPRS Primary and Secondary PDP Contexts). For simplicity we use the term *dedicated bearer* for both the GPRS Secondary PDP Context and the EPS dedicated beader.

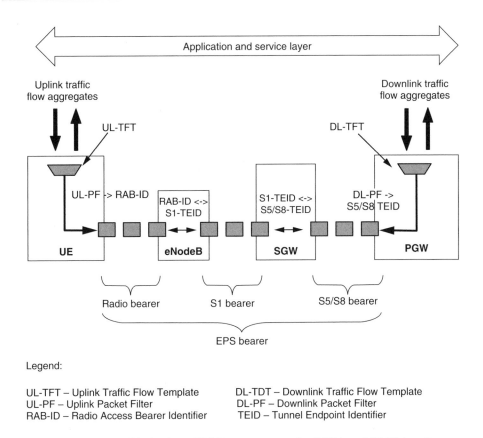

Figure 4.1 Simplified unicast EPS Bearer concept for GTP-based S5/S8-interface.

The default bearer handling is different from the EPS to the GPRS. In the GPRS, the UE can be attached to the radio network but still not have established the PDN Connection. Consequently, there is no IP address allocated to the UE and the PDN Connection either. This is a rather fundamental difference between the EPS and the GPRS regarding the consumption of IP numbering resources. A UE using EPS IP connectivity always consumes IP number resources when attached to the radio network; that is, the UE is always on IP connectivity wise. It is possible to delay the allocation of the IPv4 address for the default bearer, if the UE indicates its willingness to use Dynamic Host Configuration Protocol version 4 (DHCPv4) [10, 11] for address configuration. The address allocation for PDN Connections and the default bearer are discussed in detail in Section 4.4

There is a substantial difference between the default bearer and the dedicated bearer. Each time a new default bearer (and the founding PDN Connection) gets created, a new IPv4 address and/or IPv6 prefix is allocated and assigned for the PDN Connection. However, when a dedicated bearer gets created, it shares the existing IPv4 address and/or IPv6 prefix allocated for the PDN Connection. Dedicated bearers are differentiated using a Traffic Flow Template (TFT). A TFT is essentially an IP flow filter, and identifies an IP flow that receives a specific QoS treatment between the UE and the PGW.

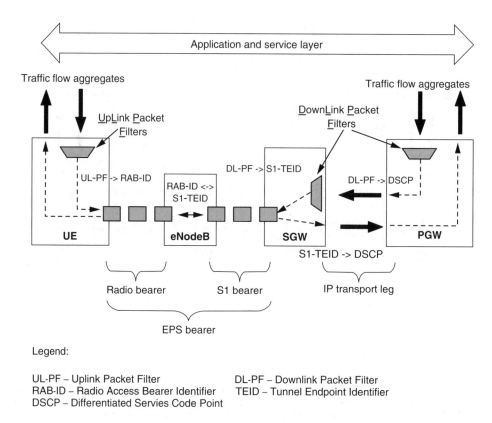

Figure 4.2 Simplified unicast EPS Bearer concept for Proxy Mobile IP (PMIP)-based S5/S8-interface.

A default bearer (and also a dedicated bearer) activation and removal has a cost. Both involve a significant amount of e.g. Non-Access Stratum (NAS) [12, 13] signaling over the radio link between the UE and the network (Mobile Management Entity (MME) or SGSN elements), a number of bearer management signaling messages, and policy and subscription data related signaling within the packet core. If bearer activations and removals are frequent, it may lead to unwanted signaling load. Additionally, each new bearer always consumes an additional Radio Access Bearer (RAB), which impacts scarce radio access resources. Finally, it is common for equipment vendors to build licensing schemes around the number of concurrent bearers, which means that operators have an incentive to optimize the number of used bearers per UE.

4.1.2 PDP and PDN Types

As discussed in Section 4.1, each PDN Connection is associated with an IPv4 address and/or IPv6 prefix. A PDN Connection can only carry the protocol type(s) that got negotiated during the default bearer establishment. (Historically, a PDP Context could also carry Point to Point Protocol (PPP) [14] frames. However, PPP has been deprecated since the introduction of EPS in 3GPP Release-8.) There are three different PDP Types (in GPRS) and PDN Types (in EPS). For simplicity we use the term *PDN Type* for both PDP Type and PDN Type, unless noted otherwise:

PDN Type IPv4 – the PDN Connection is associated with exactly one IPv4 address. This PDN Type is commonly referred as the *IPv4-only bearer* or *IPv4-only PDN Connection*. The PDN Type IPv4 has been part of the 3GPP specifications since the first release of 3GPP specifications [8].

PDN Type IPv6 – the PDN Connection is associated with exactly one IPv6 prefix with a /64 prefix length. This PDN Type is commonly referred as the *IPv6-only bearer* or *IPv6-only PDN Connection*. The PDN Type IPv6 has been part of the 3GPP specifications effectively since 3GPP Release-99 [8].

PDN Type IPv4v6 – the PDN Connection is associated with both an IPv4 address and an IPv6 prefix. This PDN Type is also commonly referred as a *dual-stack bearer* or a *dual-stack PDN Connection*. The PDN Type IPv4v6 has been part of the 3GPP specifications since the 3GPP Release-8 for EPS (i.e. for S5/S8-interfaces and S4-interfaces) [15, 16] and since Release-9 for GPRS (i.e. for Gn/Gp-interfaces) [8, 15] (For an unknown reason EPC and GPRS had an unfortunate feature disparity regarding PDN Type IPv4v6. The GPRS got to the same level only in 3GPP Release-9.)

Only the negotiated type of traffic can be transported over the PDN Connection. Furthermore, it is a common practice to apply ingress filtering to prohibit a UE from sending IP packets with a different source addresses than that associated with the PDN Connection [17]. There are, however, mobile router solutions where a complete IPv4 subnet is statically routed to a network behind the UE. Such solutions are vendor specific and not covered by 3GPP standards.

The strict separation of PDN Types and the restriction of one IPv4 address and/or IPv6 prefix per PDN Connection has caused a significant increase in the amount of concurrent PDN Connections per UE. For example, if a UE that conforms to 3GPP pre-Release-8 standards wants to have dual-stack connectivity, it basically has to create two concurrent PDN Connections: one of PDN Type IPv4 and another of PDN Type IPv6. Furthermore, when there is a need to separate traffic and provide differentiated routing in a PDN, it is common practice to create a new PDN Connection to a possibly different APN just for having an additional IPv4 address and/or IPv6 prefix configured in the UE for a new service. Since each PDN Connection (and its associated IP addressing) are separated from each other, service and PDN differentiation becomes possible. This approach actually mimics end host multi-interfacing. However, using multiple default bearers for configuring more than one IPv4 address and/or IPv6 prefix on a UE has undesired signaling and resource consumption side effects as mentioned in Section 4.1.1. We will return to this topic and possible future enhancements concerning multiple bearers and the IP addressing models in Section 6.2.2.

4.1.3 Link Models in 3GPP

Link models have a significant role from the IP functionality point of view, especially when it comes to IPv6. Originally the 3GPP GPRS architecture had a single link model: a *point-to-point link* between a UE and a GGSN, which had been modeled after a PPP link. The same link model still holds for the 3GPP EPS architecture, when GTP-based S5/S8-interfaces are used and where the point-to-point link is between a UE and a PGW. Actually, since 3GPP Release-8 there are now multiple link models due to the additional supported IP mobility and tunneling technologies in EPC [5, 18–20]. However, we only concentrate on the two IP mobility protocols and link models that have commercial deployments: a GTP-based and a PMIP-based solution for 3GPP access technologies.

The Link Model for 3GPP Access using GTP-based Interfaces

Figure 4.3 illustrates the user plane for the GTP-based S5/S8-interfaces. The Gn/Gp-based interfaces would be almost the same for the user plane except for different gateway naming. The link model for 3GPP access using GTP-based Gn-, Gp-, and S5/S8-interfaces has the following generic characteristics on the user plane:

- The 3GPP link is a point-to-point resembling link between a UE and a PGW. There can only be two nodes on the link: the UE and the first-hop router.
- The PGW is the first-hop router for the UE.
- The IPv4 address and/or IPv6 prefix are topologically anchored at the PGW.
- The link has no link-layer addresses.
- The PGW never configures an IPv4 address on its interface facing the UE.
- The lifetime of IP addresses on the link share fate of the PDN Connection lifetime.

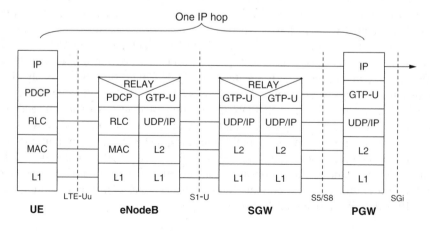

Figure 4.3   The link-model and the user plane protocol stack for EPS GTP-based interfaces.

The 3GPP link model has the following IPv6 specific properties and characteristics:

- The PGW is the only peer node with which the UE exchanges Neighbor Discovery Protocol (NDP) messages.
- The PGW has to configure a link-local IPv6 address to its UE facing interface and no other IPv6 addresses with a different scope. Specifically, the PGW does not configure any addresses out of the IPv6 prefix it advertises on the link.
- The link has exactly one, per UE unique /64 IPv6 prefix (under the same routing domain, if Unique Local Addresses (ULAs) are used) and the UE configures its non-link scoped unicast IPv6 addresses based on this IPv6 prefix.
- The lifetime of the /64 IPv6 prefix on the link shares fate of the PDN Connection lifetime. Therefore, the /64 prefix has both preferred and valid lifetimes set to infinity.
- The /64 prefix advertised on the link cannot be used for on-link determination.
- The PGW selects the IPv6 Interface IDentifier (IID) [21] for the UE. The IID is communicated to the UE over the NAS signaling and the UE is required to configure its link-local IPv6 address using the PGW selected IID. This is to make sure there can never be a Duplicate Address Detection (DAD) [22] failure on the link. (Several modem drivers and frameworks fail to deliver the IID from the radio modem and NAS signaling to the host side. Rather they generate their own IID, which opens a theoretical possibility for a link-local address collision on the 3GPP link.)
- NDP address resolution and redirection functions are not needed. There are no link-layer addresses or nodes other than the UE and the PGW on the link. However, doing for example address resolution should cause no harm other than generating unnecessary traffic.
- NDP DAD should not be needed. The `DupAddrDetectTransmits` [22] configuration variable in the host should be set to zero to turn the DAD off. However, doing DAD does no harm other than generating unnecessary traffic. The situation might change if the UE starts acting as an IPv6 tethering device using a Neighbor Discovery Proxy [23].
- NDP Neighbor Unreachability Detection (NUD) is not particularly necessary but it is recommended to enable it. The situation might change if UEs start tethering using a Neighbor Discovery Proxy [23].
- Renumbering the IPv6 prefixes on the link without re-establishing the PDN Connection is not supported.

NUD is particularly interesting on the 3GPP link referring to the specifics of the 3GPP link above (point-to-point and no link-layer addresses). The NUD algorithm using unicast Neighbor Solicitations and Advertisements as described in RFC 4861 [24] does not actually work without additional assumptions on links that have no link-layer addresses, like the 3GPP link. What there is left is then upper-layer confirmation, which specifically from the router point of view (e.g. the PGW) may not be adequate except for bidirectional and active transport layer flows. Furthermore, should the PGW ever initiate a router-host NUD using addresses configured out of the /64 prefix advertised on the link as a destination? Since the /64 prefix is routed as whole to the UE and the PGW Neighbor Discovery Protocol implementation would have no other reasons to maintain additional neighbor state for anything derived from the /64 prefix we are justified in stating that NUD should be no exception in that regard. If the PGW insists on initiating a router-host NUD using a unicast Neighbor Solicitation then it should use the network side assigned unicast link-local address of the UE as the destination. NUD over the 3GPP link, either initiated by the UE or the PGW, must handle the Neighbor Discovery Protocol state transition properly irrespective the lack of the link-layer addressing. For example, the UE and the PGW could treat the NUD originated Neighbor Discovery Protocol messages as if the appropriate link-layer addressing options were always present. It should also be understood that if NUD fails, the 3GPP specifications are silent about what should happen next and this would specifically be a concern for the router-host NUD.

Link Model for 3GPP Access using PMIP-based Interfaces

PMIP-based S5/S8-interfaces for the 3GPP access and S2a/S2b interfaces for *trusted non-3GPP access* were introduced to the 3GPP architecture starting from Release-8 [16]. Figure 4.4 illustrates the user plane for the PMIP-based S5/S8-interfaces. The link model for 3GPP access using PMIP-based S5/S8-interfaces has the following generic characteristics on the user plane:

- The link is a point-to-point resembling link between a UE and an **SGW**. There can only be two nodes on the link: the UE and the first-hop router.
- The SGW is the first-hop router for the UE. Note that this is a fundamental difference between GTP-based and PMIP-based interfaces, and there are certain known anomalies regarding the specification of the link model in the 3GPP specifications. We will discuss the PMIP-based S5/S8-interface link model further later in this section.
- The IPv4 address and/or IPv6 prefix is still topologically anchored at the PGW, in spite of the first-hop router being at the SGW.
- The 3GPP link has no link-layer addresses.
- The SGW may configure an IPv4 address on its interface facing the UE as the default gateway address. However, the NAS signaling, as of Release-11, is not able to convey the default gateway address to the UE.
- The lifetime of IP addresses on the link share fate of the PDN Connection lifetime.

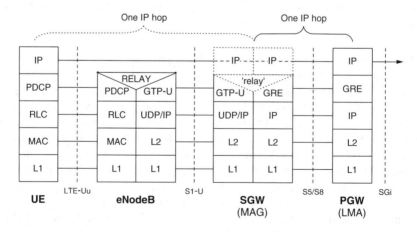

Figure 4.4   The link-model and the user plane protocol stack for EPS PMIP-based interfaces.

The 3GPP link model has the following IPv6 specific properties and characteristics:

- The SGW is the only peer node the UE exchanges NDP messages with.
- The SGW has to configure a link-local IPv6 address to its UE facing interface and no other IPv6 addresses with a different scope. Specifically, the SGW does not configure any address out of the IPv6 prefix it advertises on the link.
- The link has exactly one, per UE unique /64 IPv6 prefix and the UE configures its unicast IPv6 addresses based on this IPv6 prefix.
- The lifetime of the /64 IPv6 prefix on the link shares fate of the PDN Connection lifetime. Therefore, the /64 prefix on the link has both preferred and valid lifetimes set to infinity.
- The PGW (note, not the SGW) selects the IPv6 IID [21] for the UE. The IID is communicated to the UE over the NAS signaling and the UE is required to configure its link-local IPv6 address using the PGW selected IID. This is to ensure that there can never be a DAD [22] failure on the link. (Several modem drivers and frameworks fail to deliver the IID from the radio modem and NAS signaling to the host side. Rather they generate their own IID, which opens a theoretical possibility for a link-local address collision on the 3GPP link.)
- NDP address resolution and redirection functions are not needed. There are no link-layer addresses or nodes other than the UE and the SGW on the link. However, doing for example address resolution does no harm other than generating unnecessary traffic.
- NDP DAD should not be needed. The `DupAddrDetectTransmits` [22] configuration variable in the host should be set to zero to turn off DAD. However, doing DAD does no harm other than generating unnecessary traffic. The situation might change if the UE starts acting as an IPv6 tethering device using a Neighbor Discovery Proxy.
- NDP NUD is not particularly necessary but it is recommended to enable it. This could become important if UEs start doing tethering using a Neighbor Discovery Proxy.
- Since the first-hop router is located at the SGW, the SGW may change as a result of UE mobility and a SGW relocation. This implies that the *Idle Mode mobility* needs special care from the NDP point of view, since Router Advertisement sending intervals are not coordinated between SGWs. Once the UE becomes active the SGW should send an Router Advertisement immediately to the UE, otherwise the default router in the UE might timeout before the SGW would otherwise send its periodic unsolicited Router Advertisement.
- Renumbering the IPv6 prefixes on the link without re-establishing the PDN Connection is not supported.

According to the Internet Engineering Task Force (IETF) Proxy Mobile IPv6 (PMIPv6) specification [6, 7] a Mobile Access Gateway (MAG) is the *first-hop router* for the mobile node. This implies that the MAG, which in 3GPP architecture is located in a SGW, decrements the Hop-Limit/Time To Live (TTL) value of each IP packet that it forwards. Similarly, the Local Mobility Anchor (LMA), which in 3GPP architecture is located in a PGW, decrements the Hop-Limit/TTL value of each IP packet it forwards. Whether the SGW should decrement the Hop-Limit/TTL value is (purposely) left unspecified in the pre-Release-11 3GPP specifications [5], and therefore, there are two different interpretations and implementations around. Figure 4.4 illustrates the PMIP-based S5/S8-interfaces. The dashed lines in the SGW show the controversial part of the 3GPP defined PMIP link-model.

On the other hand, 3GPP specifications [25] are clear that the SGW, for example, originates Router Advertisements. The 3GPP specifications [16] are also clear that the SGW has to implement a Dynamic Host Configuration Protocol version 6 (DHCPv6) relay [26] function and the PGW then implements the DHCPv6 server function, which again makes it clear that the SGW is a router when PMIP-based S5/S8-interface is used. The end result is rather unfortunate functional disparity between the PMIP-based and GTP-based EPC networking nodes, which has already caused potential interoperability issues when integrating other non-3GPP access technologies into the EPC (such as – Wireless Local Area Network (WLAN)) [18] or deploying DHCPv6 Prefix Delegation (PD) [27, 28] over PMIP-based S5/S8-interface.

The 3GPP Release-11 finally corrected the link models for all non-3GPP accesses so that they follow IETF definitions on link models, including PMIP-based S5/S8-interface – now the SGW officially decreases Hop Limit/TTL.

4.2 End User IPv6 Service Impact on the 3GPP System

4.2.1 User, Control and Transport Planes

3GPP made a wise decision to separate all interfaces that are transported over IP into a *control plane*, a *user plane*, and a *transport plane*. The control plane includes signaling protocols such as Diameter [29] and GTP Control Plane (GTP-C) [3] among others. Since the IP version used to transport a control plane is independent of the actual content, the *Information Element (IE)* such as Diameter Attribute Value Pairs (AVPs), any control plane interface can be migrated to IPv6 from the information content point of view, while still transporting it over an IPv4-only transport plane in the packet core. This allows for a phased migration of the control plane to IPv6 – that is, keeping the actual transport plane in IPv4 while the control plane itself is migrated into IPv6. Figures 4.5a and 4.5b illustrate a typical common construction of the control plane and the transport plane in the 3GPP GPRS/EPC.

All end user traffic is tunneled within a packet core, typically inside a GTP User Plane (GTP-U) [4] or a PMIP [6] tunnel. Similarly to the control plane, the actual content – that is, the user plane tunneled inner IP traffic and addressing that the end user and the UE

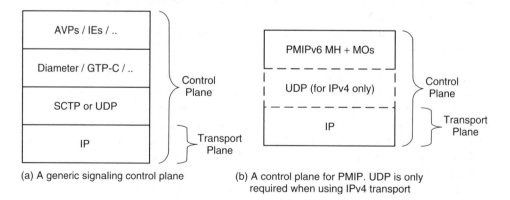

(a) A generic signaling control plane

(b) A control plane for PMIP. UDP is only required when using IPv4 transport

Figure 4.5   A control plane versus a transport plane.

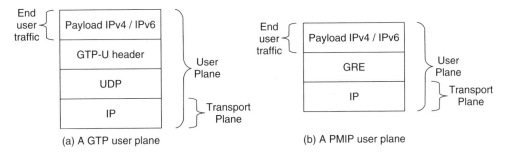

Figure 4.6 A user plane versus a transport plane.

see – is separated from the tunnel's outer IP addressing on the transport plane. Again, this allows for a phased migration of the user plane to IPv6 (and to dual-stack) while still keeping the actual transport plane in IPv4. Figures 4.6a and 4.6b illustrate a typical construction of the user plane and the transport plane in the 3GPP GPRS/EPC.

The separation of the transport plane from the user and control planes, and the fact that all user plane traffic is tunneled, would in theory allow deploying most packet core network elements (3GPP specific and pure IP routing/switching devices) without IPv6 support on their own networking connectivity level. For instance, if an operator has an existing IPv4-only Multi Protocol Label Switching (MPLS) [30] in their packet core network that can remain unchanged and be without IPv6 awareness or support [31, 32].

4.2.2 Affected Networking Elements

As discussed in Section 4.2.1, the division of 3GPP interfaces into a *control plane*, a *user plane* and a *transport plane* has made it possible to migrate different parts of the 3GPP system to IPv6 in phases. Especially, detaching the transport plane from the other two planes allows migrating the entire 3GPP system from the end user services point of view to IPv6 while still keeping the underlying operator packet core, backhaul, and radio access network IP infrastructure untouched and on IPv4. However, still at the system level most of the 3GPP networking elements are impacted when introducing IPv6 as the end user service.

Figure 4.7 shows GPRS and EPS networking elements, including the UE, which are impacted by the introduction of the end user IPv6 or dual-stack service (i.e., IPv6 delivered on the user plane). Basically everything except the IP transport plane (IP transmission) and 3GPP interfaces excluding (S)Gi is affected. The affected elements are marked with a star in Figure 4.7. The figure is far more complex (and also harder to parse) than the usual conceptual architecture pictures since we need to point out all affected elements and interfaces. We will briefly go through all the impacted elements and what is required as the entry level for the end user IPv6 service.

User Equipment

IPv6 has a significant impact on the UE on multiple layers. At a high level the following has to be verified for IPv6 support:

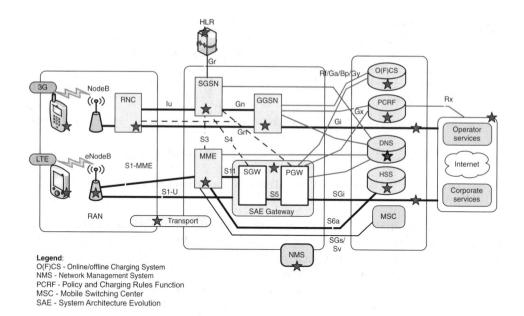

Figure 4.7   3G-GPRS and EPS networking elements affected by the IPv6 user plane service.

1. The radio modem and its firmware has to support IPv6. It is common practice not to implement every feature, even those mandatory for a specific 3GPP release, if the customer has not specifically requested for it. IPv6 and IPv4v6 PDN Type has been such for a long time.
2. The UE TCP/IP stack has to support the required IPv6 standards [33].
3. If the UE implements any kind of connection manager, it has to be IPv6 aware, for instance, to be able to request PDN Type IPv6 for IPv6 connectivity.
4. The Application Programming Interfaces (APIs) and possible middleware have to support the new IP version, IPv6.
5. The applications have to support IPv6, or in the best case the application should not need to care about IP version at all. Assuming that the APIs and middleware already provide applications interfaces that are IP version agnostic, then there is no need to port the application over to IPv6 (except when the application has to be IP-aware).

3GPP has traditionally divided UEs into two main categories depending on the integration level of the IP stack and the radio modem functions. A UE typically means a mobile phone that is a tightly *integrated* device, where the separation of the modem and the rest of the host, including the IP stack, is in practice undefined. In a standards space, the interface between the modem and the host is an ATtention (AT)-command [34] and a PPP interface. However, in reality integrated devices may have any kind of interface. Typically, an integrated UE has a tailored 'cellular aware' TCP/IP stack and APIs for applications. In this case, the IPv6 readiness – from the previous list items from 1 to 5 – is usually entirely controlled by the UE vendor, with the exception being 3rd party applications. The AT-command set and the interface has to obviously support

IPv6 extensions, and if the PPP is really used, it also has to explicitly support IPv6 and Internet Protocol version 6 Control Protocol (IPv6CP) [35].

An alternative way to implement a UE separates the end host with the IP stack (the Terminal Equipment (TE) part) and the radio modem (the Mobile Terminal (MT) part). This is a typical *dialup model* setup and is commonly referred as the *Split-UE*. The protocol between the TE and the MT has traditionally been PPP, as originally defined in 3GPP standards [25]. However, it is fairly common today to use Universal Serial Bus (USB) based connection between the TE and the MT [36, 37], and expose the cellular point-to-point link as an *Ethernet* interface. Also, the difference between an integrated and Split-UE has blurred, and many UEs that were considered as integrated are actually internally Split-UEs.

In Section 4.7 we will go into details, describing UE specific IPv6 considerations.

Radio Access Network

The Radio Access Network (RAN) ought to be completely transparent to user plane IP addressing. However, if the IP header compression is enabled and used, it can make a difference. Unfortunately there are multiple combinations of header compression algorithms, depending on what kind of RAN and which 3GPP release of it is in use. A header compression algorithm typically has explicit support for compressing IPv6 packets. The header compression takes place either on the Subnetwork Dependent Convergence Protocol (SNDCP) [38] for Gb-based 2nd Generation (2G)/GSM/Edge Radio Access Network (GERAN) or at the Packet Data Convergence Protocol (PDCP) layer [39, 40] for Iu-based 3rd Generation (3G)/UMTS Terrestrial Radio Access Network (UTRAN) or Iu-based 2G/GERAN [41, 42], and finally at the PDCP layer for the S1-based Long Term Evolution (LTE)/Evolved UMTS Terrestrial Radio Access Network (E-UTRAN).

The SNDCP layer is between the UE and the SGSN. The PDCP layer is between the UE and the RNC for the 3G/UTRAN or the Evolved Node B (eNodeB) for LTE/E-UTRAN. Should the header compression be supported, then the network has to have support for the required header compression algorithms on the network elements that terminates either SNDCP or PDCP. Table 4.1 lists the header compression algorithms that are supported by the 3GPP system, and their IPv6 support. Additionally, it is good to verify that needed IPv6 profiles of the header compression algorithms are actually implemented, since there are usually different profiles for IPv4 and IPv6.

Table 4.1 Header compression support for IPv6

Algorithm	IPv6 support	SNDCP	PDCP	Remarks
Van Jacobson [43]	No	Yes	No	Obsolete
IPHC [44]	Yes	Release-99	Release-99	Not needed by Voice over LTE (VoLTE) [45]
ROHC [46, 47, 48]	Yes	Release-6	Release-4	Profiles 0x0001 and 0x0002 are required by VoLTE [45]

If Multimedia Broadcast Multicast Service (MBMS) [49] is supported within the operator core network, then the RNC (for UTRAN) and eNodeB (for E-UTRAN) may need to have IPv6 multicast support, more specifically for Source-Specific Multicast (SSM) [50–52].

SGSN, SGW, and MME

Both a SGSN and a SGW are supposed to be transparent to the user plane traffic, since they are not regarded as routers doing forwarding but rather bridging devices. The PDN Type has a critical role during the PDN Connection creation. First, the SGSN, or in the case of EPC the MME, has to understand the requested PDN Type coming from the UE. Second, the SGSN/MME also has to understand the PDN Type downloaded as a part of the subscription profile data (downloaded over Gr- or S6d-interfaces in the case of (S4-)SGSN, and S6a-interface in the case of MME). Third, the requested and the subscribed PDN Type(s) have to match. IPv6 on the user plane also has an impact on charging, specifically the support for IPv6 and dual-stack PDN Types and address information in Charging Data Records (CDRs). We discuss charging impacts later, in Section 4.2.3.

An SGSN, for GPRS using Gn/Gp-interfaces, a S4-SGSN for EPC using S4-interface, and a SGW for EPC using S5/S8-interfaces have different support for PDN Types, especially concerning the PDN Type IPv4v6, and the handling of an unknown PDN Type varies depending on the 3GPP release. We discuss various PDN Type handling combinations and fallback scenarios in detail in Section 4.5.

In the past there were SGSNs that did not even understand the PDP Type IPv6 and just caused the PDP Context creation to fail. Such behavior can be regarded as a violation of 3GPP specifications, though.

Table 4.2 summarizes the user plane IPv6 support and its impact on the packet core SGSN, SGW, and MME elements. The user plane IPv6 (IPv6-only or dual-stack) is natively supported on EPC from 3GPP Release-8 onwards in SGW, MME, S4-SGSN, and the related signaling interfaces. In the case of GPRS, the Gn/Gp/Gr-based SGSN has IPv6-only support in virtually any release. For the dual-stack, 3GPP Release-9 is required. Specifically, Release-9 introduces the PDP Type IPv4v6 to Gn/Gp-interfaces, and `Ext-PDP-Type` and `Ext-PDP-Address` information elements for the Gr-interface [53]. In actual vendor products many of these release-specific features require the purchase of appropriate licenses to be activated and operational.

If MBMS [49] is supported within the operator core network, then the SGSN needs to have IPv6 multicast support, more specifically for SSM [50–52].

Table 4.2 SGSN, SGW, and MME support for user plane IPv6

Functionality	Gn/Gp-SGSN	SGW/S4-SGSN	MME	Remarks
IPv6-only	Release-99	Release-8	Release-8	–
Dual-stack	Release-9	Release-8	Release-8	Release-9 Gr support also required in Gn/Gp-SGSN.

GGSN and PGW

From the user plane point of view, the GGSN and the PGW are the first IP-aware packet core element for GTP-based Gn-, Gp-, and S5/S8-interfaces. In the case of the PMIP-based S5/S8-interface, the SGW is the first IP-aware packet core element.

In addition to GTP-level IPv6 impacts (see Section 4.3), the PGW supports multiple IP router and access gateway level functions that also have to be enabled for IPv6 support. These functions include the following:

- DHCPv6 server function [25, 26]. Until Release-10 only the stateless mode of DHCPv6 operation for additional parameter configuration was supported [54]. However, the introduction of the DHCPv6 PD added a stateful mode of operation to it [27, 55].
- DHCPv6 client function [25, 26].
- Remote Authentication Dial In User Service (RADIUS) and/or Diameter client function [25]. The attributes and AVPs have to have IPv6 support implemented for IPv6 values and content.
- Basic routing and respective routing protocols such as the Open Shortest Path First version 3 (OSPFv3) [56] and/or the Multi-Protocol Border Gateway Protocol (MP-BGP) [57].
- In case of MBMS, a GGSN also has to understand IPv6 SSM [50] along with the Multicast Listener Discovery version 2 (MLDv2) [51, 52].

Table 4.3 summarizes the user plane IPv6 impact of the PGW elements depending on the 3GPP release.

Additionally, the PGW may also include firewall, various flavors of Network Address Translation (NAT) and Deep Packet Inspection (DPI) functions depending on the vendor. However, these functions are not part of the 3GPP standards.

Most, if not all PGW products also support local configuration of subscription profiles (or session profiles, APN configurations depending on the vendor). This implies that the product has to understand basic subscription configurations such as: IPv6 prefix pools, IPv6 Domain Name System (DNS) server addresses [58], and Proxy Call Session Control Function (P-CSCF) addresses to name but a few.

The PGW also has Policy and Charging Control (PCC) [59] Policy and Charging Enforcement Function (PCEF) and a number of charging related interfaces. In theory,

Table 4.3 GGSN and PGW support for user plane IPv6

Functionality	GGSN	PGW	Remarks
IPv6-only	Release-99	Release-8	–
Dual-stack	Release-9	Release-8	–
DHCPv6 Client	Release-99	Release-8	–
DHCPv6 Relay	Release-99	Release-8	Relay function is 'possible'.
DHCPv6 Server	Release-99	Release-8	Stateless untill Release-10 (DHCPv6 PD).
RADIUS Client	Release-99	Release-8	–
Diameter Client	Release-99	Release-8	–

all PCC and charging related interfaces have been IPv6 aware since their introduction in Release-7. The support for dual-stack PDN Type did not cause any specific change to PCC signaling interfaces. Generally with all Diameter-based interfaces, the support for IPv6 at the Attribute Value Pair (AVP) level has to be verified for each vendor's product. There have been several cases where the IPv6 support has been left unimplemented at the AVP handling level. For example, a specific AVP even exists, but if the content is IPv6 instead of IPv4, that leads to an error or undefined behavior.

Finally, the (S)Gi interface provides connectivity to external PDNs. At minimum, the interface where (S)Gi is 'located' must understand IPv6 and be connected to a network that provides transit to some IPv6-enabled PDNs, such as the Internet. The connectivity from the (S)Gi to the external PDN can also be tunneled, for example, if the physical (transit) network directly connected to the (S)Gi is IPv4-only. We will discuss external PDN connectivity in Section 4.2.4.

As mentioned earlier, the PGW may have both a RADIUS client [60, 61] and Diameter client functions. The Authentication, Authorization, and Accounting (AAA) server located in the external PDN network is connected over the (S)Gi interface using either of the AAA protocols. Diameter is by design IPv6-aware also on the AVPs and their values level. RADIUS had IPv6 aware attributes added later [62, 63, 64].

Subscriber Management Systems, HLR and HSS

The PDN Type is part of the subscription profile data stored in an HLR or an HSS. Strictly from 3GPP specifications point of view, an HLR was replaced by an HSS starting from the 3GPP Release-5. However, in practice both systems are still developed and deployed in parallel. There is necessarily no pressing incentive for operators to deploy an HSS unless they also intend to deploy IP Multimedia Subsystem (IMS) and/or EPS.

The supported PDN Type is provisioned per each APN for every subscriber. Each subscriber can be provisioned with multiple Access Point Names. The following PDN Types are possible in the HLR/HSS for each APN:

IPv4-only PDN/PDP Type – the APN provides only IPv4 services.

IPv6-only PDN/PDP Type – the APN provides only IPv6 services.

IPv4v6 PDN/PDP Type – the APN provides dual-stack or single IP version services.

IPv4_or_IPv6 PDN Type – the APN provides either IPv4 or IPv6 services. This PDN Type is specific to an HSS and does not exist in HLR (i.e. Gr-interface [53]). Having both PDN/PDP Type IPv4 and IPv6 separately defined for the same APN equals functionally to the PDN Type IPv4_or_IPv6 but depending on the implementation may require duplication of parts of the subscription data.

The specific IPv4_or_IPv6 PDN Type brings no additional value in practice. An HLR may have multiple APN configurations of different PDN Types and even for the same name; these configurations would effectively achieve the same functionality. Furthermore, it was clarified in Release-10 that the provisioning of PDN Type IPv4v6 for a given APN means that both PDN Type IPv4 and PDN Type IPv6 are also provisioned for that APN.

It should be noted that the PDN Type IPv4v6 does not exist in an HLR, Gr-interface [53] signaling prior 3GPP Release-9. The PDP Type IPv4v6 causes the SGSN either to fail the subscription profile download (since the `Ext-PDP-Type` and `Ext-PDP-Address` information elements for Gr-interface are not understood) or to treat the PDP Type IPv4v6 as PDP Type IPv4 (see Section 4.5 for a more detailed discussion).

IP Transport Plane

The user plane IPv6 has no direct impact on the transport plane. The transport plane can be migrated to IPv6 later. However, the IPv6 packet header (minimum 40 octets) is considerably larger than the IPv4 (minimum 20 octets). Since all IP traffic is tunneled in the packet core, and the transport layer Maximum Transmission Units (MTUs) may have been optimized for typical IPv4 traffic, the undesired outcome can be increased IPv6 packet fragmentation [65] or in the worst case packets not going through at all (e.g. due the filtering of Internet Control Message Protocol (ICMP) messages). Possible introduction of IPv6 transition mechanisms is not going to make the situation any better, quite the contrary (see Section 5.2.2 for detailed discussion). Typical tunneling overhead in 3GPP networks is (assuming an IPv4-based transport plane) a minimum of 36 octets originating from the outer IPv4 header (20 octets), User Datagram Protocol (UDP) [66] header (8 octets), and GTP-U header (minimum 8 octets). Assuming a 1500-octets MTU in the packet core transport plane, this would lead to a maximum working MTU of 1464 octets on the user or control planes.

It is a common practice that 3GPP-aware UE stacks either optimize their 3GPP link default MTU value, or Transport Control Protocol (TCP) Maximum Segment Size (MSS) [67], in order to avoid unnecessary fragmentation in the packet core [68]. Again assuming 1500 octets MTU in the packet core transport plane, the maximum TCP MSS for the IPv6 traffic would be 1404 (i.e. 1500-36-40-20) octets where, as for IPv4, the maximum MSS would be 1424 (i.e. 1500-36-20-20) octets. Some Packet Data Network Gateways also try to alleviate IPv6 originated MTU issues by advertising an MTU in Router Advertisements that take the underlying GTP tunneling overhead into account or alternatively do on-wire MSS clamping. An IPv6 enabled end host [33] is recommended to perform a Path MTU Discovery procedure [69]. However, it is known that many networks unnecessarily filter Packet Too Big Internet Control Message Protocol version 6 (ICMPv6) messages [70, 71]. An alternative would be for UEs to support Packetization Layer Path MTU Discovery [72].

An enlightened reader would now be slightly confused since 3GPP specifications [8] state that the maximum safe MTU value is 1358 octets. Where does this value originate? Let us examine one of the worst case scenarios where the GTP packets are tunnel mode IPsec [73] protected between a RAN node and the core network, and IPv6 is deployed even in the transport layer. In that case the user plane packet is first encapsulated in a GTP tunnel, which results in the following overhead:

- 40 octets of IPv6 header
- 8 octets of UDP header
- 16 octets of extended GTP-U header.

This comes to 64 octets of GTP overhead. The GTP packet is further encapsulated into an IPsec tunnel. The final IPsec tunnel overhead depends on the used encryption and integrity protection algorithms. The 3GPP security specification [74] mandates the support for AES-CBC [75] with a key length of 128 bits and the use of HMAC_SHA-1 [76] for integrity protection. Therefore the overhead for IPsec with those algorithms can be calculated:

- 40 octets of IPv6 header
- 4+4 octets of Encapsulating Security Payload (ESP) [77] Security Parameters Index (SPI) and Sequence Number overhead
- 16 octets of Initialization Vector for the encryption algorithm
- Padding to make the size of the encrypted payload a multiple of 16
- Padding Length and Next Header octets (2 octets)
- 12 octets of Integrity Check Value

In order to make the user plane packet size as large as possible, a padding of 0 octets is assumed. With this zero padding assumption, the total overhead for ESP is 78 octets. In total, with 64 octets of GTP overheard, that comes to 142 octets of header overhead, which leads to $1500 - 142 = 1358$ octets for the final user plane packet.

Policy and Charging Control System

There is not much to add about PCC [59] that has not already been said earlier regarding the IPv6 support for user plane IP traffic. PCC and its Diameter based interfaces have been IPv6 compliant at the AVP level since their inclusion in the 3GPP architecture in Release-7. Essential PCC interfaces like the Gx/Rx/S9-interfaces [78–81] include both dedicated IPv6 AVPs (such as the Framed-IPv6-Prefix) or the generic any IP version AVPs (such as the CoA-IP-Address or the UE-Local-IP-Address). The generic 'any IP address' AVPs makes use of the predefined Diameter type Address, which can convey both IPv4 and IPv6 addresses with varying prefix lengths.

We have stated earlier that each PDN Connection is assigned a /64 IPv6 prefix and the prefix should be used to match traffic flows to and from the UE. Full /128 IPv6 addresses are not really supported within the 3GPP architecture. However, some interfaces like the PCC Rx-interface [80] still attempt to use full /128 IPv6 addresses. For example, the Framed-IPv6-Prefix AVP is encoded as an address by using the prefix length of /128. The PCC interface does not really make any difference between PDN Types. Whether the activated PDN Connection is IPv6-only or dual-stack can be derived from the IP address conveying AVPs.

As noted earlier, IPv6 support cannot be guaranteed although the interface supports it at the AVP level, if PCEF or Policy and Charging Rules Function (PCRF) implementations just neglect IPv6 content.

4.2.3 Charging and Billing

The 3GPP architecture has both multiple charging elements and protocols/interfaces. They do not always even coordinate too much. The charging architecture is described in [82] and specifically for the packet switched domain in [83]. The main charging interfaces of interest are:

- *Gy* – the online charging reference point between the PCEF and the Online Charging System (OCS). The Gy-interface is replaced by the Ro-interface.
- *Gz* – the offline charging reference point between the PCEF and the Charging Gateway Function (CGF)/Offline Charging System (OFCS). Gz-interface is replaced by Rf-interface.
- *Ro* – the online charging reference point between the packet core elements (i.e., SGSN, GGSN, PGW, SGW, and MME) and the OCS.
- *Rf* – the offline charging reference point between the packet core elements (i.e., SGSN, GGSN, PGW, SGW, and MME) and the Charging Data Function (CDF).

In addition to these, the packet switched domain online charging in SGSN is implemented using Customized Applications for Mobile Network Enhanced Logic (CAMEL) techniques and mechanisms [84, 85]. However, those are beyond the scope of this book. Packet core elements may also use IETF protocols for charging purposes, one example being AAA protocols over the (S)Gi interface (see Section 4.2.4 for further discussion). Generally, the charging aspects covered by this book are shallow, at best.

The main charging interfaces share the same Charging Data Record (CDR) format [86], which has evolved across 3GPP releases. There are few particular charging data record elements of interest when it comes to user plane IPv6 service:

- `PDP Type` and `PPD/PDN Type` include e.g. IPv4, IPv6 or IPv4v6.
- `Served PDP Address` and `Served PDP/PDN Address` include the IPv4 or /128 IPv6 address.
- `Served PDP/PDN Address Extension` contains the IPv4 address in case of the PDN Type IPv4v6, i.e. when dual-stack is used. In this case the IPv6 address is included in the `Served PDP Address`.
- `Served PDP/PDN Address Prefix Length` contains the IPv6 address prefix length. If this element is absent, /64 prefix length is assumed.

The support for IPv6 has been in CDRs from the beginning. The dual-stack support was introduced in parallel when it appeared in the rest of the 3GPP architecture. There are certain – typically product implementation level – details to take into account. The charging interfaces (and therefore also the CDR format) do not necessarily follow the rest of the network element release cycle. There can be situations where, for example, the SGSN/GGSN support dual-stack contexts but the charging does not, or even more mixed cases where, for instance, a combined PGW would support dual-stack charging information for EPS but not for GPRS. These combinations must be verified against each vendor's product. Fortunately, lack of dual-stack charging information does not imply that the PDN Connection activation would fail, only that the charging information received by the operator lacks information. In the case of inter-operator roaming, the lack of charging information can, of course, be a commercial-level barrier not to deploy IPv6.

In addition to 3GPP defined charging interfaces and CDR formats, the GSM Association (GSMA) has also defined its own inter-operator account record format called

TAP3 [87] for roaming billing. The TAP3 is still widely used and was updated years ago to support IPv6 and dual-stack. However, whether operators have deployed recent enough TAP3 in their roaming billing systems has to be verified case by case (most roaming contracts today are bilateral).

4.2.4 External PDN Access and the (S)Gi Interface

Access to an external PDN over the (S)Gi-interface (may) involve functions starting from a subscriber authentication and authorization, IP address allocation either from PLMN's address space or from the external PDN's address space, providing secure access to the external PDN network, and so forth. 3GPP defines two ways of accessing the external PDN such as the Internet, some corporate intranet or Internet Service Provider (ISP) network [25]. These two access methods are:

Transparent access to the Internet – IP addressing and all subscriber management belongs to the PLMN. The PGW does to need to take any part in the subscriber authentication/authorization towards the external PDN. The transparent access provides basic ISP services.

Non-transparent access to intranets or ISP networks – the UE is assigned an IP address from the corporate intranet or ISP addressing space. Furthermore, the PGW has to requests subscriber authentication/authorization and IP configuration information from servers (such as AAA, Dynamic Host Configuration Protocol (DHCP), and so forth) belonging to the corporate intranet or the ISP backend.

The non-transparent case is also frequently deployed in a way where the IP address configured to the PDN Connection is topologically anchored to a router other than the PGW. In this deployment use case the PGW effectively becomes a *bridge/relay* and all the IP level functionality is 'delegated' to this *external router* and the network behind it. So-called 'corporate APNs' or *interworking with PDN* are typically built like this.

For the GPRS and a GGSN with Gn/Gp-based interfaces, 3GPP specifications also defined *interworking with PDN using PPP* [25]. It is not supported as such by EPC, since the required PDP Type PPP was deprecated in 3GPP Release-8. In this deployment option the GGSN takes the role of Layer 2 Tunneling Protocol (L2TP) Access Concentrator (LAC) and the external router has the role of a L2TP Network Server (LNS). The PPP packets to/from the UE are either relayed by the GGSN to/from the LNS or converted to some other protocol depending on the tunneling protocol between the GGSN and the external router.

Generally speaking, there is no uniform 3GPP standard on how to accomplish the transport of the IP traffic between the UE and the external router in the case of corporate APNs, and how to make the PGW behave as a *bridge*. On the other hand, there is a description for PPP- and L2TP-based external router interworking scenario for the PDP Type PPP, which has been deprecated since 3GPP Release-8. From practical deployments and vendor implementations points of view, L2TP can be and is used with PDN Type IP. The PGW has the role of LAC and also performs the required protocol conversion from GTP to PPP, since there is no end-to-end PPP between the UE and the LNS.

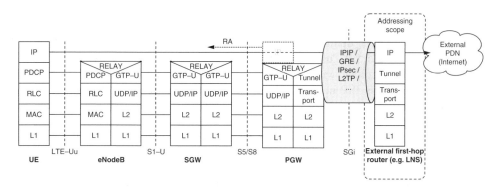

Figure 4.8   EPC and an external 'APN router' deployment.

IP traffic between the UE and the external router has to be tunneled between the PGW and the external router, since the used IP address space does not topologically belong to the operator/used PLMN. Figure 4.8 shows an example GTP-based 3GPP EPC deployment with an external to the PGW 'APN router', which should also serve as the first-hop router to the UE. There are multiple tunneling technologies that are not limited to those listed below:

- Generic Routing Encapsulation (GRE) [88]
- Layer 2 Tunneling Protocol version 3 (L2TPv3) [89]
- IPsec tunneling [73, 77]
- IP in IP tunneling (IPIP) [90].

All of the listed tunneling protocols are capable of transporting IPv6 traffic within the tunnel, if the support is implemented by both tunnel endpoints. There are some non-trivial IPv6 link model specific considerations with the *external router* deployments to take into account. There have been implementations that do not have entirely correct link model in all cases.

According to the IPv6 standard RFC 4861 [24], the first-hop router on the link is the node that can also originate the Router Advertisement. However, this is not always understood, especially when the PGW is made to behave like a bridge (or a relay) and the external router that terminates the tunneling protocols is supposed to be the first-hop router. There are deployed implementations where the PGW terminates all NDP messages (including originating the Router Advertisement) but from the IPv6 hop-limit point of view the external router is the first-hop router. This obviously is not correct. Whether it has any practical negative impact is an another issue. It could have, for example, if the PGW does not allow on-link scoped IPv6 traffic destined to multicast addresses to go through it.

Regarding PMIP on S5-interface, having an external first-hop router and still getting the IPv6 link-model correct is not trivial. This is due the fact that the SGW is supposed to be the first-hop router when PMIPv6 protocol is used instead of the PGW. Therefore, the tunnel to the external router should originate from the SGW instead of the PGW. However, this is not the case. Whether this detail has any real-life issues is another question. 3GPP has not defined how such a setup is supposed to work.

Use of AAA over the (S)Gi Interface

The (S)Gi-interface [25] defines both RADIUS [60–63, 91, 92] and Diameter [29, 93] profiles for GPRS and EPS use. A PGW assumes the role of an AAA client and the server is then located either within the operator PDN or in an external PDN (such as in a corporate or wireless ISP network). Independent of the user plane IP addressing, the AAA transport can be either IPv4 or IPv6.

The use of the AAA interface is tightly coupled with the lifetime of a PDN Connection and invoked on any event where the PDN Connection (or rather the EPS Bearer) state changes or gets modified. The (S)Gi AAA interface is frequently used for:

- Authenticating/authorizing a subscriber (and a UE) during the activation of the PDN Connection.
- Management of IP addresses for the UE; this may involve re-authorization of the subscriber.
- Informing the service infrastructure in the PDN about the IP address and subscriber/UE identity and accessed service mapping.
- Informing the service infrastructure in the PDN about the UE location, accessed PLMN information, etc.
- Accounting purposes, including periodic updates to the service infrastructure and accounting servers.
- Informing the service infrastructure about changes in the bearer (e.g. during the creation of a Secondary PDP Context or a dedicated bearer).
- Terminating the PDN Connection from the network side.

The AAA interface is available for both transparent and non-transparent access to the PDN. Our main interest is on the IP address management and specifically on the IPv6 part of it. Section 4.4 discusses IP address management and also links the use of the (S)Gi AAA interface to it. Figure 4.9 shows possible AAA interactions over the (S)Gi-interface, triggered by the management of the PDN Connection.

Use of DHCP over the (S)Gi Interface

The (S)Gi-interface [25] defines both DHCPv4 [10, 11, 94, 95] and DHCPv6 [26, 27, 54, 55, 96, 97] modes of operation for GPRS and EPS use. A PGW may assume the role of a DHCP server when it needs to serve UEs and does not need to communicate with external DHCP servers in a PDN. The PGW may assume the role of a DHCP client, with or without being a DHCP server for UEs at the same time, when it needs to interact with external DHCP servers in the PDN. Finally, the PGW may also assume the role of a DHCP relay instead of being a combined DHCP server+client. All these possible modes of operation are illustrated in Figure 4.10 and described in the 3GPP stage-3 specifications [25], but not comprehensively (or clearly) in the stage-2 architecture specifications [8, 15, 16].

The different DHCP modes in a PGW have always been confusing. When the PGW is just a server or a client, its functionality is obvious. On the other hand, what is different between *server+client* and *relay* modes of the PGW? The answer lies in how the DHCPv6 protocol behaves from a *client* point of view when it talks to a *server* or a *relay*, and from a PGW point of view for UE initiated DHCPv6 messages.

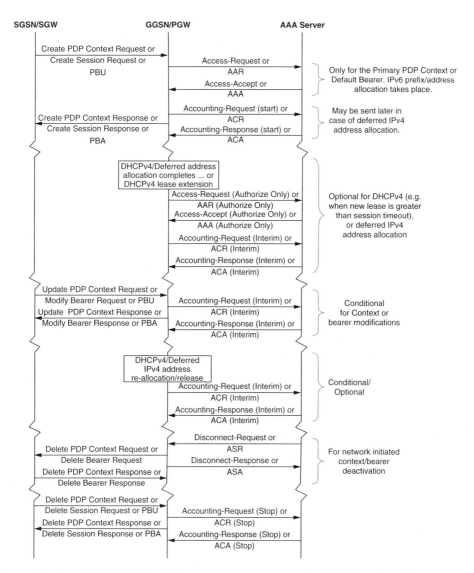

SGSN/SGW   **GGSN/PGW**   **AAA Server**

Figure 4.9 Possible RADIUS and Diameter interactions over the (S)Gi-interface with an external AAA server during the lifetime of the PDP Context/PDN Connection.

First, a DHCPv6 client can never initiate a *unicast* message to a DHCPv6 server without having talked to it earlier and having received an explicit permission for unicast communication via the Server Unicast Option sent by the DHCPv6 server [26]. Having said that, a PGW that would like, for example, to select a specific DHCPv6 server based on the APN cannot rely on the normal solicitation and multicast based DHCPv6 server discovery. Sending a unicast is, on the other hand, allowed for a DHCPv6 relay. A PGW acting as a relay could accomplish the desired DHCPv6 server selection functionality, but only in the case when the UE initiated the DHCPv6 message.

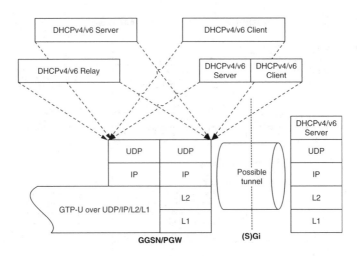

Figure 4.10 Possible DHCPv4 and DHCPv6 modes over the (S)Gi-interface with an external DHCP server.

Second, from a PGW packet processing point of view, it might be desirable that it cannot be bypassed by a UE. If the PGW has to, for any reason, inspect 'IP level control plane' such as DHCPv6, then being a relay is not optimal. If a DHCPv6 server allows unicast communication, then the UE can bypass the PGW at the user plane. Obviously IP packets still flow through the PGW, but they are not destined to it, thus the PGW has to actively inspect every user plane IP packet that goes through it. This might not be desirable from the IP forwarding efficiency point of view. If the PGW assumes the role of a DHCPv6 *server+client*, the UE would always send packets first to the PGW address, with or without unicast communication being allowed.

The use of the DHCP is tightly coupled with the establishment of a PDN Connection. The use of DHCP in 3GPP networks has been limited so far, which is visible in the lack of details learned from live deployments in relevant 3GPP standards (as in Release-11 at least). The DHCP protocol over the (S)Gi-interface is available for both transparent and non-transparent access to the PDN, and used solely for IP address management and additional end host configuration (such as configuring DNS and P-CSCF server addresses). Section 4.4 discusses IP address management and also links the use of the DHCP protocol over (S)Gi-interface to it. Figure 4.11 shows possible DHCP protocol interactions over the (S)Gi-interface triggered by the management of the PDN Connection. Currently, 3GPP specifications are silent about DHCP server initiated end host reconfiguration, but there is no reason why a vendor product could not implement such, for example, to delete a PDN Connection.

It should be noted that both AAA and DHCP can be used simultaneously over the (S)Gi-interface. One example would be using AAA for a PDN Connection authentication/authorization and then DHCP for IP address management. Another example would be using AAA for accounting purposes while DHCP is, again, used for the UE IP address management.

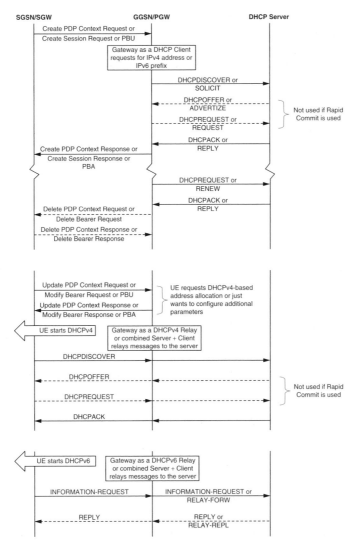

Figure 4.11 Possible DHCPv4 and DHCPv6 interactions over the (S)Gi with an external DHCP server during the lifetime of the PDP Context/PDN Connection.

4.2.5 Roaming Challenges

While there is interest in offering a roaming [98, 99] service for IPv6-enabled UEs and subscriptions, not all visited networks, such as Visited PLMNs (VPLMNs), are prepared for outbound roamers requesting IPv6 services:

- The visited network SGSN may not support the IPv6 PDP Context or IPv4v6 PDP Context types. These should mostly concern pre-Release-9 2G/3G networks without an

S4-SGSN, but there is no definitive rule, as the deployed feature sets vary depending on implementations and product licenses.

- The visited network might not be commercially ready for IPv6 outbound roamers, while everything works technically at the user plane level. This could lead to 'revenue leakage', especially from the visited operator's point of view (note that the use of a visited network PGW does not really exist today in commercial deployments for data roaming).
- The visited network might not be technically ready for billing outbound roamers using IPv6; that is, the visited operator is just not able to collect relevant and required billing and charging information from the SGSNs or the SGWs. This could again lead to a 'revenue leakage'. See Section 4.2.3 for a general discussion on charging and billing.

It might be in the interest of operators to prohibit roaming selectively from specific visited networks until IPv6 roaming is commercially in place. 3GPP does not specify a mechanism whereby IPv6 roaming is prohibited without also disabling IPv4 access and other packet services [100]. The following options for disabling IPv6 access for roaming subscribers could be available in some network deployments and vendor products:

- PCC could be used to fail the bearer authorization when a desired criterion is met. The PCRF could return, for example, the DIAMETER_ERROR_BEARER_NOT _AUTHORIZED error when a PDN Type IPv6 or IPv4v6 and a specific visited network match. The rules can be provisioned either in the home network PCRF or locally in the visited network (e.g. in a PGW or visited PCRF). The Gx(x)-interface and its AVPs are of interest. For example, the Framed-IPv6-address AVP indicates the use of IPv6, the 3GPP-SGSN-MCC-MNC AVP can be used to find out that the UE is roaming, and the Called-Station-Id AVP indicates the accessed APN. Presence of both the Framed-IPv6-address and the Framed-IP-address AVPs indicates that dual-stack PDN Connection was established.
- Some HLR and HSS subscriber databases allow prohibiting roaming in a specific (visited) network for a specified PDN Type, for example, per VPLMN granularity.
- The IP address management can be outsourced to an external PDN. Both RADIUS and Diameter protocols can be used to convey addressing information for the PDN Connection over the (S)Gi-interface from the AAA server located in the external PDN [25]. Some vendors' products allow controlling the PDN addressing over the (S)Gi-Interface using RADIUS and Diameter protocols. This functionality can also be used to restrict IPv6 roaming to a specific VPLMN. The 3GPP-PDP-Type and the 3GPP-SGSN-MCC-MNC AVPs can be used to detect that the UE is roaming within a specific VPLMN and the requested PDN Connection is either IPv6 or dual-stack. If the AAA server then responds with IPv4-only addressing information, the PGW falls back to an IPv4-only PDN Connection.
- A subscription in an HLR/HSS can be associated with up to 16 difference *Charging Characteristics* per APN [82, 101]. Although, it is not precisely described in the 3GPP specifications, the Charging Characteristics can be used to affect the PGW selection in the SGSN or MME with some vendor products. The idea would be to associate IPv6-enabled APNs and subscribers with a Charging Characteristic that can be only

put into effect when the UE is located in the Home PLMN (HPLMN). In all other cases, as when the UE is roaming, a PGW with IPv4-only APN gets selected.

- Again relying on the (S)Gi-Interface and its RADIUS or Diameter use, some vendor products allow the AAA server to override the APN during the PDN Connection establishment. Similarly as earlier, specific VPLMNs would be assigned an IPv4-only APN.
- Finally, a rather naive solution is to dedicate specific PGWs to all inbound roamers in the HPLMN. These specific PGWs would then have IPv4-only 'mirror' APNs of the IPv6-enabled counterparts configured to them.

The obvious problems are that these solutions are usually non-standard and thus, not unified across networks, and therefore also lack a well-specified fallback mechanism from the UE's point of view.

4.3   End User IPv6 Service Impact on GTP and PMIPv6 Protocols

4.3.1   GTP Control Plane Version 1

This section goes through most, if not all, GTPv1 messages and their information elements that contain user plane IPv6 addressing information. We provide a lot of information element figures and their decompositions to ease possible debugging of GTP traces.

End User Address Information Element

Figures 4.12 and 4.13 illustrate the GTPv1 End User Address information element from a Create PDP Context response, which contains the required IP addressing information. This information element is used during the PDP Context establishment between the SGSN and the GGSN to request and assign IP addresses and prefixes to the UE. From Release-9 onwards the information element can also contain two addresses at the same time [9].

	Bits							
Octets	8	7	6	5	4	3	2	1
1	Type = 0x80							
2–3	Length (6 for IPv4, 18 for IPv6, 22 for IPv4v6)							
4	Spare '1111'				PDP Type Organization (1 for IETF)			
5	PDP Type Number (0x21 for IPv4, 0x57 for IPv6, 0x8D for IPv4v6)							
6–n	PDP Address (IPv4, IPv6 or IPv4 followed by IPv6)							

Figure 4.12 The definition of the End User Address information element carrying GTPv1 IP addressing information.

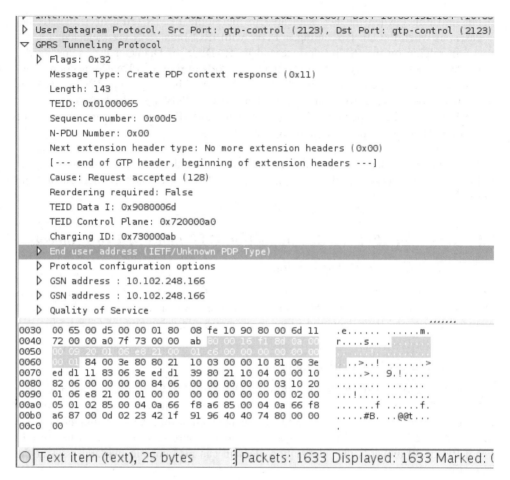

Figure 4.13 A captured End User Address information element carrying GTPv1 IP addressing information of IPv4v6 PDP Type (unknown to Wireshark version used for capturing).

The PDP Type numbering follows the PPP Data Link Layer Protocol Numbers Internet Assigned Number (IANA) registry (See *http://www.iana.org/assignments/ppp-numbers.*) for IPv4 and IPv6. However, there is no suitable value defined by the IETF for PDP Type IPv4v6, thus 3GPP has allocated its own value 0x8D (in hexadecimal) to it. The PDP Type Organization could also be the European Telecommunications Standards Institute (ETSI) in addition to IETF, but none of the protocols under the ETSI PDP Type Organization have ever been very popular, and are not even used anymore in EPS, thus we can safely always assume IETF as the organization for protocol definitions.

On GTPv1 signaling, the IID part of the IPv6 address has no meaning, other than historically. (Full IPv6 addresses are exchanged in the signaling due the early support for Legal Interception (LI) in SGSNs.) Only the prefix has significance, and it is notable that the prefix length is assumed to be /64. There is no separate field for a prefix length.

Regarding the IPv6 addresses, they are full addresses containing both prefix, which is fixed to /64, and a 64-bit IID selected by the GGSN (or under some cases by the PGW).

PDP Context Information Element

The PDP Context information element may be present in both SGSN Context Response and Forward Relocation Request GTPv1 messages that are used during the inter-SGSN Routing Area Update (RAU) and inter-Radio Access Technology (RAT) handover and the related context transfer. The PDP Context information element contains user plane IP address information within the PDP Address field. Either IPv4 or IPv6 address can be included in the PDP Address field. Prior to 3GPP Release-9 there could be only one instance of the PDP Address field. Since Release-9, the PDP Context information element header has an additional flag, the Extended End User Address flag, which indicates that a second instance of the PDP Address field can be found in the PDP Context information element. The second PDP Address was added for dual-stack purposes.

The IPv6 address is a complete /128 IPv6 address, not just a prefix.

MBMS End User Address

The MBMS End User Address information element is the same as the 'normal' End User Address information element used throughout GTPv1 signaling and shown in Figure 4.12. The only difference is that the IPv6 address is now a multicast address [21]. Note that in case of IPv6 multicast addresses, the comment on the IID part of the address and its usefulness does not apply. The format of IPv6 multicast address is different from IPv6 unicast addresses – see Section 3.1 for further descriptions of unicast and multicast address format differences.

The MBMS End User Address information element may appear in the following MBMS related GTPv1 messages: the MBMS Registration Request, the MBMS De-registration Request, the MBMS Session Start Request, the MBMS Session Stop Request, and the MBMS Session Update Request [9, 49].

MBMS UE Context

The MBMS UE Context information element may be present in both SGSN Context Response and Forward Relocation Request GTPv1 messages that are used during the inter-SGSN RAU, inter-RAT handover and the related context transfer. The MBMS UE Context information element contains user plane IP multicast address information within the PDP Address field. Either IPv4 or IPv6 address can be included in the PDP Address field. Only a single IP family is supported and 3GPP Release-9 like dual-stack operations is not allowed for MBMS Bearers.

4.3.2   GTP Control Plane Version 2

This section goes through most, if not all, GTPv2 messages and their information elements that contain user plane IPv6 addressing information. We provide a lot of information element figures and their decompositions to ease possible debugging of GTP traces.

PDN Address Allocation Information Element

Figures 4.14 and 4.15 illustrate the GTPv2 PDN Address Allocation (PAA) information element from a Create Session response, which contain the required IP addressing information. This information element is used, for example, during the PDN Connection establishment between the SGW and the PGW (S5/S8-interfaces), the S4-SGSN and the SGW (S4-interface), the MME and the SGW (S11-interface) to request and assign IP addresses and prefixes to the UE.

The PDN Type numbering follows the 3GPP's own registry defined in [3]. Furthermore, only IP packets are now allowed over the GTP tunnels, so there is no specific field for organization type like GTPv1 had.

Regarding the IPv6 addresses, they are full addresses containing both prefix and a 64-bit IID selected by the PGW. On GTPv2 signaling the prefix part of the IPv6 address has no meaning, other than being aligned with the practices of GTPv1.

There is one significant change/improvement in GTPv2 over GTPv1, namely the IPv6 prefix length is now included as part of the signaling. This small addition will better the forward compatibility of the GTPv2 if/when there are changes, for example, in IPv6 address autoconfiguration [22].

PDN Type Information Element

The PDN Type information element can only be found in the Create Session Request message during the E-UTRAN Initial Attach, UE requested PDN connectivity, or PDP Context Activation (S4 case). The information element is used only for indicating the requested PDP Type during the connection activation. Unlike in GTPv1, the requested PDN Type information is now separate from the addressing information – the PDN Address Allocation information element. Figure 4.16 shows the information element format.

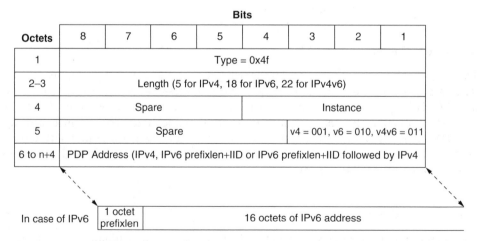

Figure 4.14 The PDN Address Allocation information element carrying GPRS Tunneling Protocol version 2 (GTPv2) IP addressing information.

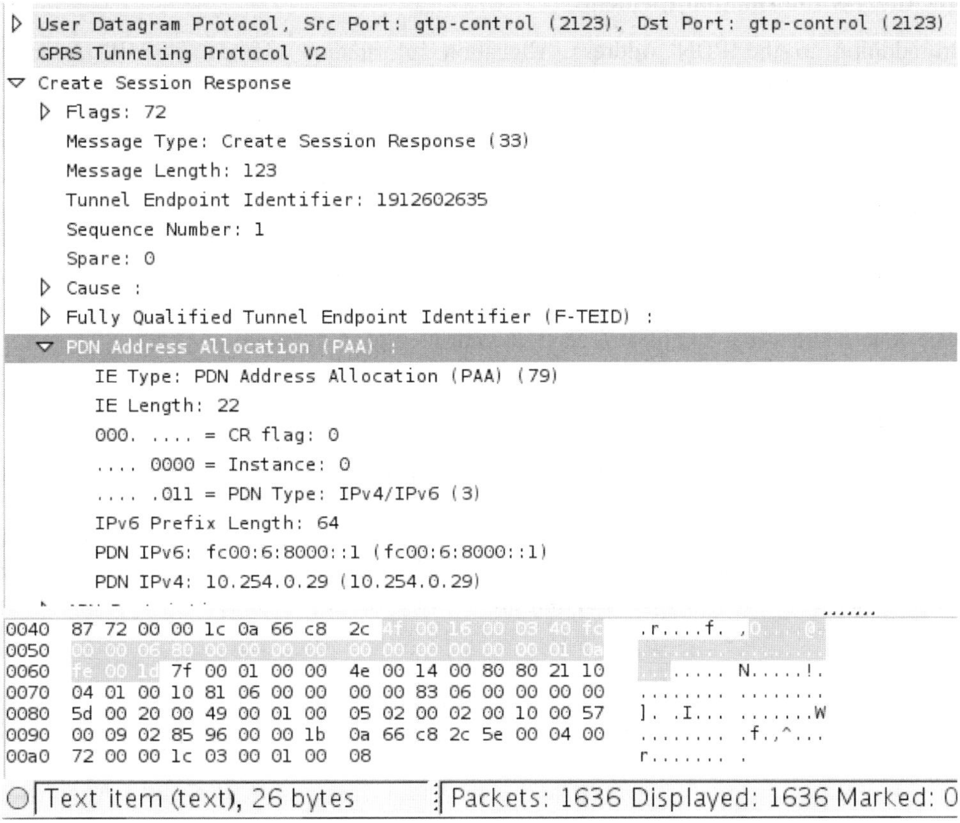

```
▷ User Datagram Protocol, Src Port: gtp-control (2123), Dst Port: gtp-control (2123)
  GPRS Tunneling Protocol V2
▽ Create Session Response
  ▷ Flags: 72
     Message Type: Create Session Response (33)
     Message Length: 123
     Tunnel Endpoint Identifier: 1912602635
     Sequence Number: 1
     Spare: 0
  ▷ Cause :
  ▷ Fully Qualified Tunnel Endpoint Identifier (F-TEID) :
  ▽ PDN Address Allocation (PAA) :
        IE Type: PDN Address Allocation (PAA) (79)
        IE Length: 22
        000. .... = CR flag: 0
        .... 0000 = Instance: 0
        .... .011 = PDN Type: IPv4/IPv6 (3)
        IPv6 Prefix Length: 64
        PDN IPv6: fc00:6:8000::1 (fc00:6:8000::1)
        PDN IPv4: 10.254.0.29 (10.254.0.29)
```

```
0040   87 72 00 00 1c 0a 66 c8  2c 4f 00 16 00 09 40 fc   .r....f. ,O....@.
0050   00 00 06 80 00 00 00 00  00 00 00 00 00 00 01 0a   ................
0060   fe 00 1d 7f 00 01 00 00  4e 00 14 00 80 80 21 10   ..... N.....!.
0070   04 01 00 10 81 06 00 00  00 00 83 06 00 00 00 00   ................
0080   5d 00 20 00 49 00 01 00  05 02 00 02 00 10 00 57   ]. .I... ......W
0090   00 09 02 85 96 00 00 1b  0a 66 c8 2c 5e 00 04 00   ........ .f.,^...
00a0   72 00 00 1c 03 00 01 00  08                        r....... .
```

○ | Text item (text), 26 bytes   ⁞ Packets: 1636 Displayed: 1636 Marked: 0

Figure 4.15 A captured PDN Address Allocation information element carrying GTPv2 IPv4 and IPv6 addressing information.

Bits

Octets	8	7	6	5	4	3	2	1
1	Type = 0x63							
2–3	Length (1 for now)							
4	Spare				Instance			
5	Spare				v4 = 001, v6 = 010, v4v6 = 011			
6 to n+4	Present if specified in the future							

Figure 4.16 The GTPv2 PDN Type information element carrying the requested PDN Type.

IP Address Information Element

In addition to the PDN Address Allocation information element, the IP Address information element is used in GTPv2 to carry around the user plane IPv4 or IPv6 address. The information element is currently (as for 3GPP Release-11) found included inside the PDN Connection Grouped Type information element used within the Forward Relocation Request and the Context Response messages. These messages are used during various handover and relocation procedures (over the S10-, S3- and S16-interfaces between MMEs and S4-SGSNs).

Figure 4.17 shows the information element format. Typically, instance value 1 is used to indicate that the PDN Address Allocation information element contains an IPv6 address. The IPv6 address is a complete /128 IPv6 address, not just a prefix.

MBMS IP Multicast Distribution for GTPv2

The MBMS IP Multicast Distribution information element may appear in the MBMS Session Start Request GTPv2 message that is sent from the MBMS Gateway to the MME/S4-SGSN [3, 49]. The information element contains both the IPv4 or IPv6 multicast destination address and the source address (for SSM [50]). Figure 4.18 shows the format of the MBMS IP Multicast Distribution information element.

4.3.3 GTP User Plane

IPv6 on the user plane has no impact on the GTP-U. The GTP-U is transparent to the IP version it carries.

4.3.4 PMIPv6

Since the PMIPv6 was originally designed and built around Mobile IPv6 (MIPv6) [102], IPv6 on the user plane has no impact to the PMIPv6, neither on the user plane nor on the control plane. However, if the transport plane remains in IPv4, then PMIPv6 has to be enhanced with support for IPv4 addressing and an IPv4 transport plane [7].

					Bits			
Octets	8	7	6	5	4	3	2	1
1	Type = 0x4a							
2–3	Length n (4 for IPv4, 16 for IPv6)							
4	Spare				Instance			
5 to n+4	IPv4 or IPv6 address (full addresses)							

Figure 4.17 The GTPv2 IP Address information element carrying either IPv4 or IPv6 address.

Bits

Octets	8	7	6	5	4	3	2	1
1	Type = 0x8e							
2–3	Length= n							
4	Spare'0000'				Instance			
5–8	Common Tunnel Endpoint Identifier							
9	v4 = 0, v6 = 1		Address Length (v4 = 4, V6 = 16)					
10–k	IP Multicast Distribution Address (IPv4 or IPv6); Multicast Address							
k+1	v4 = 0, v6 = 1		Address Length (v4 = 4, V6 = 16)					
(k+2)–m	IP Multicast Source Address (full IPv4 or IPv6 address)							
m+1	MBMS Header Compression Indicator (0 = uncomp., 1 = comp.)							
(m+2)–n	...							

Figure 4.18 The MBMS IP Multicast Distribution information element carrying either IPv4 or IPv6 address.

4.4   IP Address Assignment, Configuration, and Management

4.4.1   Addressing Assumptions

3GPP has defined multiple IP address assignment mechanisms. While our main focus is on IPv6, we briefly go through also IPv4 related address assignment and configuration aspects. After all, they are tightly linked to the behavior of a dual-stack (i.e., PDN Type IPv4v6) bearer. Table 4.4 lists all 'relevant' IPv4 and IPv6 address assignment and configuration mechanisms available in 3GPP system. 3GPP supports both dynamically or statically assigned addresses and/or prefixes for each UE.

Table 4.4   Available IPv4 and IPv6 address management and configuration methods in 3GPP

IP version	Method	Remarks
IPv4	NAS signaling	Mandatory and the most used.
IPv4	DHCPv4	Not commonly (ever) implemented in UEs or networks.
IPv4	IKEv2	For I-WLAN [103] & S2b untrusted non-3GPP access.
IPv4	MIPv4	Specification freak [20].
IPv6	SLAAC	Mandatory for IPv6.
IPv6	DSMIPv6	S2c (and H1 [104]) interfaces.
IPv6	IKEv2	I-WLAN [105] & S2b untrusted non-3GPP access [18].
IPv6	DHCPv6 PD	For mobile routers, not the UE itself.

There are certain IP addressing related peculiarities and restrictions on 3GPP bearers. Some of them are inherent from the link model as discussed in Section 4.1.3. A PDN Connection can have a single IPv4, a single /64 IPv6 prefix or both. A UE may have multiple simultaneous active PDN Connections, which would make the UE essentially multi-interfaced. Each PDN Connection is typically represented as a network interface to the host IP stack.

Prior to 3GPP Release-11, the IPv4 address that is assigned to a UE is always a /32. Since Release-11, the GTP-C signaling can also convey *default gateway* and *subnet length* information. These enhancements were needed to enable properly working DHCPv4 functionality for S2a Mobility based on GTP & WLAN access to EPC (SaMOG) [106] and unmodified end hosts with Institute of Electrical and Electronics Engineers (IEEE) 802.11 WLAN radio. In practice, /30 is the minimum workable subnet size for broadcast links such as the IEEE 802.11 WLAN. Four addresses are needed for the subnet address, default gateway on the same subnet, the end host address and the subnet broadcast address.

Similarly to IPv4, the prefix length for IPv6 is fixed to /64 on the 3GPP link. Since 3GPP Release-10, DHCPv6 PD allows for assigning an IPv6 prefixes shorter than /64 to a UE (when acting as a mobile router or a tethering device). However, these shorter prefixes are not for the use on the UE to PGW link but for the networks and devices on the UE's downstream interfaces.

Since the 3GPP link-model is a point-to-point link without native multicast capability, there is no real broadcast or multicast capabilities for IPv4 or IPv6 that would affect anything beyond the directly connected link.

When a UE decides to use DHCPv4 for its IPv4 address allocation, the UE has to transmit that decision to the network using NAS signaling (see Section 4.4.7). A PGW may have a role of a DHCPv4 server or a relay [25]. The use of DHCPv4 makes it possible for the UE to defer its address allocation to the point where the address is really needed. This so-called *deferred address allocation* procedure, in combination with PDN Type IPv4v6 bearers, could be used to conserve scarce IPv4 numbering resources. The DHCPv4 functionality also allows renewing and releasing the acquired IPv4 address at any time, which could be useful when using PDN Type IPv4v6 bearers. In the event that the UE decides to release its IPv4 address and the PDN Type is IPv4, the PGW starts a gateway initiated bearer deactivation procedure, which eventually leads to closing of the PDN Connection. The deferred IPv4 (home) address allocation is available for 3GPP accesses using DHCPv4, for trusted non-3GPP access (the S2a-interface) also using DHCPv4, and for Dual-Stack Mobile IPv6 (DSMIPv6) [19] (the S2a-interface) using Home Address (HoA) re-registration procedure.

Section 4.4.2 discusses the IPv6 and its lifetime, in more detail. However, generally speaking the lifetime of the IPv6 prefix fate-shares the PDN Connection lifetime.

The 3GPP architecture handles overlapping user plane address spaces rather well. Since the user plane traffic is always tunneled within the core network and identified with Tunnel Endpoint IDentifiers (TEIDs) or other tunneling identifiers (e.g., GRE keys), the IP numbering gets 'meaningful' only when the IP packet reaches the UE or its first-hop router, that is, a PGW in the case of GTP-based tunneling and SGW in the case of PMIPv6-based tunneling. It is quite common that multiple IPv4 enabled APNs have overlapping address spaces, especially for big operators. Naturally, when the user plane IP traffic gets routed to (S)Gi-interface, care needs to be taken if overlapping address

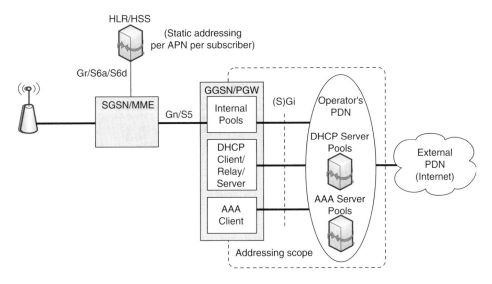

Figure 4.19 IP address management methods and places for transparent access case.

spaces are in use. Techniques vary, but most generally known techniques used in fixed IP networking to separate traffic are available on the (S)Gi-interface as well.

The IP addressing is managed either by the operator (i.e. the *transparent access to Internet*) or by an external PDN or ISP (i.e. the *non-transparent access to an intranet or ISP*). These two different deployment models were discussed in Section 4.2.4. As a reminder, Figure 4.19 shows an example architecture where the IP addressing is managed by an operator and the IP address space topologically belongs to the operator's network. Figure 4.20 shows an example architecture where the IP addressing is managed by an external PDN or an ISP, and also topologically belongs to the external network. Technically the line between the IP addressing and the UE authentication mechanisms in transparent and non-transparent cases is subtle. A thumb of rule is that in the non-transparent case, the IP address space does not topologically belong to the operator's network (where the PGW is located) and some kind of tunneling (either at layer 2 or layer 3) is involved in routing IP traffic to the PGW from the external network. Table 4.5 lists typical places where IPv4 addresses and IPv6 prefixes are managed in 3GPP deployments.

The following sections go into further details of how majority of the IP address configuration mechanisms listed in Table 4.4 work. In the majority of cases, the IP address configuration leads to a dynamic allocation of the IPv4 address or IPv6 prefix to a UE. However, the 3GPP architecture also makes it possible to assign a static IPv4 address or IPv6 prefix to the UE. Both modes of IP address assignment are described in detail.

4.4.2 Stateless IPv6 Address Autoconfiguration

IPv6 Stateless Address Autoconfiguration (SLAAC), as specified in [22, 23], is the only supported IPv6 address configuration mechanism by 3GPP specifications for a 3GPP access and a trusted non-3GPP access. These cover S5-, S8-, and S2a-interfaces. Stateful

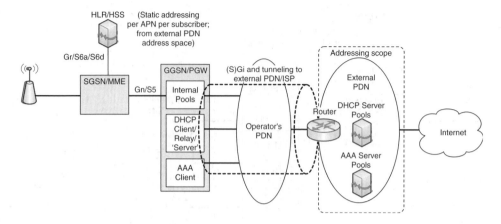

Figure 4.20 IP address management methods and places for non-transparent access case.

Table 4.5 IPv4 and IPv6 address management methods available in core network according to 3GPP standards [25]

Management method	Remarks
PGW internal pools	IP addresses and prefixes are assigned from PGW internal pools. The used pool may also be selected using for example, external AAA or DHCP server. The PGW may also implement a DHCP server function.
HLR/HSS subscription profile	IPv4 address and/or IPv6 prefix can be configured per APN and per subscriber. There can be multiple static configurations for one subscriber.
External AAA server	Multiple methods are available. An IPv4 address and/or IPv6 prefix is assigned by the AAA server, or the AAA server names the PGW internal address pool.
External DHCP server	PGW as a DHCP client or relay queries an external DHCP server for an IP address or an IPv6 prefix (including prefix delegation).
External router	The APN is terminated to an external router, which uses some mechanism to assign IPv4 address and/or IPv6 prefix. Typical setup for L2TP deployments, where a PGW is the LAC and the external router, is the LNS.

DHCPv6-based address configuration [26] is not supported by 3GPP specifications. On the other hand, stateless DHCPv6 service to obtain other configuration information is supported [54]. This implies that the M-bit is always zero and that the O-bit may be set to one in the Router Advertisement sent to the UE. The setting of the flags in the Router Advertisement complies with [107], even when it comes to DHCPv6 PD.

The 3GPP network allocates each default bearer a unique /64 IPv6 prefix, and uses layer-2 NAS signaling [12, 13] to convey an IID to the UE that is guaranteed not to conflict with gateway's IID. The UE must configure its link-local address using this IID.

The UE is allowed to use any IID it wishes for the other configured addresses with a greater scope than a link-local as long as the selected IID respects the reserved IID values [108] and IPv6 addressing architecture [21]. There is no restriction, for example, on using privacy extensions for SLAAC [109] or other similar types of mechanisms. However, there are network drivers that fail to pass the IID to the stack and instead synthesize their own IID (usually derived from a locally generated/configured IEEE 802 Media Access Control (MAC) address). If the UE skips the DAD and also has other issues with the NDP (see Section 4.9.1), then there is a small theoretical chance that the UE will configure exactly the same link-local address as the PGW. The address collision may then cause issues in IP connectivity – for instance, the UE not being able to forward any packets to the uplink.

Historically NAS signaling has conveyed a full IPv6 address from the network to the UE. This holds only for GPRS [12]. In EPS, only the IID is conveyed over the NAS signaling. Section 4.4.7 discusses further the details of NAS protocol. As an experiment, optional AT-commands *AT+CGPADDR* and *AT+CGCONTRDP* [34] can be used to query the modem for the current PDP address.

In the 3GPP link model, the /64 IPv6 prefix assigned to the UE cannot be used for on-link determination (the 'L'-flag bit in the Prefix Information Option (PIO) in the Router Advertisement must always be set to zero). If the advertised prefix is used for SLAAC, then the 'A'-flag bit in the PIO must be set to one. Details of the 3GPP link-model and address configuration are provided in Section 11.2.1.3.2a of [25]. More specifically, the PGW guarantees that the /64 prefix is unique for the UE. Therefore, there is no need to perform any DAD on addresses that the UE creates (i.e. the `DupAddrDetectTrans-mits` variable in the UE could be zero). The PGW is not allowed to generate any globally unique IPv6 addresses for itself using the /64 IPv6 prefix advertised to the UE.

The current 3GPP architecture limits the number of prefixes in each bearer to a single /64 IPv6 prefix. If the UE finds more than one prefix in the Router Advertisement, it only considers the first one and silently discards the others [25]. Therefore, multi-addressing within a single bearer is not really possible. Renumbering the previously advertised prefix without closing the layer-2 connection is also not possible. The lifetime of the /64 IPv6 prefix is bound to the lifetime of the layer-2 connection even if the advertised prefix lifetime is longer than the layer-2 connection lifetime. The advertised lifetimes (both *valid lifetime* and *preferred lifetime*) are always set to *infinite* (i.e., `0xffffffff`).

AAA Infrastructure

When RADIUS (and Diameter likewise) is used on the (S)Gi-Interface, the external AAA server can assign an IPv6 prefix for a specific APN to a UE. Either the Framed-IPv6-Prefix attribute/AVP is used to convey the IPv6 prefix or alternatively the Framed-IPv6-Pool attribute/AVP is used to name the PGW internal IPv6 prefix pool in the Access-Accept/Authentication and/or Authorization Answer message. The IID is still selected by the PGW to the UE. However, the AAA server may affect the selection of the IID and return a specific IID value in the Framed-Interface-Id attribute/AVP in the Access-Accept/Authentication and Authorization Answer message. Whether the PGW actually allows the AAA server to supply the IID is then dependent on the PGW implementation.

The PGW may also inform the AAA server about the IID it assigned to the UE. The Framed-Interface-Id in the Access-Request/Authentication and Authorization Request can be used for this purpose. The (S)Gi specification as of Release-11 [25] is not clear whether the AAA server is then allowed to override the IID selected by the PGW, thus it is safer to assume that this is not the case. Also, the specifications are not clear what to include in Framed-IPv6-* attributes/AVPs in the Access-Request/Authentication and Authorization Request when the address allocation is entirely delegated to the external AAA server that is, the PGW itself has no prefixes to propose to the AAA server.

DHCP Infrastructure

The PGW may act as a DHCPv6 client and request an IPv6 prefix for a UE from an external DHCPv6 server. Interestingly, the stateful DHCPv6 procedure as defined in RFC 3315 [26] is used and not the DHCPv6 PD procedure [27] as one might expect. The 3GPP specifications are unclear how to populate certain critical DHCPv6 parameters such as DHCP Unique IDentifier (DUID) and Identity Association IDentifier (IAID) (see Section 3.5.2). In the case of dynamic addressing, this is fortunately not that much of an issue. The DUID for the PGW should be the same for all DHCPv6 messages it exchanges.

Regarding the point that DHCPv6 is used instead of DHCPv6 PD over the (S)Gi, the 3GPP specifications are silent as to whether the IID gets conveyed over to the UE in the case of DHCPv6. The IPv6 address received from the DHCPv6 server is a /128 IPv6 address after all. Furthermore, there is no guarantee that the UE would ever respect the IID it received from the PGW when it configures its IPv6 Global Unicast Address (GUA). The details of DHCPv6 functionality fall into a vendor-specific category and UE-implementation-specific behavior.

4.4.3   Stateful IPv6 Address Configuration

The 3GPP architecture and the address allocation model does not, as of Release-11, support DHCPv6-based stateful IPv6 address allocation [26] for the PDP Context or the PDN Connection. There is, however, a stateful DHCPv6-based IPv6 prefix delegation defined in 3GPP specifications. On the other hand, DHCPv4-based stateful IPv4 address allocation is supported in the 3GPP architecture.

4.4.4   Deferred Address Allocation

Although *deferred address allocation* is specifically meant for IPv4 address management, its use is interesting in the context of PDN Type IPv4v6 as a technique to conserve IPv4 number resources. The deferred address allocation may be built on top of DHCPv4 or DSMIPv6 (the S2c-interface). In the case of DHCPv4, the UE has to specifically indicate that it wants to configure its IPv4 address using DHCPv4. The indication is done at the NAS layer using a specific Protocol Configuration Option (PCO) option (see Section 4.4.7 for further details).

As a 3GPP 'specification wise' note, the deferred address allocation as a term appears only for the non-3GPP access and DHCPv4 in stage 2 [16] (as of Release-11). The DHCPv4-based address allocation for UE is, on the other hand, possible also for 3GPP access. According to the 3GPP stage-3 specifications [25], the deferred address allocation applies also for 3GPP access. However, it is not entirely clear, since the deferred address allocation requires additional PDN Connection signaling between the SGW and the PGW after the initial PDN Connection has been established, and this signaling is currently absent from 3GPP access specifications [15].

In any case, the basic idea of deferred address allocation using DHCPv4 is the following. When the PDN Connection of PDN Type IPv4v6 gets activated, the UE would indicate its intention to configure the IPv4 address using DHCPv4. During PDN Connection activation, the PGW does not assign any IPv4 address to the UE, only the IPv6 prefix. The PDN Connection would remain IPv6-only from the UE and the network point of view to the point that the UE actually requests the IPv4 address using DHCPv4. At this point, the PGW would assign the IPv4 address to the UE, and use additional signaling between the SGW and the PGW to modify the PDN Connection with the newly assigned IPv4 address. Eventually, the UE learns the IPv4 address through DHCPv4. The UE could also release the IPv4 address later on, which would not lead to a deletion of the PDN Connection but only make it IPv6-only from the UE and the network point of view.

4.4.5 Static IPv6 Addressing

The 3GPP architecture allows assignment of a static IPv4 address and/or an IPv6 prefix to a UE. Typical to 3GPP system architecture, there are multiple provisioning and address management methods. The most straightforward method is placing the static addresses/prefixes into the per APN HLR/HSS subscription profile. In addition to the HLR/HSS based method, the *Non-Transparent access to an Intranet or ISP* [25] (see Section 4.2.4) may use external to PGW methods over the (S)Gi-Interface to allocate addresses/prefixes to a UEs. These methods include AAA (PGW being a RADIUS or Diameter client), DHCPv6 (PGW being a client), and various tunneling methods such as L2TP [89] (PGW being a LAC). In certain cases even the GTPv2 is able to differentiate whether the assigned IP address was statically assigned; more specifically the Indication information element [3] may contain static IP address information flags for the charging purposes [83] during the *modify bearer* procedure. Unfortunately, it is not clear from the specifications whether the Indication information element also applies to cases where other methods than the subscription profile were used for the static address allocation.

There are certain deployment considerations that need to be understood regarding static addressing. Static IP addresses and prefixes are always anchored to a specific PGW where the IP addresses topologically belong and eventually get routed to from a PDN. The 3GPP standards offer two basic methods for ensuring that a UE gets anchored to the correct PGW. Both methods rely on the HLR/HSS subscription profile and its per APN configuration:

- The static IPv4 address and/or IPv6 prefix are provisioned under an APN that always resolves to a specific PGW. This method works also when the subscription profile is retrieved from an HLR.
- The APN configuration profile in an HSS may contain a PGW identity (i.e., a Fully Qualified Domain Name FQDN) or an IP address of a statically assigned PGW (see the MIP6-Agent-Info AVP [110–112] for the S6a/S6b/STa-interfaces). Based on the

downloaded subscription profile, the MME or the S4-SGSN can locate the proper PGW where the statically assigned IPv4 addresses and IPv6 prefixes are also provisioned. As a side note, in this case the MME would have two sources from which to locate the proper PGW: the APN and the PGW identity. The PGW identity should take precedence in this case, since depending on the PGW 'IP architecture' the PGW may still be provisioned into the DNS as a multi-homed node and the APNs FQDN [113] could resolve to multiple IP addresses, whereas the PGW identity identifies exactly one IP address.

Camping a set of UEs always to specific PGWs can be both a scalability and a non-optimal routing issue. Some vendor products allow PGWs dynamically to advertise the address blocks that they have, using routing protocols such as Open Shortest Path First (OSPF) or Border Gateway Protocol (BGP). The use of dynamic routing could solve the issue of camping UEs (or rather subscribers) to a specific PGW, and instead let the normal gateway selection procedure select and assign a PGW to a PDN Connection. However, in the event of UEs starting to roam around making it impossible to maintain proper aggregation, there is a risk that an unbearable amount of host routes would get advertised in the routing system. Therefore, the use of static IP addressing should be carefully evaluated before being deployed in a large scale.

Subscription Profile

An HSS has per subscriber static addressing information for each APN within the subscription profile (see the APN-Configuration AVP and the Served-Party-IP-Address AVPs defined for the S6a/S6d-interfaces in [111]). Similarly for the HLR see the PDP-Address and Ext-PDP-Address parameters defined for the Gr-interface [53]. In the case of IPv6 address/prefix, the IID should be all zeros, which implies that it is not possible, according to standards, to assign a full IPv6 address to a UE. Even if a full IPv6 address were signaled over the Gr/S6a/S6d-interface to an SGSN or an MME, there is no guarantee that the IID would be delivered to the UE. In addition, even if the IID gets delivered to the UE, there is no guarantee that the UE would use it for configuring other than the link-local address.

When static addressing information comes from a subscription profile, an SGW includes the static address/prefix information into the PDN Connection activation signaling. The PGW then either authorizes or denies the use of the static address/prefix. Even in the static addressing case, the PGW should select the IID for the UE. Otherwise, there is a theoretical chance for duplicate link-local addresses on the 3GPP link.

As for 3GPP Release-11, GTPv2 specifications [3] state that the IID may be received from the HSS subscription profile. This obviously contradicts the S6a/S6d (and HSS) specifications [111].

AAA Infrastructure

In a case of RADIUS (and Diameter likewise), an external AAA server can assign a static IPv6 prefix to a UE using the Framed-IPv6-Prefix attribute in the Access-Accept message. The IID is still assigned by the PGW and included in the Framed-Interface-Id attribute in the Access-Request and Access-Accept messages. The (S)Gi-interface specification as of Release-11 [25] is not clear whether the AAA server is allowed to override the IID, thus it is safer to assume that this is not the case. The 3GPP standards are not clear either what to put into the relevant Framed-IP* attributes in the case where the address/prefix allocation is entirely delegated to an external AAA server. The attributes are mandatory in Access-Request according to the specification, but the PGW would have nothing meaningful to put into the attributes. In practice, these details are left for vendor product specifications. Similarly, as of Release-10, the Delegated-IPv6-Prefix attribute can be used to assign a static delegated IPv6 prefix to a UE. From the AAA server point of view, the UE and the requested APN identification are trivial. The AAA request messages from the PGW always contains the APN in the Called-Station-Id attribute, the Calling-Station-Id attribute contains the Mobile Station International Subscriber Directory Number (MSISDN) of the UE, the 3GPP-IMEISV contains the International Mobile Equipment Identity (IMEI) of the UE, and the 3GPP-IMSI contains the International Mobile Subscriber Identity (IMSI) of the UE.

DHCP Infrastructure

In a similar manner, a PGW may act as a DHCPv6 client and request an IPv6 prefix for a UE from an external DHCPv6 server. Interestingly, the stateful DHCPv6 procedure as defined in [26] is used and not the DHCPv6 PD procedure [27] as one might expect. The 3GPP specifications are unclear how to populate for example the IAID that would be required to identify each IPv6 address requested for a specific UE. The DUID for the PGW should obviously be the same for all DHCPv6 messages it exchanges. There is some discussion regarding the IAID and 3GPP networks in the context of DHCPv6 PD in [114]. However, there is no unambiguous 3GPP standard about it, thus enabling static IPv6 addressing using DHCPv6 would be more or less vendor specific from the DHCPv6 server provisioning point of view. Furthermore, the 3GPP specifications are silent whether the IID gets conveyed over to the UE in the case of DHCPv6. It is also known that the UE is not mandated to use the received IID for the IPv6 Global Unicast Addresss GUAs it configures. The use of DHCPv6 for static IPv6 addressing falls into a vendor specific category on both the network and the UE side.

IKEv2-based Methods

The 3GPP specifications also support Internet Key Exchange version 2 (IKEv2)-based [115, 116] Home Network Prefix (HNP) configuration for the S2c-interface (i.e. when the UE supports DSMIPv6 [19]) and full IPv6 address configuration for the (S)wu/S2b-interfaces [18, 105] (i.e. when the UE supports untrusted non-3GPP accesses). The (S)wu/S2b-interface case is probably the only solution to configure a network side selected /128 IPv6 address to the UE. Even in this case, the network has to have a deterministic mean to combine the IPv6 prefix and the IID in the PGW, which can turn out to be challenging, as discussed earlier.

L2TP-based Methods

L2TP-based 'external router' deployments were considered in Section 4.2.4. An LNS can use any method available to assign a static /64 IPv6 prefix to a UE. There is no 3GPP specification, which would have any language for combining L2TP and the rest of the '3GPP defined' static IPv6 prefix (or address) provisioning.

Manual Configuration in UE and NAS Signaling

Finally, yet another possibility is to configure a static address/prefix in the UE and use the NAS signaling to convey the address/prefix from the UE to the network. The GGSN eventually learns the UE proposed addressing during the PDP Context activation signaling. The GGSN either accepts the proposal or rejects it.

It should be noted that the static address configuration and requesting of a UE configured prefix during the PDP Context activation is only possible for GPRS and not for EPS. This is due the changes in NAS signaling. More specifically, the UE-initiated PDN Connectivity Request [13] NAS message for the creation of a default EPS Bearer cannot contain a specific IPv4 and/or IPv6 address(es) anymore. On the other hand the UE initiated PDP Activate Context Request [12] NAS message for the creation of a default (i.e. primary) PDP Context contains the PDP Address information element and possibly filled with UE requested IP addressing information.

When a UE supports DSMIPv6 (i.e. the S2c-interface) it may include a request for an IPv6 HNP at an NAS level. This is done including specific PCOs in the NAS signaling requests. The UE would request the IPv6 HNP only if the UE also requests the Home Agent (HA) IPv6 address. This feature is available for both GPRS and EPS since Release-8. Section 4.4.7 will discuss the aforementioned NAS protocol information elements in more detail.

4.4.6   IPv6 Prefix Delegation

IPv6 Prefix Delegation is a part of 3GPP Release-10 and is not covered by any earlier releases. However, the /64 prefix allocated for each default bearer (and to the UE) may be shared to the local area network by the UE implementing Neighbor Discovery (ND) proxy [23] or some similar functionality.

The Release-10 Prefix Delegation uses the DHCPv6-based PD [27]. The model defined for Release-10 requires aggregatable prefixes, which means the /64 prefix allocated for the default bearer (and to the UE) must be part of the shorter delegated prefix. This was mainly a requirement originating from the PCC architecture, where it was desirable and feasible to treat all prefixes belonging to a single PDN Connection as one shorter aggregate prefix. It is worth noting that the lifetime of the delegated prefixes fate-share the lifetime of the PDN Connection that they are associated with regardless of the lifetime values received during the DHCPv6 PD exchange. This stems from the fact that the delegated prefix and the excluded /64 prefix must aggregate. Unless the subscriber is also provisioned with a static /64 IPv6 address, there is no guarantee that the delegated prefix would remain the same across disconnections of the PDN Connection.

DHCPv6 prefix delegation has an explicit limitation, described in Section 12.1 of [27], that a prefix delegated to Requesting Routers (RRs) cannot be used by the Delegating Routers (DRs) (i.e. the PGW in this case). This implies that the shorter 'delegated prefix' cannot be given to the RRs (i.e. the UE) as such but has to be delivered by the DRs (i.e., the PGW) in such a way that the /64 prefix *allocated* to the default bearer (PDN Connection) is not part of the *delegated* prefix. An option to *exclude* a prefix from delegation [55] prevents this problem. Figure 4.21 illustrates an example deployment with an excluded prefix.

Although endorsed by 3GPP specifications, the use of [55] is optional. However, the requirement to support aggregated prefixes is still mandatory. There are two alternative ways of solving the delegation issue *without exclusions* [55].

Firstly, the DHCPv6 server could only delegate one half of the prefix space allocated to the PDN Connection, and the /64 prefix for the UE would then be taken from the other half. From this other half, the rest of the prefixes are 'lost', and handled with a *sink route*. This is obviously a very wasteful use of prefixes. See a delegation example of this approach below:

```
Prefix for the PDN Connection:  2001:db8::/56
Prefix allocated to the UE:     2001:db8::/64
Prefix delegated to the UE:     2001:db8:0:80::/57
Prefixes with a sink route:     2001:db8:0:1::/64 to 2001:db8:0:7f::/64
```

Secondly, the delegated prefix could be cut into smaller 'subnets', and then delegated as a group of prefixes. While such (complex) approach is perfectly valid from an RFC

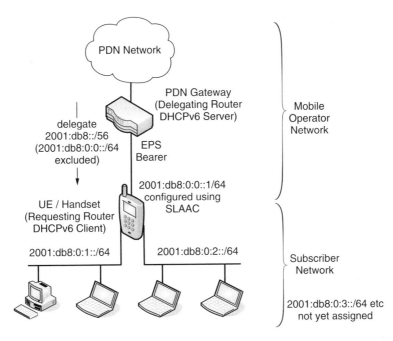

Figure 4.21 DHCPv6-based prefix delegation example and a tethering use case in 3GPP Release-10 networks.

3633 [27] point of view, it is assumed to be poorly supported by RRs. See a delegation example below:

```
Prefix for the PDN Connection:   2001:db8::/56
Prefix allocated to the UE:       2001:db8::/64
Prefix 1 delegated to the UE:    2001:db8:0:80::/57
Prefix 2 delegated to the UE:    2001:db8:0:40::/58
Prefix 3 delegated to the UE:    2001:db8:0:20::/59
Prefix 4 delegated to the UE:    2001:db8:0:10::/60
Prefix 5 delegated to the UE:    2001:db8:0:8::/61
Prefix 6 delegated to the UE:    2001:db8:0:4::/62
Prefix 7 delegated to the UE:    2001:db8:0:2::/63
Prefix 8 delegated to the UE:    2001:db8:0:1::/64
```

The third alternative would be using *'unnumbered link model'* [117], where the Requesting Router and the Delegating Router use only link-local addresses on the link they share. However, this model does not apply to the 3GPP link model at all and therefore is ruled out of possible solutions.

Due to the link model differences between PMIPv6- and GTP-based packet core, the PMIPv6 requires a slightly different handling of delegation in a SGW/MAG after a handover. The solution described in [28] makes sure that the required forwarding state for delegated prefixes is updated in a target SGW/MAG.

A notable achievement of DHCPv6 PD was introducing a stateful DHCPv6 server into the 3GPP packet core architecture and the PGW.

Prefix Management using External DHCP and AAA Servers

Delegated prefixes can be managed using the same methods that have been described earlier for /64 IPv6 prefix management. These include internal prefix pools within a PGW, external DHCPv6 server prefix pools (when the PGW implements a DHCPv6 relay or a DHCPv6 server+client combination), and external AAA server pools.

Current 3GPP Release-11 specifications [8, 15, 25] implicitly assume that either PGW internal pools or external AAA servers are used for the delegated prefix management. If the DHCPv6 were used to interface external DHCPv6 servers, both the UE /64 IPv6 address (prefix) and the delegated prefix should be assigned in one DHCPv6 protocol exchange, otherwise the handling of the excluded prefix becomes even more complicated than it already is. There are few issues regarding this in the 3GPP architecture, such a procedure not even being described in the 3GPP standards. First, the UE is explicitly prohibited from performing stateful address allocation using DHCPv6, except for requesting delegated prefixes. The 3GPP specifications [25] solve this using the combined DHCPv6 server+client mode in a PGW for PDN interworking. Furthermore, the DHCPv6 protocol itself is not entirely clear how to handle both PD (with IA_PD Option) and address assignment (with IA_NA or IA_TA Options) simultaneously. There is a good discussion of the known shortcomings and gray areas in [107, 118]. Using an external DHCPv6 server for the delegated prefix management falls into vendor specific deployment solutions as for 3GPP Release-11.

Using external AAA servers over the (S)Gi-Interface has, on the other hand, been well defined in [25] for both RADIUS and Diameter. The Delegated-IPv6-Prefix

Attribute/AVP [63] can be present in the RADIUS Access-Request or in the Diameter Authentication and Authorization Request message if the PGW assigned an IPv6 prefix from its internal pool. This is used mainly for authorization purposes. Similarly, the Delegated-IPv6-Prefix Attribute/AVP [63] can be present in the RADIUS Access-Accept or in the Diameter Authentication and Authorization Answer message when the external AAA server is used for the prefix assignment and management.

4.4.7 NAS Protocol Signaling and PCO Options

Section 4.4.5 covered the details of the NAS signaling and PCOs when related to IPv6 address configuration in UEs. The NAS protocol signaling takes place between a UE and an SGSN for GPRS [12] and between a UE and an MME for EPS [13]. The signaling is used for the call control of Circuit Switched (CS) connections, session management and mobility management for GPRS/EPS services, and radio resource management for CS and GPRS/EPS services.

The NAS protocol is also used for both PDP Context and EPS Bearer handling, and specifically for the control of the user plane, which is our concern within this section. The discussion and examples are limited to parts that involve the user plane IP configuration. The NAS protocol procedures take place before the user plane IP starts and conceptually shares similarities with PPP Network Control Protocol (NCP). Actually, in places the NAS protocol encapsulates various Internet Protocol Control Protocol (IPCP) options. The reason why we put effort into opening PCOs and NAS protocol at this level of detail is their popularity and intensive use in 3GPP networks. There is always a temptation to extend NAS protocol to every IP configuration need at the expense of IETF protocols for the same purpose. Since the introduction of non-3GPP accesses, several attempts have been made to standardize PCO delivery over DHCP [119] or PMIPv6 [120]. In the case of PMIPv6, 3GPP has already standardized a 3GPP vendor specific mobility option for encapsulating PCO [121].

Figure 4.22 shows the Packet Data Protocol Address information element [12] in detail. We have left some irrelevant (and never deployed) details out of the figure, such as the

				Bits				
Octets	8	7	6	5	4	3	2	1
1	Packet data protocol address IEI = 0x2B							
2	Length (2 for none, 6 for IPv4, 18 for IPv6, 22 for IPv4v6)							
3	Spare '0000'				PDP Type Organization 1 for IETF			
4	PDP Type 0x21 for IPv4, 0x57 for IPv6, 0x8D for IPv4v6							
5–n (24)	PDP Address (IPv4, full IPv6 or IPv4 followed by full IPv6)							

Figure 4.22 PDP Address Information Element for GPRS.

additional PDP Type organization choices and their respective PDP Type companions. Prior to Release-9 only PDP Types IPv4 and IPv6 were supported. If for any reason a UE or an SGSN does not understand the PDP Type content, it must be treated as PDP Type IPv4 (i.e. 0x21 in hexadecimal).

The Packet Data Protocol Address information element is used in GPRS, and may appear in the Activate PDP Context Request, the Activate PDP Context Accept, the Request PDP Context Activation, and the Modify PDP Context request NAS protocol messages. Unlike in EPS, the PDP Address information element may also be included in NAS messages originated by a UE (i.e. the Activate PDP Context Request message). The information element also contains the requested PDP Type, even if the actual addressing information was absent.

Even if a complete IPv6 address is included within the information element, a UE must ignore the /64 prefix when the PDP Address information element is received from the network and only extract the IID for the further configuration of the link-local address for the PDP Context (this is specifically stated in multiple 3GPP specifications [12, 13, 25]). Some SGSNs even send an IPv6 link-local address instead of the IPv6 GUA assigned to the PDP Context.

Figure 4.23 shows the PDN Address information element [13] used in EPS. It may only appear in the network originated Activate Default EPS Bearer Context Request NAS message. Unlike its GPRS companion, the information element does not carry an IPv6 prefix anymore, just the IID. Furthermore, the PDN Type for requesting a specific type of PDN Connection now has a dedicated PDN Type information element, which can be found in the UE originated PDN Connectivity Request NAS message. Figure 4.24 shows the PDN Type information element. If for any reason a MME does not understand the PDN Type content, it must be treated as PDN Type IPv6 (i.e., 010).

Another small detail concerns the EPS and the NAS signaling. Although the GTPv2 information elements (see Figure 4.14 in Section 4.3.2) allows for arbitrary length IPv6 prefix, the PDN Address information element is hardcoded to /64.

During the PDN Connection activation, a number of IP level configuration parameters gets exchanged between the UE and the PGW. The configuration information, in addition to IPv4 address and/or IPv6 prefix, is conveyed inside the PCOs [12]. The UE may request, for example, Recursive DNS Server (RDNSS) and P-CSCF addresses, or indicate whether it wants to configure its IPv4 address using DHCPv4. As a response, the PGW returns

	Bits							
Octets	8	7	6	5	4	3	2	1
1	PDN address IEI (mandatory in current NAS messages thus no code point)							
2	Length (5 for IPv4, 9 for IPv6, 13 for IPv4v6)							
3	Spare '00000'					PDN Type (v4 = 001, v6 = 010, v4v6 = 011)		
4–n (15)	PDN Address (IPv4, IPv6 IID or IPv6 IID followed by IPv4)							

Figure 4.23   PDN Address Information Element for EPS.

Bits

Octets	8	7	6	5	4	3	2	1
1	PDN Type IEI (no code point)				0	PDN Type (v4 = 001, v6 = 010, v4v6 = 011)		

Figure 4.24 PDN Type Information Element.

Bits

Octets	8	7	6	5	4	3	2	1
1	Protocol configuration options IEI = 0x27 (or 0x84 when in GTP-C)							
2	Length of option contents							
3	1	Spare '0000'				Config protocol ('000' for PPP)		
3–4	Protocol ID x							
5	Length of protocol ID x content							
6–n	Protocol ID x content							
...	Protocol ID y or Container ID z ...							

Figure 4.25 Protocol Configuration Option Information Element.

zero or more matching PCOs. Figure 4.25 shows the encoding of a PCO. The PCOs are conveyed within NAS protocol messages between the UE and the SGSN/MME, and within GTP-C messages between the MME, the SGW, and the PGW. The only difference in the encoding between NAS and GTP-C transports is the information element code number and the length field.

A number of *Protocol IDs* in a PCO are borrowed as is from PPP, and their use is indicated in the *Configuration Protocol* field (the 3rd octet in the PCO). In Release-11, the following PPP-based Protocols IDs are supported: Link Control Protocol (LCP) (0xC023), Challenge-Handshake Authentication Protocol (CHAP) (0xC223), and IPCP (0x8021). The Protocol ID content equals to a 'NCP Packet' [14] that has the 'Protocol' and the 'Padding' octets removed.

In addition to PPP-based Protocol IDs, 3GPP has defined a number of its own *Content IDs*. For instance, a Content ID 0x0003 means a *DNS Server IPv6 Address*. Containers have their own encoding described in respective 3GPP documents [12]. Figure 4.26 shows a PCO that contains both IPv4 and IPv6 DNS server addresses, and has been sent from network side to UE direction.

Tables 4.6 and 4.7 show Content IDs available in Release-11. It should be noted that Content IDs have different encoding depending on whether they were originated by a UE (requesting something) or by a network (providing configuration information).

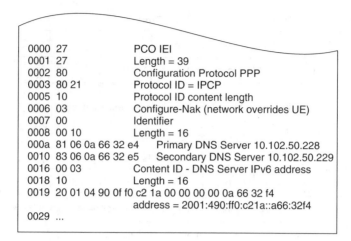

```
0000 27                    PCO IEI
0001 27                    Length = 39
0002 80                    Configuration Protocol PPP
0003 80 21                 Protocol ID = IPCP
0005 10                    Protocol ID content length
0006 03                    Configure-Nak (network overrides UE)
0007 00                    Identifier
0008 00 10                 Length = 16
000a 81 06 0a 66 32 e4        Primary DNS Server 10.102.50.228
0010 83 06 0a 66 32 e5        Secondary DNS Server 10.102.50.229
0016 00 03                 Content ID - DNS Server IPv6 address
0018 10                    Length = 16
0019 20 01 04 90 0f f0 c2 1a 00 00 00 00 0a 66 32 f4
                           address = 2001:490:ff0:c21a::a66:32f4
0029 ...
```

Figure 4.26 PCO within NAS protocol encoding example from a network to a UE, overriding the UE's proposed DNS server IPv4 addresses and also delivering a DNS server IPv6 address.

Table 4.6 Content IDs from a UE to a network direction

Content ID	Description
0x0001	P-CSCF IPv6 Address Request
0x0002	IM CN Subsystem Signaling Flag
0x0003	DNS Server IPv6 Address Request
0x0004	Not Supported
0x0005	MS Support of Network Requested Bearer Control indicator
0x0006	Reserved
0x0007	DSMIPv6 Home Agent Address Request
0x0008	DSMIPv6 Home Network Prefix Request
0x0009	DSMIPv6 IPv4 Home Agent Address Request
0x000A	IP address allocation via NAS signaling
0x000B	IPv4 address allocation via DHCPv4
0x000C	P-CSCF IPv4 Address Request
0x000D	DNS Server IPv4 Address Request
0x000E	MSISDN Request
0x000F	IFOM-Support-Request
0x0010	IPv4 Link MTU Request
0xFF00 to 0xFFFF	Reserved for operator-specific use

Last, the NAS protocol is also used to exchange TFTs between a UE and a network. TFTs are primarily IP packet filters that are specific to a PDN Connection. The TFTs may contain packet filters for the downlink direction, the uplink direction, or both directions, and they can be created, deleted, and modified dynamically. The packet filters determine the traffic mapping to PDN Connections and are used to, for example, associate IP flows to dedicated bearers for QoS enforcement purposes. The maximum length for the Traffic

Table 4.7 Content IDs from a Network to a UE direction

Content ID	Desrition
0x0001	P-CSCF IPv6 Address
0x0002	IM CN Subsystem Signaling Flag
0x0003	DNS Server IPv6 Address
0x0004	Policy Control rejection code
0x0005	Selected Bearer Control Mode
0x0006	Reserved
0x0007	DSMIPv6 Home Agent Address
0x0008	DSMIPv6 Home Network Prefix
0x0009	DSMIPv6 IPv4 Home Agent Address
0x000A	Reserved
0x000B	Reserved
0x000C	P-CSCF IPv4 Address
0x000D	DNS Server IPv4 Address
0x000E	MSISDN
0x000F	IFOM-Support
0x0010	IPv4 Link MTU
0xFF00 to 0xFFFF	Reserved for operator specific use

Flow Template information element is 257 octets, which, based on the existing encoding, can hold maximum of four full sized IPv6 packet filters per Traffic Flow Template information element. A maximum size IPv6 packet filter can be 60 octets. However, an existing TFT up to 16 filters can have added to it. There are only few IPv6 specific TFT filter components: *a full 16-octet IPv6 address*, *8-bit traffic class*, and a *20-bit flow label*.

4.4.8 Initial E-UTRAN Attach Example with IPv4 and IPv6 Address Configuration

Figure 4.27 illustrates the steps for an initial E-UTRAN attach, with the associated IPv4 and IPv6 address configuration. The example is not exhaustive from the signaling steps and individual messages point of view and therefore lacks details (see [13, 15] for more complete signaling flows). This is visible specifically in the signaling steps between the UE and the MME. Note that we have not even included the eNodeB in the example. However, we think those additional 3GPP specific details of the bearer and radio resource management are not essential for an address configuration example. The overall scenario we have selected is based on the assumption that the network and the UE support PDN Type IPv4v6, and the addresses are configured dynamically.

1. The UE initiates the attach procedure and sends the PDN Connectivity Request message via the eNodeB to the MME. The message contains PCO information element for requesting IP level configuration such as DNS server addresses and informing the network that the UE does not want to use DHCPv4 to configure its IPv4 address. The message also contains PDN Type information element requesting the PDN Type

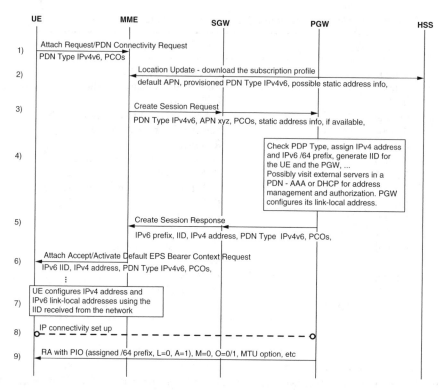

Figure 4.27 Simplified initial E-UTRAN attach example including IPv4 and IPv6 address configuration.

IPv4v6. The UE may also request a specific APN, but usually during the initial attach that is not done and the network defined default APN will be used for the PDN Connection.

2. The MME downloads the subscriber subscription profile from the HSS. The profile contains information such as the default APN, the subscribed APNs for the subscription along with the PDN Types for each APN. It is also possible that some APN is provisioned with static IPv4 address and/or IPv6 prefix, and if this is the case the static addressing information would be used during the subsequent Create Session Request messages for the PDN Connection activation.

3. The MME sends the Create Session Request GTP Control Plane version 2 (GTPv2-C) message over the S11-interface to the SGW, which further sends the corresponding Create Session Request message over the S5/S8-interface to the PGW. At this point, the MME also checks for the possible fallback cases (see Section 4.5) whether the subscribed PDN Type, the requested PDN Type, and the network wide configuration matches and based on that selects what goes into the PDN Type information element in the Create Session Request GTPv2-C message (which in this example is IPv4v6). The GTPv2-C messages also contain the PCO option, the selected APN (either from the UE or from the subscription profile), the Dual Address Bearer Flag (DAF) setting, and possibly static addressing information among other information.

4. The PGW receives the Create Session Request message and checks whether the requested PDN Connection can be established, based on the received information and what has been configured in the PGW. At this point, the PGW may also consult external AAA or DHCP servers for address allocation and/or connection authorization (see Section 4.2.4 for more information). Based on the requested PDN Type IPv4v6, the PGW has assigned an IPv4 address and an IPv6 prefix for the PDN Connection. The PGW also generates/assigns two IIDs for the UE and the PGW itself for link-local address configuration.

5. The PGW sends the Create Session Response message via the SGW to the MME. The message contains a PDN Address Allocation information element, which carries the assigned IPv6 prefix, the assigned IID for the UE, the assigned IPv4 address, and the PDN Type (IPv4v6 in this example). The message also contains the response PCO option from the PGW to the UE, which would, for example, have the DNS server information.

6. The MME initiates the PDN Connection/default bearer activation, and sends the Activate Dedicated EPS Bearer Context Request message to the UE. This NAS message contains all the previously mentioned information from the GTPv2-C message with an exception that *the IPv6 prefix is not sent to the UE*, only the PGW assigned IID. This step is followed by an additional signaling between the UE and the MME that is not shown in Figure 4.27.

7. The UE configures its IPv4 address, DNS server information, and the rest of the IP configuration. The UE also configures the IPv6 link-local address based on the IID received from the PGW.

8. At this point, the IP connectivity, that is, the PDN Connection and the default EPS Bearer, has been set up for a dual-stack. The PDN Connection appears to the UE IP stack as a dual-stack enabled network interface. The actual 'type' of the interface can be PPP, Ethernet, or something else based on the UE side implementation. The UE joins the *all-nodes multicast address* multicast group. The UE may join the *solicited-node multicast address* multicast group for all the present and future configured addresses on its interface (which would imply sending the required Multicast Listener Discovery (MLD) messages). The UE does not need to do a DAD for its link-local address since the address collision is not possible by design (assuming that the UE follows the 3GPP specifications).

9. Once the IP connection (i.e. the PDN Connection) is up, the SLAAC takes place and the UE may solicit the router by sending a Router Solicitation. The PGW sends one or more IPv6 Router Advertisement to the UE. The Router Advertisement carries at minimum the PIO option containing the /64 IPv6 prefix assigned by the network to the UE and the PDN Connection. The lifetimes of the IPv6 prefix have been set to infinite, the 'A'-flag is set to one, and the 'L'-flag is set to zero. The router lifetime setting depends of the PGW configuration but is typically close to the maximum possible value. The 'M' bit is set to zero and the 'O' bit may be set to zero or one. Other possible options in the Router Advertisement may, for example, include the MTU option. After receiving the Router Advertisement, the UE completes the SLAAC and configures a GUA for its interface. The IID of the configured GUA may be anything that the UE thinks is appropriate, for example, using privacy addresses [109]. The UE does not need to do a DAD for its GUAs since address collision is not possible by

design (assuming that the UE follows the 3GPP specifications). The UE learns the PGW link-local address from the received Router Advertisement, and since there are no link-layer addresses in the 3GPP link, the UE does not need to perform address resolution for the PGW link-local address.

4.5   Bearer Establishment and Fallback Scenarios

This section discusses the initial PDN Connection activation and handover cases and how different PDN Types behave in various UE and network configurations.

4.5.1   Initial Connection Establishment

The PDN Type that the PDN Connection eventually gets after the completion of the connection activation depends on various parameters. Since we are only concerned with the IP support on the user plane (the IP addressing delivered and seen by the end user), not every packet core network element has to understand the differences in IP addressing. Still, multiple networking elements within the 3GPP system have to support the requested PDN Type in order for the PDN Connection creation to succeed:

- The UE requested PDN Type (and the UE IP and radio modem capabilities);
- APN-specific allowed PDN Types stored in the subscription profile in the HLR/HSS;
- The SGSN release, capabilities, and settings;
- The GGSN/PGW release, capabilities, and settings;
- The DAF setting in the SGSN/MME;
- The external PDN settings and capabilities, for example in case the IP address management is delegated to an external AAA server; and
- Vendor specific means to override standard PDN Type handling logic (obviously we cannot say much about these).

The radio modem and its firmware has to support the corresponding PDN Type for the PDN Connection creation. The PDN Connection creation can be initiated by the UE or the network. In some cases the host operating system and the IP stack within the UE may not have the full knowledge of the radio modem support for different PDN Types e.g., when the end host (TE) and the modem (MT) are separate. A 3GPP Release-8 (for EPS) or a Release-9 (for GPRS) compliant UE should adhere the following rules:

If the UE is an IPv4-only, then PDN Type IPv4 PDN Connections are requested.

If the UE is an IPv6-only, then PDN Type IPv6 PDN Connections are requested.

If the UE has a dual-stack, then PDN Type IPv4v6 PDN Connection should be requested.

If the UE does not know what PDN Types the radio modem supports, then PDN Type IPv4v6 PDN Connection should be requested.

Depending on the network side support and configuration, the activated PDN Connection might have a different PDN Type than was requested. In reality, UEs do not (need to) follow these guidelines and there are typically per-APN settings provided by the UE vendor or the operator that override the 3GPP specified default behavior. There are, for example, deployment cases where it makes sense for an operator to favor single IP version PDN Connections for dual-stack capable UEs and provide dual-stack connectivity using one of the available IPv6 transition mechanisms. We discuss IPv6 transition in detail in Chapter 5.

4.5.2   Backward Compatibility with Earlier Releases

There are a few PDN Type handling rules that are important to remember. Prior to 3GPP Release-9 an SGSN treats a PDP Type IPv4v6 as a PDP Type IPv4 (like any PDP Type it does not understand) [12]. On the other hand, an SGSN starting from a Release-9 SGSN and an S4-SGSN/MME starting from Release-8 treat an unknown PDN Type as a PDN Type IPv6 [13]. The unknown PDN Type may arrive at the MME either from a UE or an HLR/HSS. The treatment should be the same.

Starting from late 3GPP Release-10, the PDN Type IPv4v6 in an HLR/HSS subscription profile implicitly means that both PDN Types IPv4 and IPv6 are supported for the given APN configuration, and a SGSN/MME can then act accordingly. Prior to Release-10 the SGSN/MME *might* have rejected the UE initiated request for the PDN Type IPv4 or IPv6 even if the PDN Type IPv4v6 was subscribed, but not IPv4 or IPv6. This was the case in some early deployments, for example, when an HLR had a PDP Type IPv4v6 provisioned for an APN and the subscription profile got downloaded into a pre-Release-10 SGSN (or even pre-Release-9). As a result, the PDP Type gets treated only as the PDP Type IPv4 and there is no way for the UE to activate a PDP Type IPv6 PDP Context. Actually, the deployment described earlier should never be allowed to happen. A network deployment must take care that the SGSN/MME versions are up to date with the HLR/HSS versions and provisioning, and there is no feature disparity.

Some early GGSNs could not handle a configuration where two APNs had the same name and different PDP Types. For their remission there was no 3GPP specification text that would have mandated such configuration prior to Release-8. However, such a limitation effectively makes fallback cases to two single IP version PDP Contexts instead of a dual-stack one harder. The UE and the network have to be configured with different APNs for each PDN Type instead of having just one.

4.5.3   Dual Address Bearer Flag

The DAF is used in GTP-C (interfaces S2b, Gn/Gp, S5/S8, S3, and S11) signaling during the activation of the PDN Connection. Its value is pre-configured by an operator and set by an SGSN/MME when a UE requests a PDN Type IPv4v6 and all SGSNs that the UE may handover to are Release-9 and PDP Type IPv4v6 capable. This is useful during a phased core network migration to fully Release-9 capable SGSNs. There is a strict requirement that when an inter-SGSN RAU or an intra-RAT handover takes place that the PDN Types must match one-to-one. During the migration phase the core network might have a mixture of Release-9 SGSNs, pre-Release-9 SGSNs, and possibly MMEs, and a handover between those should be allowed.

The DAF is signaled within the Indication information element in GTPv2 [3]. Within GTPv1 the DAF is found in the Common Flags information element [9]. The absence of the DAF is interpreted as being unset.

4.5.4 Requested PDN Type Handling in a PGW

A PGW configuration may also affect the final selection of the PDN Type for a PDN Connection. The configuration is typically specific to an APN or some PGW internal routing primitive/function construct. Table 4.8 lists the possible changes in the request PDN Type and what actually gets created. A dash in the DAF column means that the value of the DAF has no meaning.

Note 1 The PGW respons with a cause code #18 (for GTPv2) or #129 (for GTPv1) 'New PDP/PDN Type due to network preference', which a UE will eventually see as a received (E)SM cause #50 'PDN/PDP Type IPv4 only allowed' [12, 13]. The UE must not attempt to establish a second parallel PDN Connection of a different PDN Type in order to mimic dual-stack behavior.

Note 2 The PGW responds with a cause code #18 (for GTPv2) or #129 (for GTPv1) 'New PDP/PDN Type due to network preference', which a UE will eventually see as a received (E)SM cause #51 'PDN/PDP Type IPv6 only allowed' [12, 13]. The UE must not attempt to establish a second parallel PDN Connection of a different PDN Type in order to mimic dual-stack behavior.

Note 3 The PGW responds with a cause code #19 (for GTPv2) or #130 (for GTPv1) 'New PDP/PDN Type due to single address bearer only', which a UE will eventually see as a received (E)SM cause #52 'single address bearers only allowed' [12, 13]. The UE *is allowed for/should establish* a second parallel PDN Connection of a different PDN Type in order to mimic dual-stack behavior.

When a PDN Connection activation fails, for example, due to a UE requesting a PDN Type that is not supported or provisioned in the PGW for the requested APN, then the

Table 4.8 The PGW configuration and Dual Address Bearer Flag's effect on the PDN Type selection

Requested	Configured	DAF	Result	Remarks
IPv4	IPv4	–	IPv4	
IPv4	IPv6	–	Reject	
IPv4	IPv4v6	–	IPv4	
IPv6	IPv4	–	Reject	
IPv6	IPv6	–	IPv6	
IPv6	IPv4v6	–	IPv6	
IPv4v6	IPv4 and IPv6	–	IPv4 or IPv6	Note 3
IPv4v6	IPv4v6	unset	IPv4 or IPv6	Note 3
IPv4v6	IPv4	set	IPv4	Note 1
IPv4v6	IPv6	set	IPv6	Note 2
IPv4v6	IPv4v6	set	IPv4v6	

PGW can send back an error cause code #83 'Preferred PDN Type not supported'. The MME may eventually translate this to either (E)SM cause codes #50 or #51 and send those to the UE over the NAS signaling.

4.5.5 Fallback Scenarios and Rules

The general rule is that if the requested PDN Type matches one of the subscribed (provisioned) PDN Types in the HLR/HSS subscription profile, then the created PDN Connection will have that. The requested and subscribed PDN Type is first verified in an SGSN/MME during the connection setup phase after the subscription profile has been downloaded from the HLR/HSS to the SGSN/MME. The second time, the requested PDN Type gets verified against the *configured* PDN Types in a PGW. If the requested type differs from the subscribed or the configured one, then the type gets changed according to well established rules. The basic matching and conversion rules are listed in Table 4.9.

If the subscription profile has multiple PDN Types provisioned for the same APN, then the selection of the selected PDN Type is up to the implementation. For example, the subscription profile may have both PDN Types IPv4 and IPv6 for the same APN. If the PDN Type IPv4v6 is requested it is up to the SGSN/MME implementation to determine which one gets selected. Certain combinations do not make sense, such as having PDN Type IPv4 and PDN Type IPv4v6 for the same APN, since the provisioning overlaps and thus is redundant (unless there are known issues with earlier 3GPP releases in the network, as discussed earlier).

Table 4.9 summarizes all the PDN Type selection and fallback rules described so far. In the table *UE Req.* stands for the PDN Type originally requested by the UE and *NW*

Table 4.9 PDP/PDN Type matching and conversion rules non-exhaustive summary

UE Req.	Subscribed	DAF	NW Req.	PGW	Result	Remarks
IPv4	IPv4, IPv4v6, IPv4_or_IPv6	–	IPv4	IPv4, IPv4v6	IPv4	
IPv6	IPv6, IPv4v6, IPv4_or_IPv6	–	IPv6	IPv6, IPv4v6	IPv6	
IPv4	IPv4v6	–	IPv4	IPv4, IPv4v6	IPv4	
IPv6	IPv4v6	–	IPv6	IPv6, IPv4v6	IPv6	
IPv6	IPv4v6	–	IPv4	IPv4, IPv6, IPv4v6	IPv4	
IPv6	IPv4v6	–	–	–	Reject	Note 1
IPv4v6	IPv4	–	IPv4	IPv4, IPv4v6	IPv4	
IPv4v6	IPv6	–	IPv6	IPv6, IPv4v6	IPv6	
IPv4v6	IPv4, IPv4v6	–	IPv4	IPv4, IPv4v6	IPv4	Note 1
IPv4v6	IPv6	–	–	–	Reject	Note 1
IPv4v6	IPv4, IPv6, IPv4_or_IPv6	–	IPv4	IPv4, IPv6, IPv4v6	IPv4	Note 2
IPv4v6	IPv4, IPv6, IPv4_or_IPv6	–	IPv6	IPv4, IPv6, IPv4v6	IPv6	Note 3
IPv4v6	IPv4v6	–	IPv4v6	IPv4	IPv4	Note 2
IPv4v6	IPv4v6	–	IPv4v6	IPv6	IPv6	Note 3
IPv4v6	IPv4v6	–	IPv4v6	IPv4, IPv6	IPv4 or IPv6	Note 4
IPv4v6	IPv4v6	unset	IPv4v6	IPv4v6	IPv4	Note 4
IPv4v6	IPv4v6	unset	IPv4v6	IPv4v6	IPv6	Note 4
IPv4v6	IPv4v6	set	IPv4v6	IPv4v6	IPv4v6	

Req. stands for the PDN Type requested by the SGSN/MME after matching against the subscribed PDN Types has been done. The table does not include the majority of the redundant combinations such requesting a PDN Type that is not subscribed in an HLR/HSS or configured in a PGW. Neither does the table consider possible intentional restrictions on PDN Type handling such as could be the case for inbound and outbound roamers. Roaming restrictions were briefly discussed in Section 4.2.5. In Table 4.9 comma-separated lists of addressing types mean that these are the possible choices (i.e. '*or*').

Note 1 A pre-Release-9 SGSN treats an unknown PDP Type (here IPv4v6) as IPv4. This might cause an immediate PDP Context creation failure in the SGSN or conversion to PDP Type IPv4. Furthermore, the UE does not receive any Release-9 error codes, which could instruct the UE to create an IPv6 type parallel PDP Context.

Note 2 A UE will receive an (E)SM cause #50 'PDN/PDP Type IPv4 only allowed' [12, 13]. The UE must not attempt to establish a second parallel PDN Connection of a different PDN Type in order to mimic dual-stack behavior.

Note 3 A UE will receive an (E)SM cause #51 'PDN/PDP Type IPv6 only allowed' [12, 13]. The UE must not attempt to establish a second parallel PDN Connection of a different PDN Type in order to mimic dual-stack behavior.

Note 4 A UE will receive an (E)SM cause #52 'single address bearers only allowed' [12, 13]. The UE *is allowed/should* establish a second parallel PDP Context/PDN Connection of a different PDN Type in order to mimic dual-stack behavior.

In addition to the Table 4.9 selection rules, the UEs may implement their vendor-specific additional logic. One example is the case where the attempt at PDN Types IPv4v6 fails and the UE falls back to PDN Type IPv6 and eventually to Type IPv4. The logic from the UE point of view could be to vigorously try to establish any IP connectivity, since that most probably leads to a better end user experience than no connectivity at all.

4.5.6 *Inter-RAT Handovers and Inter-SGSN Routing Area Updates*

The 3GPP specifications strictly mandate that during an inter-RAT handover or an inter-SGSN RAU, the PDN Types for the activated PDN Connections in the source and target nodes must match one-to-one. It is considered to be a network planning and deployment failure if that assumption does not hold. This could build up to an issue specifically when UTRAN to E-UTRAN inter-RAT handovers are desired and PDN Type IPv4v6 is used. As a reminder, GPRS had PDP Type IPv4v6 support only since Release-9, whereas EPS had that even from Release-8. The Dual Address Bearer Flag was specifically developed to alleviate issues in such deployment scenarios. Table 4.2 in Section 4.2.2 listed the 3GPP release versions of SGSNs and an MME that have IPv6 and dual-stack support.

If the network deploys S4-SGSN then there are no 3GPP release issues with inter-RAT handovers between UTRAN and E-UTRAN since S4-SGSN uses EPS protocols and supported PDN Type IPv4v6 starting from Release-8. The issues arise if and when the GPRS side of the network still deploys SGSN for Gn/Gp-interfaces and the interworking takes place in a Gn/Gp-aware PGW.

It could be expected that misconfigurations might happen and UEs with established dual-stack connectivity attempt to move and attach to older release network segments.

GTPv1 specifications [9] prior 3GPP Release-9 state that when receiving an information element that has an invalid length in a request message, the receiving node sends an error response with a cause code set to '*Mandatory IE incorrect*'. This would actually happen when a Release-7 SGSN receives the PDP Context information element with the Extended End User Address included for dual-stack purposes in a Forward Relocation Request message. Likewise, GTPv1 specifications prior 3GPP Release-9 state that when receiving an information element that has an invalid length in a response message, the receiving node treats the signaling procedure as a failed one. This would be the case when the PDP Context information element with the Extended End User Address information element was included for dual-stack purposes in an SGSN Context Response message.

Starting from 3GPP Release-9 the handling of unexpected information element lengths is more liberal. However, regarding the specific topic discussed above, there should not be any issues since the Release-9 Gn/Gp-SGSN is supposed to understand all the required dual-stack enhancements in GTPv1.

4.6 Signaling Interfaces

This section takes a brief look at IPv6 as a transport protocol on various signaling interfaces with the 3GPP packet core. We are not going to discuss the IPv6 transport layer as such, which implies that RAN aspects are, for example, entirely neglected.

4.6.1 IPv6 as Transport

Most of the signaling and control plane interfaces within the GPRS and EPS should be IPv6 capable by definition. GTP and Session Initiation Protocol (SIP) [122] run over UDP [66], RADIUS runs over UDP and Diameter runs over Stream Control Transmission Protocol (SCTP) [123]. They are not tied to a specific IP version. Even non-IP Signaling System No. 7 (SS7) based protocols could be transported over IP using SIGTRAN [124, 125] or SIP [126].

Whether a given signaling protocol uses IPv6 transport instead of IPv4, typically depends on two factors:

- Does the core network node support IPv6 on its own network interfaces?
- Has there been a reason to upgrade the core network to IPv6?

However, within the 3GPP specifications the situation can naturally be different and not that straightforward. For instance, GTP still has an *optional* support for IPv6 transport (as of Release-11). Naturally, some systems and networking nodes were then implemented according to that. In practice, the majority of the 3GPP core network nodes are getting full IPv6 support independent of the vendor but details ought to be verified. Unpleasant surprises and feature imparity compared to IPv4 functionality is still a reality.

4.6.2 IPv6 in Information Element Level

At the information element level, the IPv6 support has been there from the beginning of GPRS, more or less throughout all signaling and control plane interfaces. This was actually the requirement to enable IPv6 support on the user plane, which has also been there, at least in the specification level, from the beginning of GPRS.

We want to highlight one detail regarding the IPv6 and signaling protocols' information elements. For some reason, there is no consistency in IPv6 address representation within 3GPP. It is somehow expected that there would be discrepancy between, for example, 3GPP and IETF defined protocols. However, 3GPP has achieved multiple encoding formats even within the same protocol.

4.7 User Equipment Specific Considerations

In this section, we discuss some of the key considerations related to IPv6 support on 3GPP enabled UEs. These UEs can range from typical mobile handsets to router-class devices to Machine-to-Machine (M2M) devices. We start the considerations by listing and categorizing the required RFCs for generic IPv6 protocol support on 3GPP UEs. Due to the properties and peculiarities of 3GPP network architecture, the list of RFCs is somewhat different from what is listed by IETF in the generic 'IPv6 Node Requirements' RFC 6434 [33]. After listing the RFCs, we describe other UE functionalities for which IPv6 support has implications.

4.7.1 IPv6 and Impacted Layers

The introduction of IPv6 support to UEs has profound implications to the used operating system and to a lesser extent to the applications and used chipsets. Figure 4.28 provides a high-level abstraction of these impacted layers and software modules, which we will discuss in more depth soon. The number of entities that IPv6 impacts is one of the main reasons why the introduction of IPv6 into UEs has been generally slow.

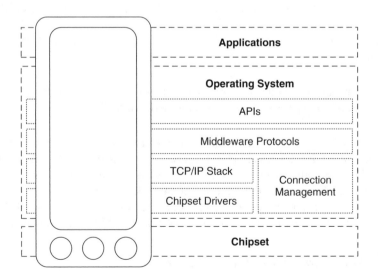

Figure 4.28   Illustration of layers needing changes due to IPv6.

As long as a UE is attached to IPv4 or dual-stack networks, it can have IPv4-only components. However, for IPv6-only deployments, which have lately been of great interest in 3GPP networks, absolutely everything must be IPv6 enabled (unless some transition scheme is used, as described in Chapter 5).

Implications for Operating Systems

The operating system, and in particular the *TCP/IP stack*, is the place that implements the IPv6 itself. Hence, it is quite understandable that the implications are significant. In addition to implementing the protocol itself, the stack's implementation may need significant restructuring due to the two parallel address families.

In order to enable higher layers to make use of IPv6, APIs need updates as well (see Section 3.9). The address family aware socket API requires changes visible to applications, in order to allow use of detailed protocol features. The APIs that provide, for the most part, address family independent API for upper layers are not without changes either: such APIs need to be able to operate on top of IPv6.

One module – perhaps not that obviously recognized as an important component to have support for IPv6 – is *Connection Management*. The Connection Management discussed herein is a high level abstraction of everything related to device provisioning, setting up of IPv6 connections including tethering, and managing interface selections and handling various possible error and fallback scenarios. An unnecessary but common problem in connection managers has been the inability to believe that IPv6-only connectivity is enough for working Internet access. Sometimes connection managers have rejected network connections that do not provide native IPv4-connectivity.

General-purpose operating systems virtually always come with a set of *middleware* protocols, such as HyperText Transfer Protocol (HTTP). These protocols, in addition to being able to use the IPv6 network layer, often need the ability to manage IPv6 address literals. Effectively all middleware protocols need updates due to IPv6.

Last of the major parts of operating systems that require IPv6 changes are *chipset drivers*. The implications may be more significant, as is the case with drivers related to interfacing 3GPP chipsets, or to a lesser extent when interfacing Ethernet-type networks. If the drivers happen to be the place that implements header compression algorithms, such as RoHC, then the implications are more significant. If nothing else, the chipset drivers need proper testing with IPv6. Even if it would seem that there are no address family dependencies, there may be: we have witnessed cases where power optimization functions caused problems for IPv6, as the optimizations assumed that only IPv4 was being used.

Implications for Applications

Depending on the API used by an *application*, and the class of the application, changes may be needed to support IPv6. If an application is using an address family aware API, it must be updated to support IPv6. Luckily, nowadays ordinary applications commonly use address family agnostic APIs (see Section 3.9.2) and hence can avoid changes.

The applications requiring changes, independent of the API they use, are the ones that pass IP address literals as part of their protocol payload (see Section 3.9.3). Often, the protocol standards used by such applications, such as SIP and File Transfer Protocol

(FTP), also need, or have needed, to be updated for IPv6. We have touched on this topic in more detail in Section 3.10.

Implications for Chipsets

For the longest time a significant hindrance for UEs' IPv6 support was the lack of support on *wireless chipsets* – on modems. For people familiar with Ethernet-based technologies this may seem strange, as in the Ethernet framing the IPv6 support has been there for such a long time already. In 3GPP the PDN Connection type and additional parameters need to be supported by protocols that are commonly implemented in wireless modems, such as SNDCP and PDCP. If a modem does not support the set up of the IPv6 type of PDN Connection, a UE is not going to be able to set up a native IPv6 3GPP connection.

Some modems supported IPv6 PDP Context type from early 2000, but most did not for the first decade of this millennium. However, at the time of writing, IPv6 support seems to be making its way to all significant chipsets, and hence the IPv6 bottleneck cannot be considered to be on modern modems.

4.7.2   Required RFCs for Host UEs

The set of required RFCs naturally depends on supported IPv6 features. The minimal set of RFCs that are needed on 3GPP UEs is listed in Table 4.10. RFCs that are optionally needed are listed in Table 4.11. Informational documents that may help build protocol and node implementations are listed in Table 4.12. Out of the listed informational documents those related to IPv6 API are particularly interesting, as they describe the basic model of how applications may be utilizing IPv6. The API RFCs are informational in nature, as the APIs are always internal to an operating system, fully dependent on the operating system's architecture, and invisible outside of a node.

It is certainly possible that a UE is also attaching to non-3GPP networks, in which case additional requirements, or priorities, may be in place. For example, while stateful DHCPv6 is not mandatory in 3GPP networks, in corporate WLAN networks it typically is. This requirement list also does not cover IPv6-related RFCs that might be required for additional non-3GPP related features. Therefore, for example, IPsec, application specific multicast use, numerous DHCPv6 options, MIPv6, or other than WLAN access type related RFCs are not listed even if they generally are optional for every implementation. We have also not included any of the IPv6 transition mechanisms in the lists we have on this section, as the basic IPv6 transition tools used in 3GPP accesses, and described on sections 5.4.2 and 5.4.3, do not require any particular transition protocol support on UEs. For those transition tools that require explicit UE support, such as double translation described in Section 5.4.4, additional RFCs would be needed. Currently it is a UE implementation specific decision what specific IPv6 transition tools, from the plethora that is available, are to be supported. The authors expect the IPv6 transition toolbox to settle during the coming years, and then the requirements will also clarify themselves.

It is worth noting that RFCs are very often large documents that contain both mandatory and optional components. Whether optional components need to be supported or not

Table 4.10   IPv6 RFCs needed by all 3GPP UEs

RFC	Title	Book Section	Reference
1981	Path MTU Discovery	3.2.3	[69]
2460	Internet Protocol Version 6	3.2	[2]
2671	Extension Mechanisms for DNS (EDNS0)	3.10.2	[127]
2710	MLD for IPv6	3.2.4	[128]
2711	IPv6 Router Alert Option	3.2.2	[129]
3590	Source Address Selection for the MLD Protocol	3.2.4	[130]
3596	DNS Extensions to Support IP Version 6	3.10.2	[131]
3986	Uniform Resource Identifier (URI): Generic Syntax	3.9.3	[132]
4191	Default Router Preferences and More-Specific Routes	3.5.6	[133]
4193	Unique Local IPv6 Unicast Addresses	3.1.2	[134]
4291	IPv6 Addressing Architecture	3.1	[21]
4443	ICMP for the IPv6	3.3	[135]
4861	Neighbor Discovery for IPv6	3.4	[24]
4862	IPv6 Stateless Address Autoconfiguration	3.5.1	[22]
4941	Privacy Extensions for SLAAC	3.5.5	[109]
5095	Deprecation of Type 0 Routing Headers in IPv6	3.2.2	[136]
5453	Reserved IPv6 Interface Identifiers	3.5.1	[108]
5722	Handling of Overlapping IPv6 Fragments	3.2.3	[137]
5942	IPv6 Subnet Model	3.1.2	[138]
5952	A Recommendation for IPv6 Address Text Representation	3.1.9	[139]
6106	IPv6 RA Options for DNS Configuration	3.4.3	[140]
6555	Happy Eyeballs: Success with Dual-stack Hosts	3.9.4	[141]
6724	Default Address Selection for IPv6	3.5.4	[142]

depends usually on what higher level features are going to be supported on a UE and sometimes on resource constraints. For example, if a 3GPP UE is strictly working on a host mode, it does not need to implement router-specific features.

4.7.3   DNS Issues

Partly due to the colorful history of RDNSS discovery, the 3GPP system includes passing of the recursive DNS server addresses as part of PCO information element signaling. Furthermore, the 3GPP specifications allow the use of stateless DHCPv6 for passing the information. At the time of writing, the 3GPP does not formally support a Router Advertisement-based approach.

For UEs it is theoretically enough to support only the PCO information element based approach, and currently that is supported in all 3GPP deployments that support IPv6. However, as UEs usually also support other access technologies where both DHCPv6 and Router Advertisement approaches are used, it is wise to also support these approaches in 3GPP interfaces – as a backup solution and for future extensibility, if for no other reason. A large UE population supporting the IP-layer approaches would enable simplification of the 3GPP architecture and hence perhaps removal of the PCO information element based method – in the distant future.

Table 4.11 IPv6 optional RFCs for 3GPP UEs

RFC	Title	Book Section	Reference
2464	Transmission of IPv6 Packets over Ethernet Networks	3.6	[143]
2675	IPv6 Jumbograms	3.2.3	[144]
3122	Extensions to IPv6 ND for Inverse Discovery Spec.	3.4.4	[145]
3315	DHCPv6 for IPv6	3.5.2	[26]
3646	DNS Configuration options for DHCPv6	3.10.2	[96]
3736	Stateless DHCPv6	3.5.2	[54]
3810	MLDv2 for IPv6	3.2.4	[52]
3971	SEcure Neighbor Discovery (SEND)	3.4.9	[146]
3972	Cryptographically Generated Addresses (CGA)	3.4.2	[147]
4242	Information Refresh Time Option for DHCPv6	3.5.2	[148]
4429	Optimistic Duplicate Address Detection (DAD) for IPv6	3.4.7	[149]
4604	Using IGMPv3 and MLDv2 for SSM	3.2.4	[51]
4607	Source-Specific Multicast for IP	3.2.4	[50]
4821	Packetization Layer Path MTU Discovery	3.2.3	[72]
4884	Extended ICMP to Support Multi-Part Messages	3.3	[150]
5072	IP Version 6 over PPP	3.6	[35]
5175	IPv6 Router Advertisement Flags Option	3.4.3	[151]
5790	Lightweight IGMPv3 and MLDv2 Protocols	3.2.4	[152]
5837	Extending ICMP for i/f and Next-Hop Identification	3.3	[153]
6085	Address Mapping of IPv6 Multicast Packets on Ethernet	3.6.2	[154]
6437	IPv6 Flow Label Specification	3.2.1	[155]
6731	Improved RDNSS Selection for MIF Nodes	3.10.2	[156]

Table 4.12 Informational RFCs for 3GPP hosts and routers

RFC	Title	Book Section	Reference
3316	IPv6 for Some Second and Third Generation Cellular Hosts	3	[157]
3493	Basic Socket Interface Extensions for IPv6	3.9.1	[158]
3542	Advanced Sockets API for IPv6	3.9.1	[159]
3678	Socket Interface Extensions for Multicast Source Filters	3.9.1	[160]
5014	IPv6 Socket API for Source Address Selection	3.9.1	[161]
6204	Basic Requirements for IPv6 Customer Edge Routers	3	[162]
6583	Operational Neighbor Discovery Problems	3.4.4	[163]

4.7.4 Provisioning

The 3GPP specifications prior to 3GPP Release-8 leave the decision whether to request IPv4 or IPv6 PDN Connection to the UE. The UEs then contain databases of operators and their settings, including address family parameters, and/or utilize provisioning system from Open Mobile Alliance (OMA) which has standardized for Device Management and Client Provisioning [164]. The OMA's specification defines parameters for Network Access Point

Definitions (NAPDEFs) (see Section 4.6.5 of [164]). Unfortunately, the NAPDEF does not support definition of the IP address type, even if it allows provisioning of IPv4 and IPv6 addresses for a UE. The lack of address family setting has led to vendor-specific extensions for defining the PDN Connection type. For example, Nokia has documented a proprietary solution, which modifies the eXtended Markup Language (XML) file used for provisioning by introducing a new setting value for NAPDEF called IFNETWORKS (see Section 2.3.4 of [165]. If the IFNETWORKS is set only to IPV4 or to IPV6, then a single address family PDN Connection will be created.

In Release-8, the 23.060 standard says that a UE, which is IPv4 and IPv6 capable, will always ask for an IPv4v6 type of PDN Connection [8]. The standard then gives a possibility for the network to downgrade, or narrow down, the UE's request and provide IPv4-only or IPv6-only PDN Connection, if that is what network operators prefers to use. However, in practice, at least at this point of IPv6 transition, cellular network operators often wish to control what type of bearer UEs request, and hence vendor specific additions to OMA's NAPDEF continue to be needed. The basis of this requirement is largely the untested nature of legacy network elements when encountering a new IPv4v6 type of request, which causes increased risks for connection establishment problems. In the Nokia's proprietary example, dual-stack connectivity can be provisioned by setting IFNETWORKS to 'IPV4,IPV6'. When both address families are set, a UE would first attempt to open an IPv4v6 dual-stack type of PDN Connection when possible, and otherwise use parallel IPv4 and IPv6 single address PDN Connections.

4.7.5 IPv6 Tethering

A tethering UE that acts as a router, as described in Section 2.6.2, needs to implement router specific features from the generic IPv6 RFCs listed in Tables 4.10 and 4.11. For example, in the case of RFC 6106, a UE acting as host may listen for the DNS configuration option in Router Advertisements, but a UE that acts as router should send the option in the Router Advertisements that the UE emits to the Local Area Network (LAN). Additionally, a routing UE node needs to implement a set of router-only RFCs listed in Table 4.13.

The exact RFCs needed depends on the supported 3GPP specification version and on the access network. The official 3GPP way for UEs to number networks below it is based on use of DHCPv6 PD, which requires RFCs 3315, 3633, and 6603. However, in 3GPP networks prior to Release-10, DHCPv6 PD is not supported as standard, and even in Release-10 based networks DHCPv6 PD support depends on the operator. If the DHCPv6 PD is not available, the UE has little choice but to implement Neighbor Discovery Proxy, as described in RFC 4389, or some variant of it. As well as DHCPv6 PD and Neighbor Discovery Proxy (or similar), there are two significant alternatives, which both come with significant issues. The first would be to implement IPv6 NAT, which should not be introduced to IPv6 as it would bring similar harm as IPv4 NAT. The second alternative is to perform stateful numbering of nodes in the local network with DHCPv6, but that approach would be applicable only if all nodes in the LAN were DHCPv6 enabled, which they are not. Hence, in practice modern implementations need to support DHCPv6 PD, and also may support the Neighbor Discovery Proxy, or similar, for the cases where the access network is not DHCPv6 PD enabled.

Table 4.13 RFCs needed for tethering use cases

RFC	Title	Book Section	Reference
2464	Transmission of IPv6 Packets over Ethernet Networks	3.6.2	[143]
3315	DHCPv6 for IPv6	3.5.2	[26]
3633	IPv6 Prefix Options for DHCPv6	3.5.2	[27]
3646	DNS Configuration options for DHCPv6	3.10.2	[96]
4389	Neighbor Discovery Proxies	3.4.10	[23]
6106	IPv6 RA Options for DNS Configuration	3.4.3	[140]
6603	Prefix Exclude Option for DHCPv6-based PD	3.5.2	[55]

Table 4.13 lists RFC 2464 as one of the required RFCs. This is because the tethering in 3GPP environments is most commonly implemented with Ethernet-based LAN – most notably, but not limited to, WLAN. Obviously, tethering can also be done with other access link types, which would then require the router to implement the corresponding RFCs. We have also included two RDNSS configuration options for serving hosts using either of those two for RDNSS discovery.

DNS Issues in the Case of Tethering

The PCO information element based RDNSS discovery comes with one significant drawback: it does not really work with Split-UEs. The problem arises as the PDN Connection is terminated on the modem, and it is the modem that receives all the information in the PCO information elements. In a traditional UE architecture where the modem is tightly integrated with the UE's operating system, this is not a problem. However, in the Split-UE (see Section 2.6.2) cases where modem and operating system are loosely integrated, sometimes even running in different devices, the problem arises of how to deliver RDNSS addresses from the modem to the operating system. One approach is to use proprietary solutions for communicating between the modem and a driver in the operating system, but it has obvious drawbacks of requiring additional software in the operating system. Another approach includes the modem modifying the Router Advertisement it proxies from GGSN and adding the RFC 6106 [140] RDNSS options into it – this modification is not detected by hosts as Secure Neighbor Discovery (SEND) is not typically used. The third approach is to run (stateless) DHCPv6 server on the modem, and serve the RDNSS option with DHCPv6. UEs that perform tethering basically have to support some of these means for delivering RDNSS addresses downstream, as 3GPP networks are not required to provide DHCPv6 services or send RDNSS information in Router Advertisements.

DHCPv6 Renew and Rebind

The DHCPv6 includes Renew and Rebind features that a requesting router can use to extend the lifetime of the delegated prefix and to request the same prefix again. In 3GPP networks, the UE's PDN Connection can drop quite easily, for example due to lost network coverage. In such cases, it is desirable that the LAN should not need to be renumbered, even if the UE's PDN Connection renumbers during the connection re-establishment phase. By ensuring that the UE remembers the prefix that it was delegated, the IAID it has used, and having constant DUID, the UE can request for the same prefix again. Hence, the delegated prefix *could*, if the network *would* allow, remain static over UE reboots or reconnections.

4.7.6 IPv6 Application Support

As long as IPv4-only applications exist for a UE, 3GPP networks cannot utilize the single-step protocol translation tools for IPv6 transition (see Section 5.4.3), but the networks have to make use of more resource consuming approaches such as dual-stack (see Section 5.4.2) or more complex setups such as double translation (see Section 5.4.4).

To pave the way for providing IPv6-only connectivity at some point in the future, and to reduce the load for IPv4 Network Address Translations, applications should be made IPv6 enabled. This, of course, is mainly the responsibility of application writers, but UEs can help by providing easy-to-use address agnostic APIs (see Section 3.9.2) and recommending and emphasizing the need to also use those APIs. When it comes to IP-aware applications, such as the generally "peer-to-peer" class of applications, those should be enhanced with the capability to understand IPv6 addressing. In this book we have provided some insights into what IPv6 means for applications in Sections 3.10.3 and 3.9.4.

It is the combined responsibility for everyone to share information and requirements to get all applications into the IPv6-world.

4.8 Multicast

IP multicast (and broadcast) delivery is realized using the MBMS service within the 3GPP architecture [49]. However, we are not going to describe MBMS in detail in this book as MBMS is not a wide spread service. Purely from the IPv6 point of view Section 4.2.2 has already discussed briefly which packet core network and RAN elements are affected by the MBMS service. In addition to radio network layer optimization, native IP multicast can be used within the 3GPP RAN and packet core as a traffic optimization mechanism to get the multicast traffic closer to the UEs that have subscribed to a multicast group or will receive the broadcast.

In order to provide MBMS services, existing 3GPP architecture functional entities – GGSN, MME, SGSN, eNodeB, RNC, Base Station Controller (BSC) – perform several MBMS related functions and procedures, some of which are specific to MBMS. The MBMS architecture differs between GPRS and EPS. Figure 4.29 illustrates the MBMS architecture for GPRS. It should be noted that a GGSN is part of the multicast delivery. Within the RAN, an RNC or a BSC may also receive native IP multicast traffic. Figure 4.30 illustrates the MBMS architecture for EPS. In EPS the PGW is bypassed and the native IP multicast traffic is delivered as far as the eNodeBs.

An MBMS specific functional entity, the Broadcast Multicast Service Centre (BM-SC), supports various MBMS user-specific services such as provisioning and delivery of multicast traffic. In EPS a functional entity called MBMS Gateway exists at the edge between the core network and the BM-SC. In the case of EPS, the PGW is bypassed regarding

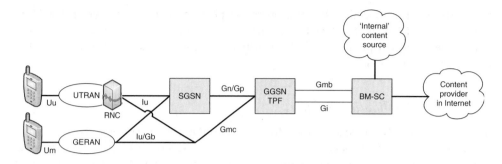

Figure 4.29   MBMS architecture for GPRS.

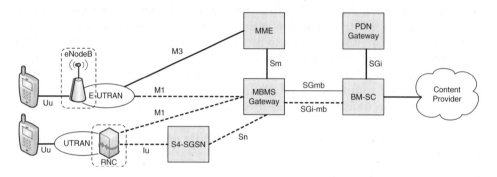

Figure 4.30   MBMS architecture for EPS.

the IP multicast traffic delivery. The BM-SC provides functions for MBMS user service provisioning and multicast traffic delivery. It may serve as an entry point for the content provider.

The BM-SC is a functional entity, which must exist for each MBMS user service. The BM-SC consists of the following sub-functions:

- membership function
- session and transmission function
- proxy and transport function
- service announcement function
- security function
- content synchronization for MBMS in UTRAN
- content synchronization for MBMS in E-UTRAN for broadcast mode
- header compression in UTRAN; header compression is not supported in E-UTRAN as of 3GPP Release-11.

4.9   Known IPv6 Issues and Anomalies

This section describes and demonstrates some of the known anomalies and issues on early IPv6 deployments in GPRS and EPS networks. The issues discussed here show how little proper testing and serious deployment trials of IPv6 in a cellular environment there has been. Some of the documented cases are indeed so trivial that they should not happen. We can partly also point at developers not fully understanding the difference between IPv4 and IPv6, specifically when it comes to link models and running IPv6 over arbitrary link technology. The documented issues in this section do not apply to all vendors and their products. The used 'sample' is more than one vendor, though.

4.9.1 IPv6 Neighbor Discovery Considerations

IPv6 NDP is a complex protocol and getting it working properly over a 3GPP link has turned out to be non-trivial. There are several known and documented shortcomings found in live GPRS networks [166]. These are typically a combination of the end host (the UE) side network interface and link abstraction framework being slightly flawed and the network side PGW being too strict on their interpretation of how IPv6 ought to work on a 3GPP defined link. The outcome is usually that no user plane traffic flows but no one is actually doing anything completely wrong.

Next we go through a couple of examples found in live networks. These are documented just for the purpose of learning and possibly to help people to identify possible error cases they face. It is expected (and hoped for) that none of these examples will be found in latest (upgraded) networks and UEs. We want to point out that these error case examples have been checked with multiple UEs vendors and PGWs vendors.

End Host Interface Treated as a Router Interface

A 3GPP link between a UE and a next-hop router (e.g. the PGW) resembles a point-to-point link, which has no link-layer addresses [157], and this has not changed from 2G/3G GPRS to EPS. The UE IP stack has to take this into consideration. When the 3GPP PDN Connection appears as a PPP interface to the UE, the IP stack is usually prepared to handle the NDP and the related Neighbor Cache state machine transitions in an appropriate way, even though NDP messages contain no link-layer address information. However, some operating systems discard Router Advertisements on their PPP interface as a default setting. This causes SLAAC to fail when the 3GPP PDN Connection gets established, thus stalling all IPv6 traffic. This kind of end host default setting has been there, for example, in Linux and early OS/X operating systems for quite some time. The reasoning has probably been that a PPP link was considered as a router-to-router link, and in that case the interface should not use SLAAC to configure its addresses.

IP Stack Incorrectly Requires Address Resolution

Currently, several operating systems and their network drivers can make the 3GPP PDN Connection appear as an IEEE 802 interface to the IP stack. Sometimes even multiple PDN Connections are bundled under the same network interface, for example, when two PDN Connections of different PDN Types are used to mimic a dual-stack connection.

One quite common approach is to connect the modem using a USB to a host, that is, between the TE and the MT. The USB based connection can be either external or internal, so it is not only limited to 'USB dongles'.

The USB-based network driver framework [36, 37] has a few known issues, especially when the IP stack is made to believe that the underlying link *has link-layer addresses*. First, the Neighbor Advertisement sent by a PGW as a response to a Neighbor Solicitation triggered by address resolution might not contain a Target Link-Layer Address option (see Section 4.4 of [157]). It is then possible that the address resolution never completes when the UE tries to resolve the link-layer address of the PGW, thus stalling all IPv6 traffic.

Partial Neighbor Discovery Protocol Implementation on the Network Side

Linked to the previous experiences of misbehaving end host driver implementations, the situation can be exacerbated. There have also been cases where the (old) GGSN simply discards all Neighbor Solicitation messages triggered by address resolution (this is because Section 2.4.1 of [157] is sometimes misinterpreted as saying that responding to address resolution and next-hop determination is not needed). As a result, the address resolution never completes when the UE tries to resolve the link-layer address of the GGSN, thus stalling all IPv6 traffic. Figure 4.31 shows a packet trace of this happening in a live network. There is little that can be done about this in the PGW, assuming that the Neighbor Discovery implementation already does the right thing. But the UE stacks must be able to handle address resolution in the manner that they have chosen to represent the

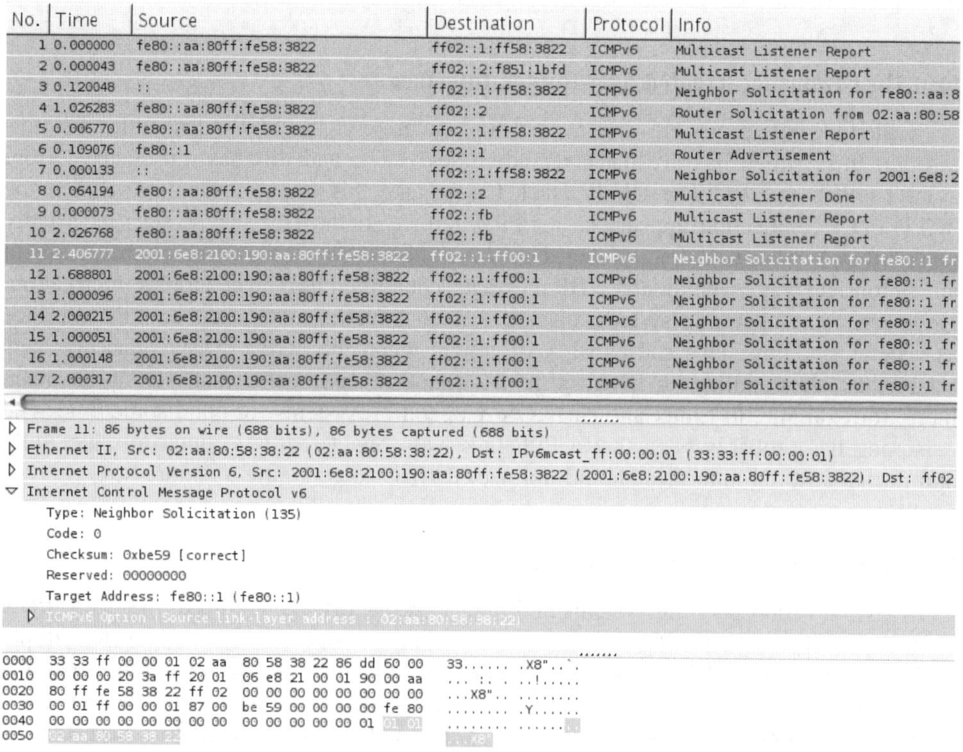

No.	Time	Source	Destination	Protocol	Info
1	0.000000	fe80::aa:80ff:fe58:3822	ff02::1:ff58:3822	ICMPv6	Multicast Listener Report
2	0.000043	fe80::aa:80ff:fe58:3822	ff02::2:f851:1bfd	ICMPv6	Multicast Listener Report
3	0.120048	::	ff02::1:ff58:3822	ICMPv6	Neighbor Solicitation for fe80::aa:8
4	1.026283	fe80::aa:80ff:fe58:3822	ff02::2	ICMPv6	Router Solicitation from 02:aa:80:58
5	0.006770	fe80::aa:80ff:fe58:3822	ff02::1:ff58:3822	ICMPv6	Multicast Listener Report
6	0.109076	fe80::1	ff02::1	ICMPv6	Router Advertisement
7	0.000133	::	ff02::1:ff58:3822	ICMPv6	Neighbor Solicitation for 2001:6e8:2
8	0.064194	fe80::aa:80ff:fe58:3822	ff02::2	ICMPv6	Multicast Listener Done
9	0.000073	fe80::aa:80ff:fe58:3822	ff02::fb	ICMPv6	Multicast Listener Report
10	2.026768	fe80::aa:80ff:fe58:3822	ff02::fb	ICMPv6	Multicast Listener Report
11	2.406777	2001:6e8:2100:190:aa:80ff:fe58:3822	ff02::1:ff00:1	ICMPv6	Neighbor Solicitation for fe80::1 fr
12	1.688801	2001:6e8:2100:190:aa:80ff:fe58:3822	ff02::1:ff00:1	ICMPv6	Neighbor Solicitation for fe80::1 fr
13	1.000096	2001:6e8:2100:190:aa:80ff:fe58:3822	ff02::1:ff00:1	ICMPv6	Neighbor Solicitation for fe80::1 fr
14	2.000215	2001:6e8:2100:190:aa:80ff:fe58:3822	ff02::1:ff00:1	ICMPv6	Neighbor Solicitation for fe80::1 fr
15	1.000051	2001:6e8:2100:190:aa:80ff:fe58:3822	ff02::1:ff00:1	ICMPv6	Neighbor Solicitation for fe80::1 fr
16	1.000148	2001:6e8:2100:190:aa:80ff:fe58:3822	ff02::1:ff00:1	ICMPv6	Neighbor Solicitation for fe80::1 fr
17	2.000317	2001:6e8:2100:190:aa:80ff:fe58:3822	ff02::1:ff00:1	ICMPv6	Neighbor Solicitation for fe80::1 fr

▷ Frame 11: 86 bytes on wire (688 bits), 86 bytes captured (688 bits)
▷ Ethernet II, Src: 02:aa:80:58:38:22 (02:aa:80:58:38:22), Dst: IPv6mcast_ff:00:00:01 (33:33:ff:00:00:01)
▷ Internet Protocol Version 6, Src: 2001:6e8:2100:190:aa:80ff:fe58:3822 (2001:6e8:2100:190:aa:80ff:fe58:3822), Dst: ff02
▽ Internet Control Message Protocol v6
    Type: Neighbor Solicitation (135)
    Code: 0
    Checksum: 0xbe59 [correct]
    Reserved: 00000000
    Target Address: fe80::1 (fe80::1)
  ▷ ICMPv6 Option (Source link-layer address : 02:aa:80:58:38:22)

```
0000   33 33 ff 00 00 01 02 aa   80 58 38 22 86 dd 60 00      33......  .X8"..`.
0010   00 00 00 20 3a ff 20 01   06 e8 21 00 01 90 00 aa      ... :. ..!.....
0020   80 ff fe 58 38 22 ff 02   00 00 00 00 00 00 00 00      ...X8"..  ........
0030   00 01 ff 00 00 01 87 00   be 59 00 00 00 00 fe 80      ........  .Y......
0040   00 00 00 00 00 00 00 00   00 00 00 00 00 00 00 01      ........  ........
0050   02 aa 80 58 38 22                                      ...X8"
```

Figure 4.31 Example of NDP failure on a 3G link, which blocks all user initiated IPv6 traffic.

interface. In other words, if they emulate IEEE 802 interfaces, they also need to process NDP messages correctly.

Duplicate Address Detection Failing

As we have already learned a PGW tells a UE the IID it is mandated to use to configure its link-local address. However, again due to misbehaving end host driver implementations or for other implementation dependent reasons, the UE IP stack may end up using an IID other than the one that the PGW told it to use. This opens the door for a theoretical DAD failure on the link-local addresses used on the link. Note that there will not be an address collision with GUAs since the PGW never configures an address for itself using the prefix advertised on the link.

People might argue that an IP stack should handle the address collision as per normal procedures defined in [22]. On the other hand, we also know that some older PGWs may skip parts of the NDP, specifically the DAD procedure, because the address collision should not happen [157]. In practice the probability of an address collision is so small that coming across one is not going to happen, except as a result of a programming error or when done on purpose.

If the PGW neglects the DAD and a duplicate link-local address gets configured on the UE (i.e. the link-local address of the PGW is configured on the UE interface), then the following error case will happen. The UE comes to the conclusion that the colliding link-local address is unique on the link because the PGW never replies to DAD generated solicited-node multicast Neighbor Solicitations with undefined source address. Any further attempt to send traffic to the PGW fails because the packets are now destined to the local interface in the UE and no user plane IP packet leaves the UE. The destination cache may still hold the old entry for the PGW link-local address that was learned when receiving the Router Advertisement.

Figure 4.32 shows an example capture where a UE is configured with a link-local address of the GGSN (i.e. `fe80::1/10`). As a side note, the (old brand) GGSN here does not answer the DAD originated Neighbor Solicitation messages. We start `ping6` to `2001:6e8:2100:100::30` at the packet #41 and a entry gets created in the destination cache for `2001:6e8:2100:100::30` pointing at `fe801::1` (the link-local address of the GGSN). On packet #48 we manually configure the address of the GGSN on the UE interface and we can also see the IPv6 stack initiating the procedures that are needed when a new address is configured, such as joining an appropriate Solicited-Node Multicast group and issuing a DAD procedure (packet #49). As there is no answer to the DAD originated Neighbor Solicitation the stack starts using the address. However, we still have the existing Destination Cache entry pointing at `fe80::1`, which is now also used by the UE. Nothing gets through over the link as all packets destined to `2001:6e8:2100:100::30` are sent to a local interface, which then fails to handle them in any way, and packets are just silently discarded (this is probably due the fact that the Neighbor Cache incorrectly points to the wrong link-layer address and the IP stack gets confused). Finally, at the packet #62 we remove `fe80::1` from the UE interface, after which the UE IPv6 stack immediately does an address resolution for the GGSN link-local address `fe80::1` and once the neighbor-cache is updated user plane packets start flowing again as normal (starting from the packet #65).

No.	Time	Source	Destination	Protocol	Info
6	5.078824	fe80::1	ff02::1	ICMPv6	Router Advertisement
7	5.078778	::	ff02::1:ffc9:35c4	ICMPv6	Neighbor Solicitation for 2001:6e8:2100:198:6
8	5.090830	fe80::6b:85ff:fec9:35c4	ff02::2	ICMPv6	Multicast Listener Done
9	5.090931	fe80::6b:85ff:fec9:35c4	ff02::fb	ICMPv6	Multicast Listener Report
10	5.108896	fe80::1	ff02::1	ICMPv6	Router Advertisement
18	8.973834	fe80::6b:85ff:fec9:35c4	ff02::fb	ICMPv6	Multicast Listener Report
34	15.191912	fe80::6b:85ff:fec9:35c4	fe80::1	ICMPv6	Neighbor Solicitation for fe80::1 from 02:6b:
35	15.195259	fe80::1	fe80::6b:85ff:fec9:35c4	ICMPv6	Neighbor Advertisement fe80::1 (rtr, sol, ovr
41	28.269187	2001:6e8:2100:198:6b:85ff:fec9:35c4	2001:6e8:2100:100::30	ICMPv6	Echo (ping) request id=0x1cd4, seq=0
42	28.558916	2001:6e8:2100:100::30	2001:6e8:2100:198:6b:85ff:fec9:35c4	ICMPv6	Echo (ping) reply id=0x1cd4, seq=0
43	29.269349	2001:6e8:2100:198:6b:85ff:fec9:35c4	2001:6e8:2100:100::30	ICMPv6	Echo (ping) request id=0x1cd4, seq=1
45	29.558793	2001:6e8:2100:100::30	2001:6e8:2100:198:6b:85ff:fec9:35c4	ICMPv6	Echo (ping) reply id=0x1cd4, seq=1
46	30.269244	2001:6e8:2100:198:6b:85ff:fec9:35c4	2001:6e8:2100:100::30	ICMPv6	Echo (ping) request id=0x1cd4, seq=2
47	30.558749	2001:6e8:2100:100::30	2001:6e8:2100:198:6b:85ff:fec9:35c4	ICMPv6	Echo (ping) reply id=0x1cd4, seq=2
48	42.898861	fe80::6b:85ff:fec9:35c4	ff02::1:ff00:1	ICMPv6	Multicast Listener Report
49	42.898895	::	ff02::1:ff00:1	ICMPv6	Neighbor Solicitation for fe80::1
50	44.790176	fe80::6b:85ff:fec9:35c4	ff02::1:ff00:1	ICMPv6	Multicast Listener Report
62	53.826088	fe80::6b:85ff:fec9:35c4	ff02::2	ICMPv6	Multicast Listener Done
63	55.194931	2001:6e8:2100:198:6b:85ff:fec9:35c4	ff02::1:ff00:1	ICMPv6	Neighbor Solicitation for fe80::1 from 02:6b:
64	55.196033	fe80::1	2001:6e8:2100:198:6b:85ff:fec9:35c4	ICMPv6	Neighbor Advertisement fe80::1 (rtr, sol, ovr
65	55.196092	2001:6e8:2100:198:6b:85ff:fec9:35c4	2001:6e8:2100:100::30	ICMPv6	Echo (ping) request id=0x1cd9, seq=7
66	55.500575	2001:6e8:2100:100::30	2001:6e8:2100:198:6b:85ff:fec9:35c4	ICMPv6	Echo (ping) reply id=0x1cd9, seq=7
67	55.557011	2001:6e8:2100:198:6b:85ff:fec9:35c4	2001:6e8:2100:100::30	ICMPv6	Echo (ping) request id=0x1cd9, seq=8

Figure 4.32 Example of NDP failure on a 3G link due to a UE configuring duplicate addresses on its interface.

Losing the Initial Router Advertisement

It is also possible that initial Router Advertisements get lost when a PDN Connection gets activated. If the initial MAX_INITIAL_RTR_ADVERTISEMENTS are lost, the UE does not have a default router or IPv6 prefix configured, which leaves the UE in a state where it has no IPv6 connectivity [167]. There are deployed (old) GGSNs that send only one Router Advertisement. Furthermore, some GGSNs may or may not respond to a Router Solicitation.

Losing a Router Advertisement on a 3GPP link should not happen. The persistent claim has been that no packet ever gets lost due to excessive retransmissions that are practiced over the 3GPP radio. In reality, Router Advertisements do get lost – not necessarily on the actual wireless link segment but *between the radio modem and the host IP stack*. Therefore, this is a real issue for *Split-UEs* and specifically on the older UEs that implement dual-stack using two parallel PDN Connections. One typical case is when a dual-stacked network interface gets up on the UE, and a number of eager IPv4 applications immediately send a flood of TCP connection attempts, DNS queries, and various service discovery queries to the modem; this then gets overloaded while the PDN Connection activation is still in progress and as a result the Router Advertisement can just get lost. A decent end host IP stack implementation should send a Router Solicitation after a while but this seems not to be the case in all situations.

Solutions to Fix the Situation

There are multiple approaches for fixing possible NDP issues regarding the 3GPP link model and IPv6. Some IPv6-enabled radio modems and PGWs are known to do the following hacks/tricks:

- A PGW adds a forged Source Link-Layer Address option into the Router Advertisement message, which makes the UE ND cache transition from the INCOMPLETE state to the STALE state immediately on receiving the Router Advertisement message.
- A PGW adds a forged Source Link-Layer Address option into the address resolution solicited Neighbor Advertisement message, which makes the address resolution complete and the UE ND cache transitions from the INCOMPLETE state to the STALE state immediately on receiving the Neighbor Advertisement message.
- A radio modem firmware adds a forged Source Link-Layer Address option into the Router Advertisement message on behalf of the PGW.
- A radio modem firmware *unicasts* the Neighbor Solicitation to the PGW when a host attempts the address resolution. This looks like an NUD to the PGW, and since all implementations are mandated to respond to the NUD [25], there is a far better chance that the PGW responds with something meaningful in its Neighbor Advertisement (some PGW implementations add a forged Source Link-Layer Address option into the Neighbor Advertisement message).

All the above are still hacks/tricks to mostly alleviate issues caused by broken UE implementations. The UE and the PGW implementations should just understand the 3GPP link model properly and then implement the NDP state changes accordingly.

4.9.2 PDN Connection Model and Multiple IPv6 Prefixes

When 3GPP was adding IPv6 into their specifications, IETF recommended that the 3GPP link should allow multiple IPv6 prefixes on the link – see Section 2 of [168]. As we know, this recommendation never materialized in 3GPP. Instead when multiple prefixes are desired then multiple PDN Connections (default bearers) get activated. There are certain benefits to this approach such as the possibility for a UE to be connected to different PGWs simultaneously that may be dedicated to different purposes. On the other hand, using multiple PDN Connections just to enable multi-addressing sounds a bit wasteful. Each connection requires a handful of resources on the network side. The connection activation signaling is not the most lightweight one and involves multiple core network elements. On the UE side, multiple connections imply multiple Radio Access Bearers (RABs) over the radio interface, which are also considered as an expensive resource.

However, a single /64 prefix on a PDN Connection is such a fundamental part of the 3GPP link model that changing it is not straightforward. We still consider the current model too restrictive. Changes to the 3GPP link model and the PDN Connection addressing assumptions are definitely something worth rethinking for future 3GPP releases. We will offer some initial discussions about the possible evolution of the 3GPP architecture in Chapter 6.

4.10 IPv6 Specific Security Considerations

IPv6 has been advocated as a more secure protocol than IPv4. Unfortunately, this not true and was most likely inspired by the early IPv6 node requirements [169] stating that every

IPv6 stack must support ESP [77] and Authentication Header (AH) [170]. However, over the years it became clear that the industry does not want to implement Security Architecture for the Internet Protocol [73] for every IPv6 stack. Therefore, the IPv6 node requirements specifically on the security part got loosened [32] and now support for Security Architecture is optional, though highly recommended.

There have been numerous attempts in recent years to correct the identified IPv6 specific security vulnerabilities. We can categorize them and the related attack vectors to several groups based on their nature and the exploited vulnerability:

- IPv6 addressing, which covers host scanning, poisoning the router ND cache with a flood of packets that causes an address resolution to take place, and privacy breaches made possible by easily predictable addresses or stable IIDs.
- IPv6 fragmentation and reassembly defects used for forging IPv6 fragments in order to, for example, bypass firewalls.
- IPv6 first-hop and/or on-link security vulnerabilities which cover, for example, NDP message spoofing, forged redirects, traffic stealing, and rogue Router Advertisements.
- IPv6 firewall bypassing made possible by oversized IPv6 extension header chains or overlapping fragments.
- IPv6 extension header and option defects used for smurf attacks and traffic amplification using ICMPv6 and unknown options or Type 0 Routing Header.
- Router load generation using the router alert IPv6 hop-by-hop extensions.

Some of the listed vulnerabilities do not apply as such to 3GPP architecture. More specifically, the 3GPP link model and the fact that there can be exactly one UE and the first-hop router (i.e. the PGW or in case of PMIPv6 the SGW) on the same link, make certain on-link attacks less critical. Several documented vulnerabilities are specifically targeted to shared links like Ethernet. On the other hand, 3GPP architecture from the UE point of view is as vulnerable as any IP network for off-link attacks. In the following sections we will discuss the earlier listed security threats from a 3GPP architecture perspective and concentrate only on those security topics that concern the user plane. A good summary of generic operational IPv6 security concerns can be found in [171].

The attacks, vulnerabilities and security threats discussed in the following sections do not form a definitive list. They represent the set that the authors were aware of and have also been documented in IETF for the purpose of fixing the vulnerabilities found.

4.10.1 IPv6 Addressing Threats

This section discusses several known IPv6 addressing exploiting vulnerabilities in the context of the 3GPP architecture. We only consider user plane addressing which concerns the UE, thus we are not going to look into attacks and vulnerabilities that directly target the GPRS or EPC core network nodes on their transport and control planes. We also concentrate only on remote network attacks since there can be exactly one end host on the 3GPP link, which makes on-link local area network attacks less important.

IPv6 address architecture has been criticized for its privacy issues. Specifically, SLAAC configured addresses tend to have a *permanent cookie* in the form of the EUI-64 based IID. If the EUI-64 is derived from a MAC address, or from any fixed identifier, then independent of changing prefix the IID remains the same, allowing tracking of the device/user globally. Privacy addresses [109], which are supported today in practically every modern operating system, solve the issue of the IID being a *permanent cookie*. Privacy addresses then have a new set of issues such as the address changing too frequently, for example, from the network access control and its resource consumption point of view. The recent work on stable privacy addresses [172] makes the IID remain the same for a longer time, practically as long as the link gets renumbered or the end host changes to another link that has a different prefix. In 3GPP networks, the generation of a new stable privacy address would take place during each PDN Connection activation.

Scanning of the address range, or rather a /64 IPv6 prefix, for possible victims is another concern. It is often naively bypassed, stating that the IPv6 address's IID number space is so huge (2^{64} possibilities) that off-link initiated scanning attempts are doomed to fail. However, in practice the IID space is much smaller [173]. If the scanning attack is targeted at a specific brand of devices, then the Organizationally Unique Identifier (OUI) part of the EUI-64 can be fixed (that is 24 bits less to scan). If the target is expected to have an Ethernet network interface, then another 16 bits of the IEEE 48 bit MAC address derived from modified EUI-64 is fixed (the `0xfffe` part in the middle of the EUI-64). Now we are down to 24 bits of number space to scan, which does not sound that bad anymore. Similarly, if the address is manually configured or the end host is possibly dual-stacked, guessing the address has higher probability due the laziness of network management personnel. It is surprisingly common to just embed small numbers that fit into the last octet of the IID (like 69) or the service port number as the IPv6 IID. Another common practice is to use the IPv4 address of the end host or some hexadecimal combination forming a readable word as the IPv6 IID (take `2001:db8::192.0.2.1` and `2001:db8::abad:cafe` as examples). For the latter, manually configured IIDs, stable privacy addresses [172] or 'human-safe' Cryptographically Generated Addresses (CGAs) [174] work as a good solution to avoid predictable addresses.

Another remote IPv6 address scanning side effect is the unnecessary address resolution performed by the first-hop router on the link trying to resolve the link-layer address of the (bogus) IPv6 destination address. This is known as a Denial of Service (DoS) attack, which attempts to fill the ND cache of the router [163]. Rate limiting of address resolution or filtering incoming packets with 'impossible' IID ranges in their destination address are a few of the approaches to mitigate this attack.

Are these threats and vulnerabilities issues in the 3GPP architecture? Naturally some are but not all. As we saw in Section 4.1.3 the 3GPP link is a point-to-point link without link-layer addresses, and the UE is the only end host on the link. Also the /64 prefix advertised to the 3GPP UE is never on-link (the PIO 'L'-flag is set to 0). Therefore, a PGW can skip address resolution on the 3GPP link and just forward the whole /64 to the UE. As a result, the PGW is not vulnerable, for example, to an ND cache poisoning attack.

The EUI-64 that the end host comes up with should never be derived from an IEEE 48-bit MAC address as there in no kind of link-layer address available in the UE. As noted in Section 4.9.1, this is not always the case. Furthermore, there have also been concerns that the UE IP stack might use permanent subscriber identities, such as an IMSI, as the source for the IPv6 address IID. This would be a privacy threat and would allow tracking of subscribers. Therefore, the use of an IMSI (or any 3GPP specific identity [113]) as the IID is prohibited [15]. The use of privacy addresses [109] is encouraged in the UE implementations. However, there is no standardized method of blocking misbehaving UEs that use a 'wrong' type of IID. Most network connections are expected to be relatively short-lived, and static IPv6 addressing is probably not going to be too popular in the consumer mass market. As a result the UEs (and subscribers) are likely to have frequently changing prefix, which combined with any decent IPv6 IID generation method should provide adequate privacy from the IPv6 addressing point of view.

4.10.2 IPv6 First-hop Security

This section discusses known IPv6 NDP exploiting vulnerabilities in the context of 3GPP architecture. We only consider user plane addressing which concerns the UE, thus we are not going to look into attacks and vulnerabilities that directly target the GPRS/EPC core network nodes on their transport and control planes. We have already dealt with the ND cache poisoning attack in Section 4.10.1, so that will not be reintroduced in this section.

There are a number of ingenious ways to attack hosts and routers on a multi-access link [175] if the attacker is able to gain access to the link and also spoof its NDP packet source address. Redirecting traffic to on-link or even off-link victim hosts, preventing a host from sending traffic to off-link destinations, removing legitimate default routers from end hosts' default router list, preventing hosts completing SLAAC by purposely failing a DAD procedure and advertising bogus prefixes on the link are just some well-known attacks making use of NDP's vulnerabilities [176]. All these culminate in the lack of proving the ownership of the claimed source address of the NDP message.

Rogue Router Advertisements can also be an annoying issue and prevent the end host reaching its desired off-link destinations [177]. In many cases rogue Router Advertisements seen in the wild are unintentional: the byproduct of a badly configured host or a flawed implementation in the host operating system.

Secure Neighbor Discovery SEND [146] has a mechanism that is used for showing address ownership on individual nodes and is based on CGAs [147]. SEND utilizes X.509 certificates that include the extension for IPv6 addresses [178, 179, 180] to prove that, for example, the on-link router is authorized to advertize a given IPv6 prefix. The challenge of SEND is that it requires support from both sender and receiver of the SEND protected NDP message. There are also further network and device management challenges. The required certificate provisioning, and the Resource Public Key Infrastructure (RPKI) [181] for RPKI certificates [182] handling might be an even bigger issue, preventing wider adoption than the technology itself.

Another approach to addressing on-link threats is using dedicated (layer-2) devices or functions incorporated into other required networking devices that do the required tracking of source addresses, NDP messages, and address assignment on the link. Technologies like Source Address Validation Improvements (SAVI) [183, 184] and RA Guard [185] fit into this category. These mechanisms are intended to complement ingress filtering techniques to help detect and prevent source address spoofing, and also filter bogus Router Advertisement messages.

Are the above threats and vulnerabilities issues in the 3GPP architecture? This is not likely, due to the 3GPP link model described in Section 4.1.3. A PGW router and a UE can be the only nodes on the same 3GPP point-to-point link. Furthermore, as discussed in Section 4.10.1 most of the NDP procedures can just be neglected in the UE or in the PGW. For any of the on-link threats to be possible, the PGW should more or less be compromised and its software modified.

4.10.3 IPv6 Extension Header Exploits

IPv6 is extended using a number of different extension headers and their options – see Section 3.2. The attacks we are discussing in this section are based on a misuse of legitimate extension headers.

Option Processing Exploits

Standard IPv6 hop-by-hop and destination option header handling can be used for a smurf and traffic amplification attack [186]. The attack requires that the hostile host is able to spoof the source address of the sent packet, and multicast forwarding is supported in the attack sourcing network. The used exploit is rather simple. The IPv6 extension header has a one-octet-long option type code and its highest-order two bits specify the action that must be taken if the IPv6 node does not recognize the option type. One of the option types, namely 10xxxxxx, causes the processing node to discard the packet *and* send an ICMPv6 Parameter Problem, Code 2, message to the original packet's source address regardless of the original destination address being a multicast address. As a result, a hostile host can send any spoofed packet with a bogus option type in any hop-by-hop or destination option to a known multicast group or address, which would then cause all the multicast receiving nodes to flood the victim host with ICMPv6 messages.

Another traffic amplification attack is available using the now deprecated Type 0 Routing Header (RH0) [187, 136]. The RH0 allows source routing of IPv6 packets via a list of predefined destination nodes. A single 1280-octet IPv6 packet with a single RH0 can contain over 70 intermediate node addresses. A cleverly constructed packet with an RH0 option can be made to oscillate between two IPv6 nodes or routers as long as all entries (destinations) in the RH0 option have been consumed. This allows for an off-link attacker a rather cheap way to cause congestion on a path between two nodes.

All hop-by-hop options must be processed or at least examined by all intermediate nodes between the source and the destination. IPv6 Router Alert Option is an example [129] of a hop-by-hop option, which is currently used by MLD and MLDv2, for example. A router on a path would examine the packet more closely, which typically means it gets pulled out of the fast path for the Central Processing Unit (CPU) to process. Obviously, this is more costly from the router perspective and when a router is flooded with such packets, the IPv6 Router Alert option can be used as a DoS mechanism.

Are the above attacks and vulnerabilities issues in the 3GPP architecture? They potentially are, and could make a mobile system as a good resource pool for an attacker to source distributed attacks. However, we could speculate that the 3GPP link model and the lack of native IP level multicast and non-existing deployments of MBMS could mitigate somewhat the described attacks. At least orchestrating such remotely would be slightly more challenging.

Bypassing Firewalls

IPv6 allows arbitrary extension header size that can even be fragmented into several packets. Use of unusually long extension header chains [188, 189] opens possibilities for a couple of attacks and possible vulnerabilities in firewall products. Firewall filtering rules are typically based on the following five-tuple:

- source address
- destination address
- next header
- source port number (when available)
- destination port number (when available).

Now if the extension header chain is so long that the chain and the actual payload (with a potential port information) will not fit into the first fragment, then the firewall has no other choices than to:

- reassemble the whole packet, which can be a considerable scalability issue for the firewall product and a potential threat for DoS attack against the firewall
- let fragmented packets through, which can then be a severe security breach
- just drop all fragments.

A straightforward fix for the above would be mandating that the whole extension header chain and the final payload transport protocol header must fit into the first fragment. This would allow building meaningful filtering rules against the five-tuple without the need to reassemble packets in the firewall.

Uncommonly long header chains have additional non-security-related issues. Take IPv6 transition tools as an example. A stateful IPv4/IPv6 Network Address Translation (NAT64) [190] needs to have the five-tuple again in order to build the translation state. If the entire header is not present, the NAT64 fails to establish the state and the packets are dropped (assuming that the NAT64 does not reassemble the packets).

While we are at IPv6 transition, various tunneling and translation mechanisms are used for transporting IPv6 packets over IPv4 networks [191]. Briefly, when IPv6 is transported over the IPv4, then the specific IPv6 access control and filtering rules may be completely bypassed [192]. A naive solution would be just blocking the known protocol types for IPv6 in IPv4 tunnels (such as 41 for Connection of IPv6 domains via IPv4 clouds (6to4)) but that could block old-fashioned legitimate configured tunnels as well. A better approach would be looking deeper into the packet and verifying from the inner IPv6 packet headers what is actually going on. That is, however, more computationally expensive. IPv6 filtering [193] in general is easily done wrong and may cause unwanted side effects such as hindering NDP.

Are these threats and vulnerabilities issues in 3GPP architecture? They potentially are, and there is really no difference compared to any other (fixed network) IPv6 deployment.

Atomic and Overlapping Fragments

Last we look into two fragmentation-specific security threats; overlapping fragments [137] and using *atomic fragments* as a tool to launch other fragmentation and predictable identifier based attacks [194, 195].

The IPv6 specification [2] did not specifically prohibit fragments overlapping each other when reassembling packets. That is, the start of a fragment indicated by the fragment offset makes the new fragment overlap parts of the other fragment parts. This can be used to fool firewalls and intrusion detection systems into assuming that the received packet fragments are legitimate where as the final reassembled packet contains some attack vector such as a specially crafted TCP packet. Later it was clarified that overlapping fragments cause all previously received fragments and those still on the fly to be discarded [137].

Atomic fragments are IPv6 packets that contain the fragment header with both *fragment offset* and '*M*'-*flag* set to zero, that is, there are no other fragments following and the packet, even if it contains a fragment header, is not fragmented. IPv6 allows such packets because of potential IPv6 to IPv4 translation, where the IPv4 side of the network may require fragmenting packets to less than the minimum allowed IPv6 packet size of 1280 octets. An atomic fragment is triggered by an attacker sending a spoofed ICMPv6 Packet Too Big to a victim host containing a next-hop MTU value less than 1280 octets and the spoofed source address set to some third-party host. Now the victim host will (assuming it does not have other counter measures to mimic the spoofing) add a fragment header to every packet it sends to the third-party host. Now the attacker can launch other fragment (and possible predictable identifier) based attack towards the victim and third-party host. One example is 'simple' polluting of the fragment identifier space of the third-party host by the attacker sending spoofed overlapping fragments to the third-party host and effectively causing all communication from the victim host to be discarded at the third-party host.

Are the above attacks and vulnerabilities issues in the 3GPP architecture? They potentially are, and there is really no difference compared to any other (fixed network) IPv6 deployment.

4.11 Chapter Summary

This chapter described how the *user plane* IPv6 is integrated into the 3GPP architecture and which packet core networking nodes are affected by the introduction of the user plane IPv6. We explain how the 3GPP point-to-point link model works and the specifics of IPv6 on it. The IPv6 address assignment, management, and provisioning are discussed in detail. The static addressing has not been forgotten either. We also looked into the lower layers and control plane protocols, in order to point out places where user plane IPv6 information gets exchanged between the packet core network elements and also to the UE. The impact of the user plane IPv6 in mobile devices was also discussed in detail.

Known issues in existing IPv6 deployments were also discussed and their root causes were revealed to the extent that the authors are aware of them. We also discussed in many places where 3GPP specifications are 'not there yet' or they contradict each other. Finally we took a look at a number of known IPv6 security issues and reflect them to the 3GPP architecture.

In summary, we can say that there is still a lot that could be done better regarding IPv6 in 3GPP Networks. This comes from the fact that deployments using IPv6 in real live 3GPP networks are few and operational aspects are still in their infancy. Needless to say, the packet core networking products still have a lot to catch up, mainly due to the lack of feature enhancements originating in operational experience.

References

1. Postel, J. *Internet Protocol*. RFC 0791, Internet Engineering Task Force, September 1981.
2. Deering, S. and Hinden, R. *Internet Protocol, Version 6 (IPv6) Specification*. RFC 2460, Internet Engineering Task Force, December 1998.
3. 3GPP. *3GPP Evolved Packet System (EPS); Evolved General Packet Radio Service (GPRS) Tunnelling Protocol for Control plane (GTPv2-C); Stage 3*. TS 29.274, 3rd Generation Partnership Project (3GPP), March 2012.
4. 3GPP. *General Packet Radio System (GPRS) Tunnelling Protocol User Plane (GTPv1-U)*. TS 29.281, 3rd Generation Partnership Project (3GPP), June 2010.
5. 3GPP. *Proxy Mobile IPv6 (PMIPv6) based Mobility and Tunnelling protocols; Stage 3*. TS 29.275, 3rd Generation Partnership Project (3GPP), March 2012.
6. Gundavelli, S., Leung, K., Devarapalli, V., Chowdhury, K., and Patil, B. *Proxy Mobile IPv6*. RFC 5213, Internet Engineering Task Force, August 2008.
7. Wakikawa, R. and Gundavelli, S. *IPv4 Support for Proxy Mobile IPv6*. RFC 5844, Internet Engineering Task Force, May 2010.
8. 3GPP. *General Packet Radio Service (GPRS); Service description; Stage 2*. TS 23.060, 3rd Generation Partnership Project (3GPP), March 2012.
9. 3GPP. *General Packet Radio Service (GPRS); GPRS Tunnelling Protocol (GTP) across the Gn and Gp interface*. TS 29.060, 3rd Generation Partnership Project (3GPP), March 2012.
10. Droms, R. *Dynamic Host Configuration Protocol*. RFC 2131, Internet Engineering Task Force, March 1997.
11. Park, S., Kim, P., and Volz, B. *Rapid Commit Option for the Dynamic Host Configuration Protocol version 4 (DHCPv4)*. RFC 4039, Internet Engineering Task Force, March 2005.
12. 3GPP. *Mobile radio interface Layer 3 specification; Core network protocols; Stage 3*. TS 24.008, 3rd Generation Partnership Project (3GPP), March 2012.
13. 3GPP. *Non-Access-Stratum (NAS) protocol for Evolved Packet System (EPS); Stage 3*. TS 24.301, 3rd Generation Partnership Project (3GPP), March 2012.
14. Simpson, W. *The Point-to-Point Protocol (PPP)*. RFC 1661, Internet Engineering Task Force, July 1994.
15. 3GPP. *General Packet Radio Service (GPRS) enhancements for Evolved Universal Terrestrial Radio Access Network (E-UTRAN) access*. TS 23.401, 3rd Generation Partnership Project (3GPP), March 2012.
16. 3GPP. *Architecture enhancements for non-3GPP accesses*. TS 23.402, 3rd Generation Partnership Project (3GPP), March 2012.
17. Baker, F. and Savola, P. *Ingress Filtering for Multihomed Networks*. RFC 3704, Internet Engineering Task Force, March 2004.
18. 3GPP. *Access to the 3GPP Evolved Packet Core (EPC) via non-3GPP access networks; Stage 3*. TS 24.302, 3rd Generation Partnership Project (3GPP), September 2011.
19. 3GPP. *Mobility management based on Dual-Stack Mobile IPv6; Stage3*. TS 24.303, 3rd Generation Partnership Project (3GPP), September 2011.
20. 3GPP. *Mobility management based on Mobile IPv4; User Equipment (UE) – foreign agent interface; Stage 3*. TS 24.304, 3rd Generation Partnership Project (3GPP), April 2011.
21. Hinden, R. and Deering, S. *IP Version 6 Addressing Architecture*. RFC 4291, Internet Engineering Task Force, February 2006.
22. Thomson, S., Narten, T., and Jinmei, T. *IPv6 Stateless Address Autoconfiguration*. RFC 4862, Internet Engineering Task Force, September 2007.
23. Thaler, D., Talwar, M., and Patel, C. *Neighbor Discovery Proxies (ND Proxy)*. RFC 4389, Internet Engineering Task Force, April 2006.
24. Narten, T., Nordmark, E., Simpson, W., and Soliman, H. *Neighbor Discovery for IP version 6 (IPv6)*. RFC 4861, Internet Engineering Task Force, September 2007.
25. 3GPP. *Interworking between the Public Land Mobile Network (PLMN) supporting packet based services and Packet Data Networks (PDN)*. TS 29.061, 3rd Generation Partnership Project (3GPP), December 2011.
26. Droms, R., Bound, J., Volz, B., Lemon, T., Perkins, C., and Carney, M. *Dynamic Host Configuration Protocol for IPv6 (DHCPv6)*. RFC 3315, Internet Engineering Task Force, July 2003.
27. Troan, O. and Droms, R. *IPv6 Prefix Options for Dynamic Host Configuration Protocol (DHCP) version 6*. RFC 3633, Internet Engineering Task Force, December 2003.

28. Zhou, X., Korhonen, J., Williams, C., Gundavelli, S., and Bernardos, C. J. *Prefix Delegation for Proxy Mobile IPv6*. Internet-Draft draft-ietf-netext-pd-pmip-05, Internet Engineering Task Force, October 2012. Work in progress.

29. Calhoun, P., Loughney, J., Guttman, E., Zorn, G., and Arkko, J. *Diameter Base Protocol*. RFC 3588, Internet Engineering Task Force, September 2003.

30. Rosen, E., Viswanathan, A., and Callon, R. *Multiprotocol Label Switching Architecture*. RFC 3031, Internet Engineering Task Force, January 2001.

31. Clercq, J. De, Ooms, D., Prevost, S., and Faucheur, F. Le. *Connecting IPv6 Islands over IPv4 MPLS Using IPv6 Provider Edge Routers (6PE)*. RFC 4798, Internet Engineering Task Force, February 2007.

32. Asati, R., Manral, V., Papneja, R., and Pignataro, C. *Updates to LDP for IPv6*. Internet-Draft draft-ietf-mpls-ldp-ipv6-07, Internet Engineering Task Force, June 2012. Work in progress.

33. Jankiewicz, E., Loughney, J., and Narten, T. *IPv6 Node Requirements*. RFC 6434, Internet Engineering Task Force, December 2011.

34. 3GPP. *AT command set for User Equipment (UE)*. TS 27.007, 3rd Generation Partnership Project (3GPP), December 2011.

35. S. Varada, Haskins, D., and Allen, E. *IP Version 6 over PPP*. RFC 5072, Internet Engineering Task Force, September 2007.

36. USB,. *Universal Serial Bus Communications Class Subclass Specification for Mobile Broadband Interface Model*. CDC MBIM Revision 1.0, USB Implementers Forum, Inc. (USB-IF), November 2011.

37. USB,. *Universal Serial Bus Communications Class Subclass Specification for Ethernet Control Model Devices*. CDC ECM Revision 1.2, USB Implementers Forum, Inc. (USB-IF), February 2007.

38. 3GPP. *Mobile Station (MS) – Serving GPRS Support Node (SGSN); Subnetwork Dependent Convergence Protocol (SNDCP)*. TS 44.065, 3rd Generation Partnership Project (3GPP), December 2009.

39. 3GPP. *Packet Data Convergence Protocol (PDCP) specification*. TS 25.323, 3rd Generation Partnership Project (3GPP), October 2010.

40. 3GPP. *Evolved Universal Terrestrial Radio Access (E-UTRA); Packet Data Convergence Protocol (PDCP) specification*. TS 36.323, 3rd Generation Partnership Project (3GPP), January 2010.

41. 3GPP. *GSM/EDGE Radio Access Network (GERAN) overall description; Stage 2*. TS 43.051, 3rd Generation Partnership Project (3GPP), December 2009.

42. 3GPP. *UTRAN Iu interface data transport and transport signalling*. TS 25.414, 3rd Generation Partnership Project (3GPP), April 2011.

43. Jacobson, V. *Compressing TCP/IP Headers for Low-Speed Serial Links*. RFC 1144, Internet Engineering Task Force, February 1990.

44. Degermark, M., Nordgren, B., and Pink, S. *IP Header Compression*. RFC 2507, Internet Engineering Task Force, February 1999.

45. GSMA. *IMS Profile for Voice and SMS. PRD IR.92* 5.0, GSM Association (GSMA), December 2011.

46. Jonsson, L-E., Pelletier, G., and Sandlund, K. *The RObust Header Compression (ROHC) Framework*. RFC 4995, Internet Engineering Task Force, July 2007.

47. Bormann, C., Burmeister, C., Degermark, M., Fukushima, H., Hannu, H., Jonsson, L-E., Hakenberg, R., Koren, T., Le, K., Liu, Z., Martensson, A., Miyazaki, A., Svanbro, K., Wiebke, T., Yoshimura, T., and Zheng, H. *RObust Header Compression (ROHC): Framework and four profiles: RTP, UDP, ESP, and uncompressed*. RFC 3095, Internet Engineering Task Force, July 2001.

48. Jonsson, L-E., Sandlund, K., Pelletier, G., and Kremer, P. *RObust Header Compression (ROHC): Corrections and Clarifications to RFC 3095*. RFC 4815, Internet Engineering Task Force, February 2007.

49. 3GPP. *Multimedia Broadcast/Multicast Service (MBMS); Architecture and functional description*. TS 23.246, 3rd Generation Partnership Project (3GPP), December 2011.

50. Holbrook, H. and Cain, B. *Source-Specific Multicast for IP*. RFC 4607, Internet Engineering Task Force, August 2006.

51. Holbrook, H., Cain, B., and Haberman, B. *Using Internet Group Management Protocol Version 3 (IGMPv3) and Multicast Listener Discovery Protocol Version 2 (MLDv2) for Source-Specific Multicast*. RFC 4604, Internet Engineering Task Force, August 2006.

52. Vida, R. and Costa, L. *Multicast Listener Discovery Version 2 (MLDv2) for IPv6*. RFC 3810, Internet Engineering Task Force, June 2004.

53. 3GPP. *Mobile Application Part (MAP) specification*. TS 29.002, 3rd Generation Partnership Project (3GPP), March 2012.

54. Droms, R. *Stateless Dynamic Host Configuration Protocol (DHCP) Service for IPv6*. RFC 3736, Internet Engineering Task Force, April 2004.

55. Korhonen, J., Savolainen, T., Krishnan, S., and Troan, O. *Prefix Exclude Option for DHCPv6-based Prefix Delegation*. RFC 6603, Internet Engineering Task Force, May 2012.

56. Coltun, R., Ferguson, D., Moy, J., and Lindem, A. *OSPF for IPv6*. RFC 5340, Internet Engineering Task Force, July 2008.

57. Bates, T., Chandra, R., Katz, D., and Rekhter, Y. *Multiprotocol Extensions for BGP-4*. RFC 4760, Internet Engineering Task Force, January 2007.

58. Mockapetris, P. V. *Domain names – concepts and facilities*. RFC 1034, Internet Engineering Task Force, November 1987.

59. 3GPP. *Policy and charging control architecture*. TS 23.203, 3rd Generation Partnership Project (3GPP), March 2012.

60. Rigney, C., Willens, S., Rubens, A., and Simpson, W. *Remote Authentication Dial In User Service (RADIUS)*. RFC 2865, Internet Engineering Task Force, June 2000.

61. Rigney, C. *RADIUS Accounting*. RFC 2866, Internet Engineering Task Force, June 2000.

62. Aboba, B., Zorn, G., and Mitton, D. *RADIUS and IPv6*. RFC 3162, Internet Engineering Task Force, August 2001.

63. Salowey, J. and Droms, R. *RADIUS Delegated-IPv6-Prefix Attribute*. RFC 4818, Internet Engineering Task Force, April 2007.

64. Dec, W., Sarikaya, B., Zorn, G., Miles, D., and Lourdelet, B. *RADIUS attributes for IPv6 Access Networks*. Internet-Draft draft-ietf-radext-ipv6-access-15, Internet Engineering Task Force, January 2013. Work in progress.

65. Savola, P. *MTU and Fragmentation Issues with In-the-Network Tunneling*. RFC 4459, Internet Engineering Task Force, April 2006.

66. Postel, J. *User Datagram Protocol*. RFC 0768, Internet Engineering Task Force, August 1980.

67. Postel, J. *Transmission Control Protocol*. RFC 0793, Internet Engineering Task Force, September 1981.

68. Postel, J. *TCP maximum segment size and related topics*. RFC 0879, Internet Engineering Task Force, November 1983.

69. McCann, J., Deering, S., and Mogul, J. *Path MTU Discovery for IP version 6*. RFC 1981, Internet Engineering Task Force, August 1996.

70. Jaeggli, J., Colitti, L., Kumari, W., Vyncke, E., Kaeo, M., and Taylor, T. *Why Operators Filter Fragments and What It Implies*. Internet-Draft draft-taylor-v6ops-fragdrop-00, Internet Engineering Task Force, October 2012. Work in progress.

71. deBoer, M. and Bosma, J. Discovering Path MTU black holes on the Internet using RIPE Atlas. Masters thesis, University of Amsterdam, MSc. Systems & Network Engineering, July 2012.

72. Mathis, M. and Heffner, J. *Packetization Layer Path MTU Discovery*. RFC 4821, Internet Engineering Task Force, March 2007.

73. Kent, S. and Seo, K. *Security Architecture for the Internet Protocol*. RFC 4301, Internet Engineering Task Force, December 2005.

74. 3GPP. *3G security; Network Domain Security (NDS); IP network layer security*. TS 33.210, 3rd Generation Partnership Project (3GPP), June 2010.

75. Frankel, S., Glenn, R., and Kelly, S. *The AES-CBC Cipher Algorithm and Its Use with IPsec*. RFC 3602, Internet Engineering Task Force, September 2003.

76. Krawczyk, H., Bellare, M., and Canetti, R. *HMAC: Keyed-Hashing for Message Authentication*. RFC 2104, Internet Engineering Task Force, February 1997.

77. Kent, S. *IP Encapsulating Security Payload (ESP)*. RFC 4303, Internet Engineering Task Force, December 2005.

78. 3GPP. *Policy and Charging Control (PCC) over Gx/Sd reference point*. TS 29.212, 3rd Generation Partnership Project (3GPP), March 2012.

79. 3GPP. *Policy and charging control signalling flows and Quality of Service (QoS) parameter mapping*. TS 29.213, 3rd Generation Partnership Project (3GPP), December 2011.

80. 3GPP. *Policy and charging control over Rx reference point*. TS 29.214, 3rd Generation Partnership Project (3GPP), December 2011.

81. 3GPP. *Policy and Charging Control (PCC) over S9 reference point; Stage 3*. TS 29.215, 3rd Generation Partnership Project (3GPP), March 2012.

82. 3GPP. *Telecommunication management; Charging management; Charging architecture and principles*. TS 32.240, 3rd Generation Partnership Project (3GPP), April 2011.

83. 3GPP. *Telecommunication management; Charging management; Packet Switched (PS) domain charging*. TS 32.251, 3rd Generation Partnership Project (3GPP), December 2011.

84. 3GPP. *Customised Applications for Mobile network Enhanced Logic (CAMEL) Phase 4; Stage 2*. TS 23.078, 3rd Generation Partnership Project (3GPP), December 2011.

85. 3GPP. *Customised Applications for Mobile network Enhanced Logic (CAMEL) Phase X; CAMEL Application Part (CAP) specification*. TS 29.078, 3rd Generation Partnership Project (3GPP), June 2010.

86. 3GPP. *Telecommunication management; Charging management; Charging Data Record (CDR) parameter description*. TS 32.298, 3rd Generation Partnership Project (3GPP), March 2012.

87. GSMA,. *TAP3 Format Specification*. PRD TD.57 30.3, GSM Association (GSMA), May 2012.

88. Farinacci, D., Li, T., Hanks, S., Meyer, D., and Traina, P. *Generic Routing Encapsulation (GRE)*. RFC 2784, Internet Engineering Task Force, March 2000.

89. Lau, J., Townsley, M., and Goyret, I. *Layer Two Tunneling Protocol – Version 3 (L2TPv3)*. RFC 3931, Internet Engineering Task Force, March 2005.

90. Perkins, C. *IP Encapsulation within IP*. RFC 2003, Internet Engineering Task Force, October 1996.

91. Chiba, M., Dommety, G., Eklund, M., Mitton, D., and Aboba, B. *Dynamic Authorization Extensions to Remote Authentication Dial In User Service (RADIUS)*. RFC 5176, Internet Engineering Task Force, January 2008.

92. Zorn, G. *Microsoft Vendor-specific RADIUS Attributes*. RFC 2548, Internet Engineering Task Force, March 1999.

93. Calhoun, P., Zorn, G., Spence, D., and Mitton, D. *Diameter Network Access Server Application*. RFC 4005, Internet Engineering Task Force, August 2005.

94. Alexander, S. and Droms, R. *DHCP Options and BOOTP Vendor Extensions*. RFC 2132, Internet Engineering Task Force, March 1997.

95. Schulzrinne, H. *Dynamic Host Configuration Protocol (DHCP-for-IPv4) Option for Session Initiation Protocol (SIP) Servers*. RFC 3361, Internet Engineering Task Force, August 2002.

96. Droms, R. *DNS Configuration options for Dynamic Host Configuration Protocol for IPv6 (DHCPv6)*. RFC 3646, Internet Engineering Task Force, December 2003.

97. Schulzrinne, H. and Volz, B. *Dynamic Host Configuration Protocol (DHCPv6) Options for Session Initiation Protocol (SIP) Servers*. RFC 3319, Internet Engineering Task Force, July 2003.

98. GSMA,. *Inter-Service Provider IP Backbone Guidelines*. PRD IR.34 5.0, GSM Association (GSMA), December 2010.

99. GSMA,. *GSMA PRD IR.88 'LTE Roaming Guidelines'*. PRD IR.88 6.0, GSM Association (GSMA), August 2011.

100. 3GPP. *Technical realization of Operator Determined Barring (ODB)*. TS 23.015, 3rd Generation Partnership Project (3GPP), December 2009.

101. 3GPP. *Telecommunication management; Charging management; Charging data description for the Packet Switched (PS) domain*. TS 32.215, 3rd Generation Partnership Project (3GPP), June 2005.

102. Perkins, C., Johnson, D., and Arkko, J. *Mobility Support in IPv6*. RFC 6275, Internet Engineering Task Force, July 2011.

103. 3GPP. *3GPP system to Wireless Local Area Network (WLAN) interworking; System description*. TS 23.234, 3rd Generation Partnership Project (3GPP), December 2009.

104. 3GPP. *Mobility between 3GPP Wireless Local Area Network (WLAN) interworking (I-WLAN) and 3GPP systems; General Packet Radio System (GPRS) and 3GPP I-WLAN aspects; Stage 3*. TS 24.327, 3rd Generation Partnership Project (3GPP), June 2010.

105. 3GPP. *3GPP system to Wireless Local Area Network (WLAN) interworking; WLAN User Equipment (WLAN UE) to network protocols; Stage 3*. TS 24.234, 3rd Generation Partnership Project (3GPP), September 2011.

106. 3GPP. *Study on S2a Mobility based on GTP & WLAN access to EPC*. TR 23.852, 3rd Generation Partnership Project (3GPP), December 2011.

107. Singh, H., Beebee, W., Donley, C., and Stark, B. *Basic Requirements for IPv6 Customer Edge Routers*. Internet-Draft draft-ietf-v6ops-6204bis-12, Internet Engineering Task Force, October 2012. Work in progress.

108. Krishnan, S. *Reserved IPv6 Interface Identifiers*. RFC 5453, Internet Engineering Task Force, February 2009.

109. Narten, T., Draves, R., and Krishnan, S. *Privacy Extensions for Stateless Address Autoconfiguration in IPv6*. RFC 4941, Internet Engineering Task Force, September 2007.

110. Korhonen, J., Bournelle, J., Tschofenig, H., Perkins, C., and Chowdhury, K. *Diameter Mobile IPv6: Support for Network Access Server to Diameter Server Interaction*. RFC 5447, Internet Engineering Task Force, February 2009.

111. 3GPP. *Evolved Packet System (EPS); Mobility Management Entity (MME) and Serving GPRS Support Node (SGSN) related interfaces based on Diameter protocol*. TS 29.272, 3rd Generation Partnership Project (3GPP), March 2012.

112. 3GPP. *Evolved Packet System (EPS); 3GPP EPS AAA interfaces*. TS 29.273, 3rd Generation Partnership Project (3GPP), March 2012.

113. 3GPP. *Numbering, addressing and identification*. TS 23.003, 3rd Generation Partnership Project (3GPP), March 2012.

114. Sarikaya, B., Xia, F., and Lemon, T. *DHCPv6 Prefix Delegation in Long-Term Evolution (LTE) Networks*. RFC 6653, Internet Engineering Task Force, July 2012.

115. Kaufman, C., Hoffman, P., Nir, Y., and Eronen, P. *Internet Key Exchange Protocol Version 2 (IKEv2)*. RFC 5996, Internet Engineering Task Force, September 2010.

116. Giaretta, G., Kempf, J., and Devarapalli, V. *Mobile IPv6 Bootstrapping in Split Scenario*. RFC 5026, Internet Engineering Task Force, October 2007.

117. Shen, N. and Zinin, A. *Point-to-Point Operation over LAN in Link State Routing Protocols*. RFC 5309, Internet Engineering Task Force, October 2008.

118. Troan, O. and Volz, B. *Issues with multiple stateful DHCPv6 options*. Internet-Draft draft-ietf-dhc-dhcpv6-stateful-issues-03, Internet Engineering Task Force, November 2012. Work in progress.

119. Melia, T. and Mghazli, Y. El. *DHCP option to transport Protocol Configuration Options*. Internet-Draft draft-melia-dhc-pco-00, Internet Engineering Task Force, February 2009. Work in progress.

120. Zhu, C. and Zhou, X. *The interworking between IKEv2 and PMIPv6*. Internet-Draft draft-zhu-netext-ikev2-interworking-pmip-00, Internet Engineering Task Force, April 2011. Work in progress.

121. 3GPP. *Mobile IPv6 vendor specific option format and usage within 3GPP*. TS 29.282, 3rd Generation Partnership Project (3GPP), December 2009.

122. Rosenberg, J., Schulzrinne, H., Camarillo, G., Johnston, A., Peterson, J., Sparks, R., Handley, M., and Schooler, E. *SIP: Session Initiation Protocol*. RFC 3261, Internet Engineering Task Force, June 2002.

123. Stewart, R. *Stream Control Transmission Protocol*. RFC 4960, Internet Engineering Task Force, September 2007.

124. Ong, L., Rytina, I., Garcia, M., Schwarzbauer, H., Coene, L., Lin, H., Juhasz, I., Holdrege, M., and Sharp, C. *Framework Architecture for Signaling Transport*. RFC 2719, Internet Engineering Task Force, October 1999.

125. Loughney, J., Sidebottom, G., Coene, L., Verwimp, G., Keller, J., and Bidulock, B. *Signalling Connection Control Part User Adaptation Layer (SUA)*. RFC 3868, Internet Engineering Task Force, October 2004.

126. Camarillo, G., Roach, A. B., Peterson, J., and Ong, L. *Integrated Services Digital Network (ISDN) User Part (ISUP) to Session Initiation Protocol (SIP) Mapping*. RFC 3398, Internet Engineering Task Force, December 2002.

127. Vixie, P. *Extension Mechanisms for DNS (EDNS0)*. RFC 2671, Internet Engineering Task Force, August 1999.

128. Deering, S., Fenner, W., and Haberman, B. *Multicast Listener Discovery (MLD) for IPv6*. RFC 2710, Internet Engineering Task Force, October 1999.

129. Partridge, C. and Jackson, A. *IPv6 Router Alert Option*. RFC 2711, Internet Engineering Task Force, October 1999.

130. Haberman, B. *Source Address Selection for the Multicast Listener Discovery (MLD) Protocol*. RFC 3590, Internet Engineering Task Force, September 2003.

131. Thomson, S., Huitema, C., Ksinant, V., and Souissi, M. *DNS Extensions to Support IP Version 6*. RFC 3596, Internet Engineering Task Force, October 2003.

132. Berners-Lee, T., Fielding, R., and Masinter, L. *Uniform Resource Identifier (URI): Generic Syntax*. RFC 3986, Internet Engineering Task Force, January 2005.

133. Draves, R. and Thaler, D. *Default Router Preferences and More-Specific Routes*. RFC 4191, Internet Engineering Task Force, November 2005.

134. Hinden, R. and Haberman, B. *Unique Local IPv6 Unicast Addresses*. RFC 4193, Internet Engineering Task Force, October 2005.

135. Conta, A., Deering, S., and Gupta, M. *Internet Control Message Protocol (ICMPv6) for the Internet Protocol Version 6 (IPv6) Specification*. RFC 4443, Internet Engineering Task Force, March 2006.

136. Abley, J., Savola, P., and Neville-Neil, G. *Deprecation of Type 0 Routing Headers in IPv6*. RFC 5095, Internet Engineering Task Force, December 2007.

137. Krishnan, S. *Handling of Overlapping IPv6 Fragments*. RFC 5722, Internet Engineering Task Force, December 2009.

138. Singh, H., Beebee, W., and Nordmark, E. *IPv6 Subnet Model: The Relationship between Links and Subnet Prefixes*. RFC 5942, Internet Engineering Task Force, July 2010.

139. Kawamura, S. and Kawashima, M. *A Recommendation for IPv6 Address Text Representation*. RFC 5952, Internet Engineering Task Force, August 2010.

140. Jeong, J., Park, S., Beloeil, L., and Madanapalli, S. *IPv6 Router Advertisement Options for DNS Configuration*. RFC 6106, Internet Engineering Task Force, November 2010.

141. Wing, D. and Yourtchenko, A. *Happy Eyeballs: Success with Dual-Stack Hosts*. RFC 6555, Internet Engineering Task Force, April 2012.

142. Thaler, D., Draves, R., Matsumoto, A., and Chown, T. *Default Address Selection for Internet Protocol version 6 (IPv6)*. RFC 6724, Internet Engineering Task Force, September 2012.

143. Crawford, M. *Transmission of IPv6 Packets over Ethernet Networks*. RFC 2464, Internet Engineering Task Force, December 1998.

144. Borman, D., Deering, S., and Hinden, R. *IPv6 Jumbograms*. RFC 2675, Internet Engineering Task Force, August 1999.

145. Conta, A. *Extensions to IPv6 Neighbor Discovery for Inverse Discovery Specification*. RFC 3122, Internet Engineering Task Force, June 2001.

146. Arkko, J., Kempf, J., Zill, B., and Nikander, P. *SEcure Neighbor Discovery (SEND)*. RFC 3971, Internet Engineering Task Force, March 2005.

147. Aura, T. *Cryptographically Generated Addresses (CGA)*. RFC 3972, Internet Engineering Task Force, March 2005.

148. Venaas, S., Chown, T., and Volz, B. *Information Refresh Time Option for Dynamic Host Configuration Protocol for IPv6 (DHCPv6)*. RFC 4242, Internet Engineering Task Force, November 2005.

149. Moore, N. *Optimistic Duplicate Address Detection (DAD) for IPv6*. RFC 4429, Internet Engineering Task Force, April 2006.

150. Bonica, R., Gan, D., Tappan, D., and Pignataro, C. *Extended ICMP to Support Multi-Part Messages*. RFC 4884, Internet Engineering Task Force, April 2007.

151. Haberman, B. and Hinden, R. *IPv6 Router Advertisement Flags Option*. RFC 5175, Internet Engineering Task Force, March 2008.

152. Liu, H., Cao, W., and Asaeda, H. *Lightweight Internet Group Management Protocol Version 3 (IGMPv3) and Multicast Listener Discovery Version 2 (MLDv2) Protocols*. RFC 5790, Internet Engineering Task Force, February 2010.

153. Atlas, A., Bonica, R., Pignataro, C., Shen, N., and Rivers, JR. *Extending ICMP for Interface and Next-Hop Identification*. RFC 5837, Internet Engineering Task Force, April 2010.

154. Gundavelli, S., Townsley, M., Troan, O., and Dec, W. *Address Mapping of IPv6 Multicast Packets on Ethernet*. RFC 6085, Internet Engineering Task Force, January 2011.

155. Amante, S., Carpenter, B., Jiang, S., and Rajahalme, J. *IPv6 Flow Label Specification*. RFC 6437, Internet Engineering Task Force, November 2011.

156. Savolainen, T., Kato, J., and Lemon, T. *Improved Recursive DNS Server Selection for Multi-Interfaced Nodes*. RFC 6731, Internet Engineering Task Force, December 2012.

157. Arkko, J., Kuijpers, G., Soliman, H., Loughney, J., and Wiljakka, J. *Internet Protocol Version 6 (IPv6) for Some Second and Third Generation Cellular Hosts*. RFC 3316, Internet Engineering Task Force, April 2003.

158. Gilligan, R., Thomson, S., Bound, J., McCann, J., and Stevens, W. *Basic Socket Interface Extensions for IPv6*. RFC 3493, Internet Engineering Task Force, February 2003.

159. Stevens, W., Thomas, M., Nordmark, E., and Jinmei, T. *Advanced Sockets Application Program Interface (API) for IPv6*. RFC 3542, Internet Engineering Task Force, May 2003.

160. Thaler, D., Fenner, B., and Quinn, B. *Socket Interface Extensions for Multicast Source Filters*. RFC 3678, Internet Engineering Task Force, January 2004.

161. Nordmark, E., Chakrabarti, S., and Laganier, J. *IPv6 Socket API for Source Address Selection*. RFC 5014, Internet Engineering Task Force, September 2007.

162. Singh, H., Beebee, W., Donley, C., Stark, B., and Troan, O. *Basic Requirements for IPv6 Customer Edge Routers*. RFC 6204, Internet Engineering Task Force, April 2011.

163. Gashinsky, I., Jaeggli, J., and Kumari, W. *Operational Neighbor Discovery Problems*. RFC 6583, Internet Engineering Task Force, March 2012.

164. OMA. *Provisioning Content*. TS 1.1, Open Mobile Alliance (OMA), July 2009.

165. Nokia. *Nokia Views on IPv6 Transition*. White Paper 2.5, Nokia, December 2011.

166. Korhonen, J., Soininen, J., Patil, B., Savolainen, T., Bajko, G., and Iisakkila, K. *IPv6 in 3rd Generation Partnership Project (3GPP) Evolved Packet System (EPS)*. RFC 6459, Internet Engineering Task Force, January 2012.

167. Krishnan, S., Anipko, D., and Thaler, D. *Packet loss resiliency for Router Solicitations*. Internet-Draft draft-krishnan-6man-resilient-rs-01, Internet Engineering Task Force, July 2012. Work in progress.

168. Wasserman, M. *Recommendations for IPv6 in Third Generation Partnership Project (3GPP) Standards*. RFC 3314, Internet Engineering Task Force, September 2002.

169. Loughney, J. *IPv6 Node Requirements*. RFC 4294, Internet Engineering Task Force, April 2006.

170. Kent, S. *IP Authentication Header*. RFC 4302, Internet Engineering Task Force, December 2005.

171. Chittimaneni, K., Kaeo, M., and Vyncke, E. *Operational Security Considerations for IPv6 Networks*. Internet-Draft draft-ietf-opsec-v6-01, Internet Engineering Task Force, November 2012. Work in progress.

172. Gont, F. *A method for Generating Stable Privacy-Enhanced Addresses with IPv6 Stateless Address Auto-configuration (SLAAC)*. Internet-Draft draft-ietf-6man-stable-privacy-addresses-03, Internet Engineering Task Force, January 2013. Work in progress.

173. Gont, F. *Network Reconnaissance in IPv6 Networks*. Internet-Draft draft-ietf-opsec-ipv6-host-scanning, Internet Engineering Task Force, December 2012. Work in progress.

174. Yourtchenko, A., Asadullah, S., and Pisica, M. *Human-safe IPv6: Cryptographic transformation of host-names as a base for secure and manageable addressing*. Internet-Draft draft-yourtchenko-humansafe-ipv6-00, Internet Engineering Task Force, February 2012. Work in progress.

175. Nikander, P., Kempf, J., and Nordmark, E. *IPv6 Neighbor Discovery (ND) Trust Models and Threats*. RFC 3756, Internet Engineering Task Force, May 2004.

176. Gont, F. *Neighbor Discovery Shield (ND-Shield): Protecting against NeighborDiscovery Attacks*. Internet-Draft draft-gont-opsec-ipv6-nd-shield-00, Internet Engineering Task Force, June 2012. Work in progress.

177. Chown, T. and Venaas, S. *Rogue IPv6 Router Advertisement Problem Statement*. RFC 6104, Internet Engineering Task Force, February 2011.

178. Lynn, C., Kent, S., and Seo, K. *X.509 Extensions for IP Addresses and AS Identifiers*. RFC 3779, Internet Engineering Task Force, June 2004.

179. Gagliano, R., Krishnan, S., and Kukec, A. *Certificate Profile and Certificate Management for SEcure Neighbor Discovery (SEND)*. RFC 6494, Internet Engineering Task Force, February 2012.

180. Gagliano, R., Krishnan, S., and Kukec, A. *Subject Key Identifier (SKI) SEcure Neighbor Discovery (SEND) Name Type Fields*. RFC 6495, Internet Engineering Task Force, February 2012.

181. Lepinski, M. and Kent, S. *An Infrastructure to Support Secure Internet Routing*. RFC 6480, Internet Engineering Task Force, February 2012.

182. Huston, G., Michaelson, G., and Loomans, R. *A Profile for X.509 PKIX Resource Certificates*. RFC 6487, Internet Engineering Task Force, February 2012.

183. Wu, J., Bi, J., Bagnulo, M., Baker, F., and Vogt, C. *Source Address Validation Improvement Framework*. Internet-Draft draft-ietf-savi-framework-06, Internet Engineering Task Force, January 2012. Work in progress.

184. Nordmark, E., Bagnulo, M., and Levy-Abegnoli, E. *FCFS SAVI: First-Come, First-Served Source Address Validation Improvement for Locally Assigned IPv6 Addresses*. RFC 6620, Internet Engineering Task Force, May 2012.

185. Levy-Abegnoli, E., deVelde, G. Van, Popoviciu, C., and Mohacsi, J. *IPv6 Router Advertisement Guard*. RFC 6105, Internet Engineering Task Force, February 2011.

186. Gont, F. *Security Implications of IPv6 options of Type 10xxxxxx*. Internet-Draft draft-gont-6man-ipv6-smurf-amplifier-02, Internet Engineering Task Force, January 2013. Work in progress.

187. Biondi, P. and Ebalard, A. *IPv6 Routing Header Security*. Pdf, CanSecWest Security Conference 2007, April 2007.

188. Gont, F. and Manral, V. *Security and Interoperability Implications of Oversized IPv6 Header Chains*. Internet-Draft draft-ietf-6man-oversized-header-chain-02, Internet Engineering Task Force, November 2012. Work in progress.

189. Gont, F. *Security Implications of the Use of IPv6 Extension Headers with IPv6 Neighbor Discovery*. Internet-Draft draft-gont-6man-nd-extension-headers-03, Internet Engineering Task Force, June 2012. Work in progress.
190. Bagnulo, M., Matthews, P., and van Beijnum, I. *Stateful NAT64: Network Address and Protocol Translation from IPv6 Clients to IPv4 Servers*. RFC 6146, Internet Engineering Task Force, April 2011.
191. Gont, F., and Liu, W. *Security Implications of IPv6 on IPv4 Networks*. Internet-Draft draft-ietf-opsec-ipv6-implications-on-ipv4-nets-02, Internet Engineering Task Force, December 2012. Work in progress.
192. Giobbi, R. *Bypassing firewalls with IPv6 tunnels*. html, CERT, April 2009.
193. Davies, E. and Mohacsi, J. *Recommendations for Filtering ICMPv6 Messages in Firewalls*. RFC 4890, Internet Engineering Task Force, May 2007.
194. Gont, F. *Processing of IPv6 "atomic" fragments*. Internet-Draft draft-ietf-6man-ipv6-atomic-fragments-03, Internet Engineering Task Force, December 2012. Work in progress.
195. Gont, F. *Security Implications of Predictable Fragment Identification Values*. Internet-Draft draft-gont-6man-predictable-fragment-id-03, Internet Engineering Task Force, January 2013. Work in progress.

5

IPv6 Transition Mechanisms for 3GPP Networks

This chapter starts with the motivation for the Internet Protocol version 6 (IPv6) transition and then follows with a high level overview of the technical solution approaches and their benefits and drawbacks. With the motivation and set of solution approaches described, we will list the detailed toolbox and scenarios for the IPv6 transition for 3rd Generation Partnership Project (3GPP) networks. We conclude the chapter by listing the impacts that IPv6 transition has on the 3GPP architecture and how application writers, phone vendors, and network operators can plan the transition.

5.1 Motivation for Transition Mechanisms

The main motivation to move to IPv6 is the Internet Protocol version 4 (IPv4) address shortage and the implications it causes: need for address translation, overlapping private address spaces, and so forth. All of those result in increased costs for operating IPv4 networks, and they hinder innovation. The IPv6 removes the address shortage, and the side-effects it has caused, thus promising operational cost savings in the long run. All the other reasons for deploying IPv6 are secondary, and often also result from the increased address space. A large set of tools and techniques have had to be developed to make the long transition from the IPv4-only world to the world dominated by IPv6 smooth and seamless – for the users of the network. The transition will have succeeded if the world one day is predominantly (fully is too far in the future to consider) using IPv6, and the average user has not noticed a thing. Definitely for the network administrators, software developers, and hardware vendors the change is not as transparent, and hard work is required. For the rest of this chapter we look at how this journey is planned to be made in 3GPP networks.

Early Days

In the early days of IPv6 the thinking was that IPv4 and IPv6 would live in parallel until all nodes would be IPv6 enabled and then IPv4 could be gradually disbanded. This

Deploying IPv6 in 3GPP Networks: Evolving Mobile Broadband from 2G to LTE and Beyond, First Edition.
Jouni Korhonen, Teemu Savolainen and Jonne Soininen.
© 2013 John Wiley & Sons, Ltd. Published 2013 by John Wiley & Sons, Ltd.

approach is the dual-stack transition, as was documented in RFC 1933 [1] back in 1996 and then later updated in RFC 4213 [2]. RFC 1933 also defined the means for setting up manual and automatic IPv6 over IPv4 tunnels in order to allow initial deployments of IPv6 pockets, or clouds, which were to be connected over IPv4 infrastructure.

As often happens with great plans, the world did not really start deploying IPv6 on a large scale until IPv4 address pools were effectively dry. Therefore, the idea of getting everybody to IPv6 before IPv4 addresses ran out did not succeed. Of course many IPv6 deployments were made, and backbone networks were updated to a large extent, but the gap to the target was huge. Nevertheless, the dual-stack transition mechanism is still the most favored tool for transition in 3GPP, as we will see in Section 5.4.

For a long time the focus on transition mechanisms were how to get IPv6 connectivity in the face of IPv4 infrastructures. This resulted in standardization of many IPv6-over-IPv4 tunneling technologies, and even more proposals about how the tunneling could be done. Besides manually and statically configured tunnels [2], maybe the most significant tools for automatically tunneling of IPv6 over IPv4 were, and to some extent still are, *Connection of IPv6 domains via IPv4 clouds (6to4)* [3], *IPv6 over IPv4 without explicit tunnels (6over4)* [4], *IPv6 Rapid Deployment on IPv4 infrastructures (6RD)* [5], *Intra-Site Automatic Tunnel Addressing Protocol (ISATAP)* [6], and *Tunneling IPv6 over UDP through NATs (Teredo)* [7]. We want to emphasize that there really are large numbers of other mechanisms, but these are the most famous.

Advances in Tool Development

As time passed and IPv4 addresses grew short, but no IPv6-only killer application had emerged, it started to be clear that there is not that much need for connecting hosts to IPv6 clouds via tunnels. Furthermore, it was widely understood that tunneling IPv6 over IPv4 does nothing to alleviate the IPv4 address shortage. This situation made people interested in actually replacing IPv4 with IPv6 in the access networks. Since the world is still mostly IPv4, it also meant that a need arose for connecting to the IPv4 Internet over IPv6 networks.

A harbinger of the coming IPv4-over-IPv6 transition tools work was perhaps the *Dual Stack Transition Mechanism (DSTM)* proposal by Laurent Toutain, Hossam Afifi, and Jim Bound in 1999 (documented in an expired Internet-Draft 'draft-toutain-ngtrans-dstm-00.txt'). These gentlemen saw the need for IPv4 connectivity over IPv6-only access networks, and they proposed a tool where a host could ask for a temporary allocation of an IPv4 address, and then tunnel the IPv4 packets over IPv6. Almost a decade after the introduction of DSTM, the work really started flying, and after lots of discussions the Internet Engineering Task Force (IETF) ended up specifying *Dual-Stack Lite* [8] and the IETF has also additional tools in the pipeline.

Maybe No Tunneling for Hosts After All

At the start of the IPv6 transition, as of the writing of this book, it began to seem that perhaps it is not necessarily a good idea to deploy *host-based* transition solutions that utilize tunneling after all. In particular, for 3GPP accesses tunneling does not seem to be the right way to go, due to reasons described in depth in Section 5.2.2. That said, *network-based* tunneling solutions do seem to be playing an important role: the tunneling over single address family infrastructures can be performed, for example by the Consumer Premises Equipment (CPE) or by an access concentrator (such as Packet Data Network Gateway – PGW).

For more or less the same reasons there were for developing tunneling-based transition tools, protocol translation based tools have also been developed. The toolset for translation is much smaller, and essentially is about IPv4/IPv6 Network Address Translation (NAT64) (described in Section 5.3.3). The protocol translation between incompatible IPv4 and IPv6 causes issues described in Section 5.2.1, and hence has been disfavored by many. However, if the issues are accepted or worked around, the translation can be a powerful tool. And because of that, the IETF is currently spending a significant number of cycles in perfecting the protocol translation technologies (or to put it another way, to mitigate and minimize the issues).

5.1.1   Phasing the Transition

The original idea of dual-stack transition has been watered down by the industry not actively deploying IPv6 early enough to avoid the need for more complex tools – not an uncommon problem for any resource exhausting situation that mankind faces. Now, as the limited resource of IPv4 addresses is really in short supply, work to deploy IPv6 has begun. In part to solve the chicken and egg problem, several major content providers have gone to the trouble of making their content available in IPv6. These deployments got publicity due to events such as *World IPv6 Day* on 8th June in 2011, see http://www.worldipv6day.org, and *World IPv6 Launch* on 6th June in 2012, see http://www.worldipv6launch.org. Following these significant deployments, it is now possible for network operators to actually get traffic into IPv6 just by providing IPv6 connectivity for the hosts they serve, and therefore, operators can reduce Network Address Translation (NAT) costs by deploying IPv6. This is significant, as it concretely allows the first cost savings of IPv6.

The operating systems used by hosts, especially those of personal computers (desktops, laptops), are generally IPv6 enabled already. Hence if provided with IPv6 connectivity, these hosts actually start using IPv6. Unfortunately, the situation with 3GPP User Equipments (UEs) is not as good. The UEs at the time of writing this book are generally (but with exceptions) unable to use IPv6 on cellular access. Not all of the components on the UEs that need updating (as shown in Figure 4.28) are done, yet. Thus, we can see that even the equipment implementation part of the transition is still ongoing.

Due to the end equipment being increasingly IPv6 enabled (hosts, content), the burden of progressing the transition is now mostly on the access network providers but partly also on the UE vendors, whose devices are not yet IPv6 enabled, and obviously on the remaining content providers. We have written this book to help people advocate and add IPv6 support to the components they are responsible for.

We expect to see an increase in access network IPv6 support during the coming few years, in addition to increased support by hosts and content providers. Especially in 3GPP, the IPv6 deployments are expected to become more widespread through deployment of Long Term Evolution (LTE). The vast majority of the deployments will look like dual-stack to the hosts, hence the dual-stack transition continues to be dominant – even if tunneling or double translation would be internally used by access networks.

The need to provide connectivity for IPv4-only hosts will remain for a long time. It is not justifiable to disconnect the huge number of IPv4-only computers and entertainment systems that people have in their homes. Perhaps a decade after the last IPv4-only device is sold it might become possible to turn IPv4 off, but we do not even see the day yet when the last such device would be made. Hence, for the foreseeable future, we are going to witness IPv4 and IPv6 coexistence with occasional pockets of IPv6-only deployments (with connectivity to IPv4 domain e.g., protocol translation).

5.2 Technology Overview

In this section, we will give a more detailed overview of IPv6 transition technologies and their pros and cons. A common benefit for both encapsulation and translation based solutions is that they avoid the need to deploy native dual-stack connectivity all the way.

5.2.1 Translation

Protocol translation approach is attractive as it effectively avoids the issues related to tunneling (see Section 5.2.2) – except Maximum Transmission Unit (MTU) challenges remain due to the bigger IPv6 header size. The translation also provides access to the IPv4 Internet even if the host itself is in an IPv6-only domain.

Unfortunately the translation comes with several major challenges. These challenges are so major that they caused the original Network Address Translation - Protocol Translation (NAT-PT) technology, defined in 2000 in RFC 2766 [9], to be even declared 'historic' in 2007 by RFC 4966 [10]. However, since the NAT-PT, IETF has worked on to improve the translation technologies, and they have indeed got better, although they are still not perfect.

The biggest current issues related to protocol translation, from 3GPP perspective when applicable, are listed below.

Application Compatibility

Application protocols that transport Internet Protocol (IP)-address literals in their payloads require explicit support to work through protocol translation. This means that the protocol translator would need to have an Application-Level Gateway (ALG) for each such protocol that has IP address family specific content in its protocol payload. In the worst case, the application protocol does not support change of the content by a protocol translator (encrypted, or protocol not supported). This means that effectively the protocol translator cannot support 100 percent of unhelpful applications. The applications of course can be modified to work with protocol translators.

However, a special case of protocol translation, called double translation and discussed in Section 5.4.4, can work with problematic applications. This is because for the applications the translation looks like IPv4 NAT: the first protocol translator translates only the outermost header from IPv4 to IPv6, while the second protocol translator translates the header back from IPv6 to IPv4. Effectively the IPv4 address and port changes, and hence applications can cope with the same mechanisms that they use for IPv4 NAT traversal.

Lying with Domain Name System

The whole idea of protocol translation is founded on the idea that the destination is IPv4-only, and hence does not have an IPv6 address (AAAA) resource record in the Domain Name System (DNS). On the other hand, the host on the IPv6-only network definitely requires to get AAAA records for its DNS queries. Therefore, an entity called DNS Extensions for Network Address Translation (DNS64) synthesizes AAAA resource records having IPv6 addresses from real IPv4 address (A) resource records. The problems that a host will have with these synthesized addresses are the following:

1. Synthetic addresses cannot be verified with Domain Name System Security Extensions (DNSSEC) [11], as by design they are altered by a middlebox.
2. Synthetic addresses cannot be given as referrals to other nodes.
3. Synthetic addresses are only valid in the network where the address was received.

Lost Header Information

The IP and Internet Control Message Protocol (ICMP) header contents are different from IPv4 to IPv6. After a translation between address families there is no possibility of reconstructing the original header anymore. The fragmentation related *identification field* serves as an example. In an IPv4 header, the identification field is 16 bits whereas in IPv6 it is 32 bits. A class of Internet Control Message Protocol version 4 (ICMPv4) headers is supported for IPv6 by translation algorithms such as IP/ICMP Translation Algorithm [12]. Specifically it is stated for IP/ICMP Translation Algorithm and NAT64 that there is no attempt to translate IPv4 options. They are silently ignored except when the IPv4 packet contains unexpired source routing options, which then causes the entire packet to be discarded (and an ICMPv4 Destination Unreachable, Source Route Failed may be sent).

State on the Network

The traditional Internet Service Provider (ISP) model assumes the core network has as little per customer state as possible and the required NAT state is pushed down to the customer network. The less state the better the network equipment scales. CPEs equipped with a NAT from IPv4 to IPv4 (NAT44) made this model possible. Also various logging functions are easier to handle since every customer could still be identified uniquely by the CPE's public IPv4 address.

Now there is a possible threat from the ISPs point of view that transition technologies change existing practices and the NAT function has to be placed inside the ISP core network. This would be the outcome when ISPs have to, for example, provide two levels of NAT when they run out of public addresses to number the CPEs [13]. Another example is the stateful NAT64 function [14]. These stateful NAT functions may face real scaling issues, not to mention public authority requirements for address translation logging [15].

From a mobile operator point of view the situation is similar. There is the slight difference as 3GPP has systematically refused to standardize and deploy transition/migration solutions that would have UE impact in a sense of new software requirements. The IPv6 transition comes with an additional cost: in order to provide UEs with a connectivity to

IPv4 Internet from an IPv6-only access technology (such as the Packet Data Network (PDN) Type IPv6 would provide) some migration/translation solution has to be provided. If that is a stateful NAT64 (see Section 5.3.3), then the state has to be in the mobile operator core network, which implies investment in a powerful NAT farm. On the other hand, the mobile operator may seek to deploy dual-stack access that solves the issue of accessing the IPv4 Internet. However, dual-stack does not make the IPv4 address shortage to go away. On a contrary, every dual-stack UE must eventually be configured with an IPv4 address, which typically increases the pressure for deploying even bigger NAT44 farms within the mobile operator packet core. Since these NAT solutions are usually stateful, scalability concerns are real.

5.2.2 Encapsulation

The approach of encapsulating packets belonging to one address family to packets of the other address family comes with a set of benefits and drawbacks.

The main benefit is that the encapsulated packets do not need to be modified, and hence can be forwarded as is. This provides the best compatibility and great transparency of the tunneling technologies for the nodes whose packets are tunneled.

Due to the strong benefits, the IETF community generally prefers the tunneling based approaches when native dual-stack is not available.

However, in particular for 3GPP networks, a set of significant drawbacks for host-based tunneling approaches has been identified. We discuss these drawbacks below.

GPRS Tunneling Protocol (GTP) is Already There

The GTP [16] used in 3GPP already is a tunneling protocol that provides mobiles with IP connectivity over various types of network infrastructures – it can tunnel over IPv4 as well as IPv6, and hence any kind of tunneling protocol on top of GTP is essentially just overhead. It has been shown from practical experience that, excluding rare Serving Gateway Support Nodes (SGSNs) that do not support the IPv6 type of Packet Data Protocol (PDP) context, the IPv6 PDP context can be set up almost anywhere on this planet where 3GPP packet access is provided.

Quality of Service (QoS) Cannot Be Provided

A non-obvious reason against tunneling based approaches is the lack of 3GPP's bearers to differentiate tunneled traffic. This restriction comes from the Traffic Flow Templates (TFTs), which define rules for filtering traffic to different connections (e.g., between Primary and Secondary PDP Contexts). The TFT has rules only for the outermost headers (see Sections 4.1.1 and 4.4.7), and hence QoS differentiation cannot be provided based on the content of the tunneled header.

Deep Packet Inspection (DPI) Costs

A feature that 3GPP networks sometimes need for various reasons, such as for service enrichment or for traffic forwarding reasons, is a DPI. The DPI is not part of the 3GPP standards, but its existence is well recognized. Tunneling significant amounts of data would require more resources from DPI systems, and hence is unfavored – even if technically DPI can look at the tunneled packets.

MTU Implications

Even today the MTU that can be provided for the PDN Connections is usually less than 1500 bytes, as the 3GPP network tunnels the user-plane packets over network segments that typically have 1500-byte MTUs. Adding any tunneling protocol would make the MTU smaller, and while that is not significant, it is still a waste. In some cases the tunneling also causes fragmentation and reassembly at the tunnel endpoints, and hence increases resource consumption and the risk of packet losses. In other cases, tunnel endpoints must generate ICMP Packet Too Big messages of the tunneled protocol address family type, when receiving Packet Too Big messages within the address family used for tunneling (e.g. in the IPv6-over-IPv4 case, if the tunnel endpoint receives an ICMPv4 Packet Too Big message, it may need to generate Internet Control Message Protocol version 6 (ICMPv6) Packet Too Big message) [2].

Added Complexities and Deployment Challenges

Last but not least, getting tunneling implementations to a large number of different UEs, including the numerous Terminal Equipments (TEs) of Split-UE scenarios, is very difficult to do. And even if done, it would require tools for tunnel endpoint discovery, tunnel setups, and tunnel maintenance. This all further increases the signaling, which in general should be decreased rather than increased (existing networks are already suffering from large amounts of signaling traffic).

5.2.3   Mesh or Hub-and-spoke

When selecting a suitable IPv6 transition and migration solution, either translation or encapsulation based, one architectural decision aspect is whether the solution supports *mesh* or *hub-and-spoke* [17] topologies. Some solutions may support both and the decision is then up to the network administrator.

In hub-and-spoke topologies, several 'clients' (spokes) contact a smaller number of gateways (hubs) that typically perform the translation function or are just tunnel/traffic concentrator points. Spokes do not communicate directly with each other. Within the translation/migration context, the core network connecting spokes and hubs needs to support only one address family and the other families are then either translated or tunneled between spokes and hubs. A group of spokes typically gets locked to a specific hub. Dual-Stack Lite (DS-Lite) [8] is a textbook example of a hub-and-spoke solution.

In mesh topologies, the 'clients' may communicate directly with each other, if allowed by the administration. The edge devices (routers) provide connectivity to core network external networks. Within the translation/migration context, the core network connecting clients and edge devices needs to support only one address family, and the other families are then either translated or tunneled over it. The transition/migration function itself may be done at the client or the edge device side depending on the technology. Since mesh supports any-to-any connectivity, the transition/migration solution is implicitly stateless, at least on the edge device side. 6RD [5] using IPv4 anycast to reach the nearest Border Relay (BR) is an example of mesh topology.

5.2.4 Scalability Concerns

The transition/migration technology selection should pay close attention to various scalability concerns. We have already mentioned possible unwanted core network side translation state increase in Section 5.2.1 as one of the scalability concerns. However, scalability is not only about the processing and maintaining the translation state. The following other scalability affecting concerns should also be considered:

- Does the transition/migration solution have wasteful requirements on the address space? It is common that transition/migration solutions make use of embedding all kinds of transition specific information into the IPv6 address. For example, a typical 6RD deployment requires at minimum /32 IPv6 prefix to be dedicated just for transition purposes for each 6RD domain that the operator deploys. It is so much easier, from a management and rollout point of view, to embed the whole IPv4 address into the *6RD delegated prefix* rather than starting to design prefixes and routing based on the scattered IPv4 subnets. The popularity of 6RD even affected the IPv6 initial allocation and assignment size policies at Regional Internet Registry (RIR) level [18]. Another example is 464XLAT [19], which basically promotes assignment of two /64 IPv6 prefixes for each UE (one for native IPv6 traffic and the other for the translated IPv6 traffic, though it *can* also work with a single /64 prefix). While one can argue that there are enough IPv6 addresses for all possible imaginable uses, the wasteful IPv6 address use might come back to haunt us in few years' time.
- Does the transition/migration solution have an impact on the external and/or internal routing infrastructure? For example, a transition solution might require advertising anycast routes or host routes in the Interior Gateway Protocol (IGP).
- What kind of logging impact does the transition/migration solution have? The solution might not have a good way to deterministically build, for example, the translation state. This might lead to an excessive amount of log information being produced, since each translation binding may need to be logged due to local governmental regulation.
- Does the transition/migration solution have address or port number sharing ratio concerns? This is topical in deployments where public IPv4 addresses still have to be used. Since the days of NAT44 port sharing the technology and deployment requirements have also expanded into sharing one address across multiple end devices. The *A+P* or *address and port sharing* [20] is here to stay and driven by recent transition technologies like Mapping of Address and Port with Encapsulation or Translation (MAP) [21].

5.3 Transition Toolbox

From the plethora of IPv6 transition tools that the IETF has produced, we have included the ones that are the most relevant for 3GPP networks. This does not mean that the other tools could not be used. They can be vendor or network specific choices, but are not part

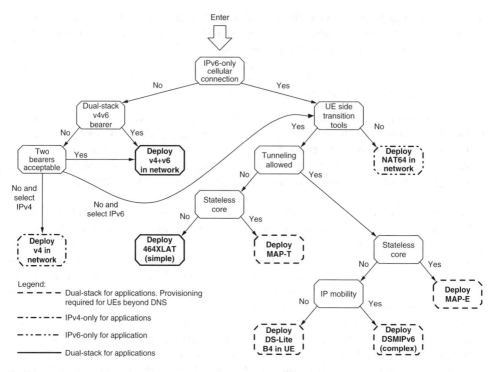

Figure 5.1 Transition technology solution selection flowchart when deploying IPv6-enabled PDN Connectivity that also provides access to IPv4 PDNs.

of the big picture of IPv6 transition in 3GPP access. Furthermore, it is good to realize that as the UEs can also access non-3GPP access networks, they may find it useful to implement some tunneling based transition solutions for such networks (if they wish to obtain IPv6 connectivity).

Figure 5.1 shows a flowchart that aims to recommend a transition tool to use when certain requirements and capabilities in the UE and the network side have to be met. The flowchart only includes those transition tools that are either part of the 3GPP system architecture or otherwise seen useful in this problem space. Transition tools defined in IETF that aim to transport IPv6 over IPv4 are all excluded. Here we also consider dual-stack [2] as a transition tool. It should be noted that several of the flowchart's transitions reflect the UE capabilities for activating a specific type of a PDN Connection or supporting a specific transition solution.

5.3.1 Transition Solutions Not Included

Effectively all tunneling based tools that would be terminated on a UE, be that IPv6-over-IPv4 or IPv4-over-IPv6, are effectively considered irrelevant for the 3GPP access for the reasons listed in Section 5.2.2. That said, it is certainly possible for the UEs to implement tunneling solutions if, for example, they need to always have access to IPv6 or they utilize IP mobility solution not provided by the access network operators. For example, UE can have Dual-Stack Mobile IPv6 (DSMIPv6) Mobile Node (MN) software installed [22, 23] and access to a Home Agent (HA) somewhere. It is thus recognized that for special cases and needs, the benefits of tunneling can outweigh the drawbacks. In addition, the increase of sharing, tethering, the Internet connection via a broadband mobile network over for instance Wireless Local Area Network (WLAN) means that the operating system might not even know that its Internet connectivity is provided by a 3GPP network. Some operating systems automatically try to get access to IPv6 using a multitude of transition tools.

Additionally, we do not pay attention to various (some very ingenious) solutions that are designed to prolong the lifetime of IPv4. On the other hand, any means of keeping IPv4 alive and functional would be in the scope of dual-stack. As we know, dual-stack does not solve the IPv4 exhaustion problem in a short term or actually ever. The only long-term solution for IPv4 exhaustion is IPv6-only networking.

Special Considerations Regarding 6to4

The 6to4 protocol is designed to configure itself automatically in an environment where IPv6 is not present. The principle of the protocol is to configure itself, in a stateless manner, an IPv6 address out of the public IPv4 address it already has. The IPv6 address is created by using the 6to4 prefix `2002::/16` (see also Figure 3.7), and adding the 32-bit IPv4 address behind it. This creates a 6to4 prefix of /48. The 6to4 operation is also made in a stateless manner: the packets from a 6to4 network or a 6to4 host are sent upstream with destination address set to the 6to4 router IPv4 anycast address `192.88.99.1`. On downstream the packets are routed using the 6to4 network's, or host's, public IPv4 address that is embedded in the 6to4 IPv6 prefix.

There are a number of disadvantages in the 6to4 technology in the 3GPP environment. The first is the use of public addresses. Because one 6to4 IPv6 prefix requires one public IPv4 address, it is an ill fit as a UE based transition tool. 6to4 automatic tunneling is prone to routing errors, and therefore, it does not provide a production grade tool for a gateway (Gateway GPRS Support Node (GGSN), or PGW) based transition solution. In addition, some early implementations needed one public address per UE also in the gateway, which wasted addresses.

While the 6to4 technology is not adopted or recommended to be used in 3GPP networks, it is still beneficial for 3GPP operators to host 6to4 relays. This is due to the fact that an IPv6-enabled UE may be communicating with a destination that is using 6to4. The 6to4 is known for its vulnerability to issues caused by asymmetric routing and ingress filtering dropping packets coming via the 'wrong' path. To help avoid these issues, a network operator can deploy local 6to4 relay through which the packets traverse in both directions. This issue and the requirements for 6to4 relays are discussed in depth in RFC 6343 [24].

5.3.2   Dual-stack

The main transition tool for IPv6 transition is the dual-stack as defined in RFC 4213 [2]. This approach, when referred to as a transition tool, not only means that the IP stack has implementation for both IPv4 and IPv6, but that the stack is actually using both of them at the same time and in parallel (instead of, for example, tunneling IPv4 over IPv6). Both address families use independent means for address configuration, for example IPv4 can use Dynamic Host Configuration Protocol (DHCP) or link-layer means, and IPv6 can use Stateless Address Autoconfiguration (SLAAC) or Dynamic Host Configuration Protocol version 6 (DHCPv6), when supported.

In dual-stack cases hosts ask for both A and AAAA DNS resource records when performing name resolution (see more in Section 3.10.2). We highlight this because performing both queries efficiently is an important feature to save time in connection establishments. Furthermore, in the case of single address family connections, a host might optimize and only send out DNS queries for the addresses that can actually be used for communications.

IPv4 Addressing Issues

Nowadays the dual-stack approach is often – and will be even more in the future – challenged by the shortage of IPv4 addresses. Therefore, the IPv4 addresses are increasingly network address translated, i.e. assigned to hosts from the private IPv4 address space allocated in RFC 1918 [25] (i.e. addresses from spaces 10/8, 172.16/12, or 192.168/16). In 3GPP networks it is also possible to use an approach called deferred IPv4 address allocation to mitigate load for an operator's IPv4 address pools (see Sections 4.4 and 4.4.4).

As a result of NAT'ed IPv4 address, the hosts using dual-stack access do not typically get the same level of service with both address families. The IPv4 connections may need more frequent keep-alive traffic than IPv6, and IPv4 is more likely to not allow incoming connections due to NAT, while IPv6 does not always come with a firewall.

Address Family Performance Differences

While the host has both IPv4 and IPv6 in use in parallel, it does not mean that they are both equally good. The IPv4 may suffer from the restrictions mentioned above, but also for communicating to Internet destinations different address families may be providing different characteristics, such as latency, bandwidth, and even lack of connectivity, when communicating towards a destination. This is due to fact that as IPv4 and IPv6 practically form different networks, the routing of IPv4 and IPv6 packets may use different routes. To tackle these sorts of issues hosts can utilize the Happy Eyeballs approach to find out which address family yields better results – generally or per destination – see more in Section 3.9.4.

5.3.3 NAT64 and DNS64

NAT64 [12, 14] and DNS64 [26] are based on the framework for IPv4/IPv6 translation [27]. For translation purposes a new set of *IPv4-embedded* IPv6 address formats has been defined [28]. The NAT64 comes in two flavors: stateless [12] and stateful [14]. The stateless NAT64 (IP/ICMP translation algorithm) replaces the Stateless IP/ICMP

Translator (SIIT) [29], which was used as the foundation for the now deprecated NAT-PT [9, 10].

Although NAT64 itself as a technology does not require DNS64 to work, these two usually go hand in hand, specifically for the stateful version of NAT64. Therefore, whenever we mention NAT64 we actually assume that there is a DNS64 deployed somewhere in the same network unless specifically otherwise noted.

The IPv4/IPv6 translation framework often refers to various translation scenarios. In the context of this book and 3GPP architecture we are only going to cover Scenario 1 (an IPv6 network to the IPv4 Internet) and Scenario 5 (an IPv6 network to IPv4 network). Both stateless and stateful NAT64/DNS64 solutions can support Scenario 1 and Scenario 5. While the stateful flavor is the most suitable fit into the 3GPP architecture, we also describe a potential deployment architecture using a nearly stateless NAT64 within mobile operators' packet core. NAT64 as a technology has received positive feedback among 3GPP operators for two main reasons: it requires no changes in UEs and it does not tunnel anything over the wireless part of the 3GPP network.

The basic principle of NAT64 and DNS64, as already described in general terms in Section 5.2.1, is that a DNS64 returns a synthesized IPv6 address to the end host and packets sent to the said synthesized IPv6 destination get routed through a NAT64 device that knows how to handle the synthesized IPv6 address. Figure 5.2 illustrates a complete signaling flow where an IPv6-only UE sends ICMPv6 Echo Requests to an IPv4-only web server. Terms like Network Specific Prefix (NSP) are discussed in the following sections.

For address translation to work, the NAT64 (or a pool of NAT64s) and the DNS64 need to have a common understanding of the used IPv6 topology and used addressing plan. Specific prefixes used to enable the translation must be routed through the NAT64 devices, whereas the native IPv6 traffic can, and actually should, bypass the NAT64 devices using alternative routes in the network.

Address Formats

In order to enable NAT64 translation, IETF has defined an IPv4-embedded IPv6 address format for the IPv6/IPv4 translation purposes [28]. The address format is shown in Figure 5.3. The `Pref64::/n` is an IPv6 prefix assigned by an operator for use in algorithmic mapping of addresses. The 'u' octet contains the familiar 'g' and 'u' bit positions from the modified EUI-64 based Interface IDentifier (IID) (see Section 3.1). The 'u' octet should always be set to zero. Depending on the `Pref64::/n` length, which is always in steps of 8, the embedded IPv4 address may need to be split into two parts around the 'u' octet. The *suffix* part fills the remaining bits of the IPv4-embedded IPv6 address to reach 128 bits of total length. The content of the suffix is currently all zeros but implementations should be prepared for the fact that it may contain meaningful information in the future.

The Well-Known Prefix (WKP) is a special reserved prefix: `64:ff9b::/96`. It can be used within an operator's own network instead of the NSP. The WKP has certain limitations such as that it cannot be advertised into the global Internet routing system. The NSP is a 'normal' IPv6 prefix that the operator has reserved for NAT64 translation purposes. Whether the NSP is a Unique Local Address (ULA) or a globally routable prefix is for the operator to decide.

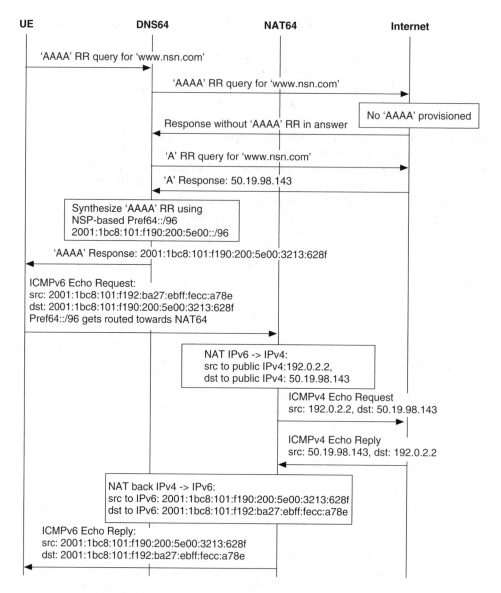

Figure 5.2   An example of DNS64 and a NAT64 operation.

IPv4-mapped IPv6 addresses are further divided into two categories:

- *IPv4-converted IPv6 addresses*: IPv6 addresses used to represent IPv4 nodes in an IPv6 network. These would, for example, be the synthesized IPv6 addresses received from DNS64.
- *IPv4-translatable IPv6 addresses*: IPv6 addresses assigned to IPv6 nodes for use with stateless translation. Nodes with IPv4-translatable IPv6 addresses are reachable from the IPv4 internet side (at least in theory). For each IPv4-translatable IPv6 address

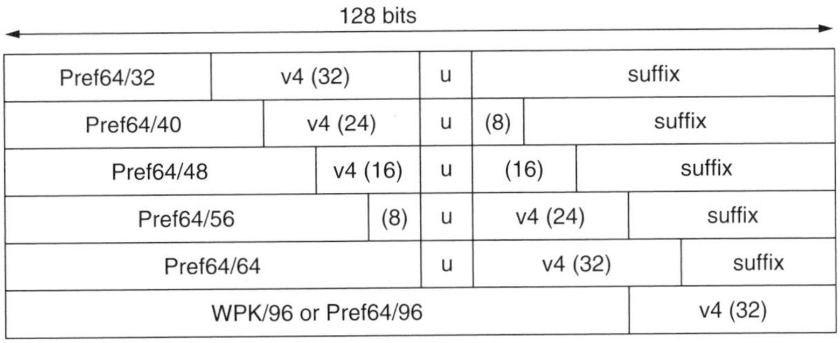

128 bits

Pref64/32	v4 (32)	u	suffix		
Pref64/40	v4 (24)	u	(8)	suffix	
Pref64/48	v4 (16)	u	(16)	suffix	
Pref64/56	(8)	u	v4 (24)	suffix	
Pref64/64		u	v4 (32)	suffix	
WPK/96 or Pref64/96					v4 (32)

Figure 5.3 IPv4-embedded IPv6 address format for translation purposes.

Table 5.1 IPv4-embedded IPv6 address synthesis examples

Network-specific prefix	IPv4 address	IPv4-embedded IPv6 address
2001:db8::/32	192.0.2.1	2001:db8:c000:201::
2001:db8:1100::/40	192.0.2.2	2001:db8:11c0:2:2::
2001:db8:1122::/48	192.0.2.3	2001:db8:1122:c000:2:300::
2001:db8:1122:3300::/56	192.0.2.4	2001:db8:1122:33c0:0:204::
2001:db8:1122:3344::/64	192.0.2.5	2001:db8:1122:3344:c0:2:500::
2001:db8:1122:3344::/96	192.0.2.6	2001:db8:1122:3344::192.0.2.6
Well-known prefix	IPv4 address	IPv4-embedded IPv6 address
64:ff9b::/96	192.0.2.7	64:ff9b::192.0.2.7

that should also be reachable from the IPv4 Internet, there must be a matching public IPv4 address. Syntactically IPv4-translatable IPv6 address does not differ from IPv4-converted IPv6 address.

Table 5.1 illustrates how to synthesize IPv6-embedded addresses out of both NSP and WKP. The textual format follows the standard IPv6 text representation of IPv6 addresses [30].

Stateful NAT64 Deployment

RFC 6146 [14] describes a stateful version of NAT64. It is probably the most popular flavor of deployed NAT64s, because it can provide IPv4 connectivity to IPv6 hosts using far less public IPv4 addresses than stateless NAT64. In Figure 5.4 we show an example architecture of a NAT64/DNS64 deployment within a mobile operator network, where NAT64 function could be embedded inside a PGW. Alternatively, the translation function could be a separate node. It does not really matter from the translation outcome point of view. There are certain benefits in combining the mobile network gateway and the translation function. Those usually boil down to more fine-grained subscription awareness within a combined

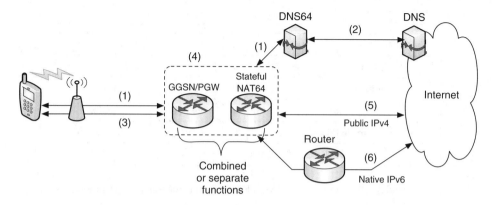

Figure 5.4 An example of DNS64 and a stateful NAT64 that can be integral to a GGSN/PGW or a separate function.

No.	Time	Source	Destination	Protocol	Info
34	0.002056	2001:1bc8:101:f192:ba27:ebff:fecc:a78e	2001:1bc8:101:f190:200:5e00:3213:628f	ICMPv6	Echo (ping) request id=0x0a3f, seq=1
35	0.145955	2001:1bc8:101:f190:200:5e00:3213:628f	2001:1bc8:101:f192:ba27:ebff:fecc:a78e	ICMPv6	Echo (ping) reply id=0x0a3f, seq=1
89	0.126968	2001:4830:120:1::2	2001:1bc8:101:f192:ba27:ebff:fecc:a78e	ICMPv6	Echo (ping) reply id=0x0a41, seq=1
120	0.006888	2001:1bc8:101:f192:ba27:ebff:fecc:a78e	2001:1bc8:101:f190:200:5e00:5c7a:5b50	ICMPv6	Echo (ping) request id=0x0a42, seq=1
121	0.051375	2001:1bc8:101:f190:200:5e00:5c7a:5b50	2001:1bc8:101:f192:ba27:ebff:fecc:a78e	ICMPv6	Echo (ping) reply id=0x0a42, seq=1
153	0.005606	2001:1bc8:101:f192:ba27:ebff:fecc:a78e	2001:1bc8:101:f190:200:5e00:5b79:4527	ICMPv6	Echo (ping) request id=0x0a49, seq=1
154	0.059112	2001:1bc8:101:f190:200:5e00:5b79:4527	2001:1bc8:101:f192:ba27:ebff:fecc:a78e	ICMPv6	Echo (ping) reply id=0x0a49, seq=1

```
▷ Frame 88: 118 bytes on wire (944 bits), 118 bytes captured (944 bits)
▷ Ethernet II, Src: Raspberr_cc:a7:8e (b8:27:eb:cc:a7:8e), Dst: Routerbo_f7:ea:43 (00:0c:42:f7:ea:43)
▷ Internet Protocol Version 6, Src: 2001:1bc8:101:f192:ba27:ebff:fecc:a78e (2001:1bc8:101:f192:ba27:ebff:fecc:a78e), Dst: 2001:4830:120:1::2
▽ Internet Control Message Protocol v6
      Type: Echo (ping) request (128)
      Code: 0
      Checksum: 0x01a7 [correct]
```

```
0000  00 0c 42 f7 ea 43 b8 27  eb cc a7 8e 86 dd 60 00    ..B..C.'......`.
0010  00 00 00 40 3a 40 20 01  1b c8 01 01 f1 92 ba 27    ...@:@ ........'
0020  eb ff fe cc a7 8e 20 01  48 30 01 20 00 01 00 00    ...... .H0. ....
0030  00 00 00 00 00 02 80 00  01 a7 0a 41 00 01 c8 ac    ...........A....
0040  43 50 98 67 00 00 08 09  0a 0b 0c 0d 0e 0f 10 11    CP.g............
0050  12 13 14 15 16 17 18 19  1a 1b 1c 1d 1e 1f 20 21    .............. !
0060  22 23 24 25 26 27 28 29  2a 2b 2c 2d 2e 2f 30 31    "#$%&'() *+,-./01
0070  32 33 34 35 36 37                                   234567
```

| ⊙ File: "/Volumes/work/resear... | Packets: 182 Displayed: 8 Marked: 0 Ignored: 80 Load time: 0:... | Profile: Default |

Figure 5.5 An IPv6-only host ping6ing some infamous websites.

node, easier means to combine Access Point Names (APNs) or session profile configuration with translation, and the rest of the functions that the PGW does to the user plane traffic (DPI, policy control, differentiated charging, layer-7 filtering, to name but a few).

Figures 5.5 and 5.6 show a short captured packet trace where a client in an IPv6-only network pings a few IPv4-only web servers in the Internet.

Depending on the implementation of the combined PGW and NAT64, the native IPv6 traffic should bypass the translation function entirely. Scalability is one of the main concerns of the combined NAT64 and PGW. The PGW already aggregates a lot of traffic and especially sessions. The scaling up traffic-wise typically means redundant nodes. Scaling up the number of sessions usually means (naively) more memory for session state information and either bigger IPv4 pools on the egress side of the NAT64 or smaller

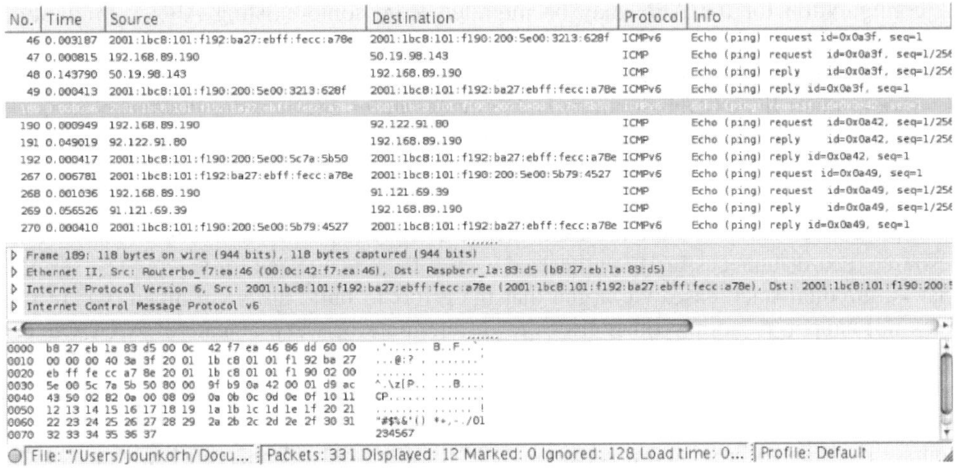

No.	Time	Source	Destination	Protocol	Info
46	0.003187	2001:1bc8:101:f192:ba27:ebff:fecc:a78e	2001:1bc8:101:f190:200:5e00:3213:628f	ICMPv6	Echo (ping) request id=0x0a3f, seq=1
47	0.000815	192.168.89.190	50.19.98.143	ICMP	Echo (ping) request id=0x0a3f, seq=1/256
48	0.143790	50.19.98.143	192.168.89.190	ICMP	Echo (ping) reply id=0x0a3f, seq=1/256
49	0.000413	2001:1bc8:101:f190:200:5e00:3213:628f	2001:1bc8:101:f192:ba27:ebff:fecc:a78e	ICMPv6	Echo (ping) reply id=0x0a3f, seq=1
190	0.000949	192.168.89.190	92.122.91.80	ICMP	Echo (ping) request id=0x0a42, seq=1/256
191	0.049019	92.122.91.80	192.168.89.190	ICMP	Echo (ping) reply id=0x0a42, seq=1/256
192	0.000417	2001:1bc8:101:f190:200:5e00:5c7a:5b50	2001:1bc8:101:f192:ba27:ebff:fecc:a78e	ICMPv6	Echo (ping) reply id=0x0a42, seq=1
267	0.006781	2001:1bc8:101:f192:ba27:ebff:fecc:a78e	2001:1bc8:101:f190:200:5e00:5b79:4527	ICMPv6	Echo (ping) request id=0x0a49, seq=1
268	0.001036	192.168.89.190	91.121.69.39	ICMP	Echo (ping) request id=0x0a49, seq=1/256
269	0.056526	91.121.69.39	192.168.89.190	ICMP	Echo (ping) reply id=0x0a49, seq=1/256
270	0.000410	2001:1bc8:101:f190:200:5e00:5b79:4527	2001:1bc8:101:f192:ba27:ebff:fecc:a78e	ICMPv6	Echo (ping) reply id=0x0a49, seq=1

```
▷ Frame 189: 118 bytes on wire (944 bits), 118 bytes captured (944 bits)
▷ Ethernet II, Src: Routerbo_f7:ea:46 (00:0c:42:f7:ea:46), Dst: Raspberr_1a:83:d5 (b8:27:eb:1a:83:d5)
▷ Internet Protocol Version 6, Src: 2001:1bc8:101:f192:ba27:ebff:fecc:a78e (2001:1bc8:101:f192:ba27:ebff:fecc:a78e), Dst: 2001:1bc8:101:f190:200:...
▷ Internet Control Message Protocol v6

0000  b8 27 eb 1a 83 d5 00 0c  42 f7 ea 46 86 dd 60 00   .'......B..F..`.
0010  00 00 00 40 3a 3f 20 01  1b c8 01 01 f1 92 ba 27   ...@:? ........'
0020  eb ff fe cc a7 8e 20 01  1b c8 01 01 f1 90 02 00   ...... .........
0030  5e 00 5c 7a 5b 50 80 00  9f b9 0a 42 00 01 d9 ac   ^.\z[P.....B....
0040  43 50 02 82 0a 00 08 09  0a 0b 0c 0d 0e 0f 10 11   CP..............
0050  12 13 14 15 16 17 18 19  1a 1b 1c 1d 1e 1f 20 21   .............. !
0060  22 23 24 25 26 27 28 29  2a 2b 2c 2d 2e 2f 30 31   "#$%&'()*+,-./01
0070  32 33 34 35 36 37                                  234567

○ File: "/Users/jounkorh/Docu...  Packets: 331 Displayed: 12 Marked: 0 Ignored: 128 Load time: 0...  Profile: Default
```

Figure 5.6 A NAT64 doing translation.

port ranges for each source IPv6 address. Also, NAT64 translation with stateful sessions does need processing beyond swapping IP header and possible re-computation of IP and transport protocol checksums. When the session density is high, NAT64 processing also becomes computationally expensive and also something to consider.

A standalone NAT64 is actually rather easy to scale up by increasing the number of functional NAT64 nodes. Assuming that each NAT64 function has its own dedicated NSP `Pref64::/n` prefix, a simple DNS64 round-robin of multiple `Pref64::/n` prefixes distributes the load among a cluster of NAT64 nodes. It should be noted that a 'standalone' NAT64 function can be within the same physical appliance as the rest of the PGW. The level of integration into the rest of the 3GPP system is very low.

How NAT64 is then provisioned and deployed in the mobile operator network is another issue. The key information to provision to subscribers is the address of the DNS64-enabled DNS server. This information is typically part of the generic APN or session profile within the PGW. Another issue is then how the subscription profile gets provisioned. Will there be a separate NAT64-enabled APN among others and how is that then provisioned in a Home Location Register (HLR)/Home Subscriber Server (HSS)? One potential provisioning example is as follows:

1. The PGW is configured with a dual-stack capable APN. The APN configuration points to a DNS64 server, which gets assigned to any UE accessing this very APN, independent of whether the PDN Connection is IPv4-only, IPv6-only (the reason for all this), or IPv4v6.
2. The DNS64 server is used as a caching Recursive DNS Server (RDNSS). The DNS64 functionality only gets triggered when the server is queried for AAAA resource records for destinations not having AAAA records. The A resource record queries are always handled in a 'traditional' way.
3. The subscriber is provisioned with a dual-stack APN for all its needs, that is, whether the UE activates an IPv4-only, IPv6-only, or IPv4v6 PDN Connection. The other

configuration for the APN may be migrated from some existing APN configuration just by conducting a mass update for the PDN type. There is not necessarily a need to create a new APN and corresponding configuration.

4. Should plain IPv4 PDN Connection use be de-preferred, then the PGW can be made to reject PDN Type IPv4 context activation attempts. Such configuration would favor both IPv6 and dual-stack (IPv4 would only be possible when using dual-stack).

In the above example, there are cases where the DNS64 server is accessed and AAAA resource records unnecessarily synthesized even if there is no need for any NAT64 service, that is, the dual-stack case. We argue that the effort to build an environment where this does not happen or adding other tricks to differentiate between a 'native' AAAA query versus IPv6-only query may not be worth it. Depending on the terminal capabilities and provisioning, there may not be any need to push new APN configuration into devices. There is probably a need to update an existing configuration for IPv6, though. It is, of course, possible to provision a separate APN for IPv6-only connectivity with a NAT64 service, and then push that into the mobile devices.

Nearly Stateless NAT64 Deployment Reusing NAT44

Typically mobile operators have already deployed a highly scalable NAT44 service using mature technologies for their ever-increasing IPv4 demand when the numbering resources have in practice been already exhausted. There might be little incentive to invest in another flavor of high end NAT cluster – the NAT64 cluster. When the percentage of IPv6 traffic increases, the investment put into NAT64 probably cannot be that easily justified, as at the same time there is increasing amount of freed capacity in the existing NAT44 deployment.

The authors have been thinking that it would be possible to capitalize the existing NAT44 by introducing a (lightweight) *nearly stateless NAT64* function into the network. The driver behind this is that the fewer states a device has, the better it scales. If the translation were entirely algorithmic, it would enable it to have similar capability to MAP and 6RD anycast based reachability of NAT64 functions (see Section 5.3.6). Such a stateless NAT64 function could easily be implemented as an integrated feature into the PGW.

The idea of (nearly) stateless NAT64 solution in a nutshell is the following. During the PDN Connection activation the UE is assigned a /64 IPv6 prefix that embeds both *base IPv6 prefix* and the lower bits of a private RFC 1918 address – see Table 5.2 for addressing examples. The APN where the stateless NAT64 has been enabled must be provisioned with the base IPv6 prefix, a private IPv4 address pool and prefix lengths for both. When using `10.0.0.0/8` range a maximum of 16 million mobile devices can be

Table 5.2 IPv6 address for NAT64 purposes

APN base prefix	Private IPv4	Constructed IPv6 prefix	Private IPv4 pool
2001:db8:1100::/40	10.0.32.1	2001:db8:1100:2001::/64	10.0.0.0/8
2001:db8:1120::/44	172.16.64.3	2001:db8:1120:4003::/64	172.16.0.0/12
2001:db8:1122::/48	192.168.128.2	2001:db8:1122:8002::/64	192.168.0.0/16

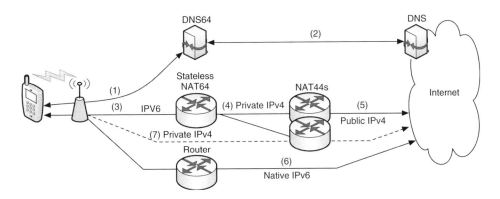

Figure 5.7 An example of DNS64 and a stateless NAT64 with a reused NAT44 function.

numbered within one routing domain (since there can be overlapping private IPv4 address spaces, care must be taken that these areas are separated somehow).

When each UE in a specific APN has a unique 'IPv4-translatable IPv6 address'-*like* address (not exactly as such according to RFC 6052, though), the stateless NAT64 can extract the private IPv4 address out of the IPv6 source address on the fast path and concatenate it with the provisioned upper bits of the private IPv4 address pool. The resulting IPv4 address is then used as the source address of the translated IPv4 packet. The rest of the NAT64/DNS64 operation for resolving the IPv4-converted IPv6 address for the destination node in the IPv4 Internet goes as with any NAT64/DNS64 deployment. The translated IPv4 packet is then routed towards the existing NAT44 – just as possible native IPv4 traffic UE might originate. Moreover, if the existing NAT44 device already performs other operator-specific functions (e.g. access control, charging, policy control) then this extra function remains there without any change and does not add any load to the PGW where the NAT64 is located. Figure 5.7 illustrates a possible (nearly) stateless NAT64 deployment that reuses the existing NAT44 function.

Why then do we refer to the above idea as 'nearly stateless'? In 3GPP the network has no control over the IID that the UE is using beyond link-local addresses. And what we described earlier has no means of preserving the information of the IID that the UE was using. That is where the 'nearly stateless' originates. The NAT64 (preferably in the PGW) has to record the used IID for each translated IPv6 and IPv4 address pairs. That is still fairly resource conservative but adds some level of undesired connection tracking. Moreover, if the UE is using privacy addresses [31], there can potentially be multiple overlapping entries when the UE changes the source IPv6 address. That would mean that the NAT64 needs to start looking into additional information beyond the IP header, which basically destroys the whole idea of statelessness. If 3GPP architecture would allow using DHCPv6 for the address assignment, we could achieve the stateless NAT64 function.

Another enhancement to the (nearly) stateless idea is to assign private IPv4 addresses from the pool, say, in increments of four addresses: `10.0.0.0, 10.0.0.4, 10.0.0.8,` etc. When a new IID is seen for one IPv6 prefix constructed out of the assigned private IPv4 address, then up to four simultaneous IIDs can be mapped back and forth to the same /64 IPv6 prefix and to four different private IPv4 addresses with minimal

Table 5.3 IPv6 address for NAT64 and interface identifier mapping purposes

UE IPv6 address	Assigned IPv4 address	Mapped IPv4 address for NAT44
2001:db8:1100:4::e/64	10.0.0.4 /30	10.0.0.4
2001:db8:1100:4::2/64	10.0.0.4 /30	10.0.0.5
2001:db8:1100:4::a/64	10.0.0.4 /30	10.0.0.6
2001:db8:1100:4::9/64	10.0.0.4 /30	10.0.0.7
2001:db8:1100:8::e/64	10.0.0.8 /30	10.0.0.8
2001:db8:1100:8::5/64	10.0.0.8 /30	10.0.0.9
2001:db8:1100:8::1/64	10.0.0.8 /30	10.0.0.10
2001:db8:1100:8::b/64	10.0.0.8 /30	10.0.0.11

state maintenance. The approach is slightly wasteful on IPv6 prefixes (which should not be an issue in practice) and still has some session state in the NAT64. Table 5.3 shows how the mapping works for a base IPv6 prefix 2001:db8:1100::/40, a private IPv4 pool 10.0.0.0/8 with four address increments, and two UEs with IPv6 prefixes 2001:db8:1100:4::/64 and 2001:db8:1100:8::/64. Four reserved IPv4 addresses actually equal a /30 subnet for one /64 IPv6 prefix.

NAT64 and DNS64 Issues

NAT64 (and also DNS64) has a number of issues that cast shadows on prominent IPv6 deployments. Two known problems relate to the use of IP literals as referrals or inside payload protocols, and how to deal with verifying DNSSEC [11] resolvers will be discussed in more detail in the following section.

There is a class of applications that do not work through NAT64 [32]. Some of these applications may, on the other hand, work just fine through a normal NAT44. The problem is the assumption that the IP address family does not change, and therefore the NAT-traversal solution within the, typically closed, protocol and implementation just does not work. The inherent cause of these problems is embedding IP literals inside the payload protocol. The obvious correction to such applications and protocols is to make them independent of IP literals or deploy protocol-specific ALGs on NAT64 [33].

Another class of failing applications are IPv4-only applications. There is not much else you can do to correct the situation other than implementing IPv6 support for those when used in IPv6-only end host and networking environment. If the failing IPv4-only application is closed source, there is little that can be done other than wait for the application maintainer to provide an update (if ever). Recently, IETF has worked on a protocol solution for running IPv4-only legacy applications in a dual-stack end host that has IPv6-only network access. We will present a solution called 464XLAT [19] in Section 5.3.4. Another solution approach for IPv4-only legacy applications is called Bump-In-the-Host (BIH) [34], which is discussed further in Section 5.3.5. Both 464XLAT and BIH require end host modifications.

Currently there is no protocol level means to distinguish between a 'vanilla' DNS server and a DNS64 capable DNS server. At the same time there could be network operations driven desires to use the same DNS server for all IPv4-only, IPv6-only, and dual-stack end hosts. And why is this of any concern? When the DNS64 server receives an AAAA query from an end host, it does not know whether it has to apply normal DNS server operations or act as a DNS64 server (and therefore be prepared to do the address synthesis when needed). An unfortunate packet reordering or a glitch in the network can cause the DNS64 server to receive just an A response instead of an AAAA response. Unnecessarily synthesized AAAA records cause dual-stacked end hosts to use synthesized IPv6 addresses instead of native IPv4. While preferring IPv6 has nothing wrong with it, it causes unnecessary load on the NAT64 cluster. Also a certain class of applications may work better through a NAT44 than a NAT64. There was a proposal in IETF that had explicit end host to DNS(64) server signaling using EDNS0 option [35].

In the absence of the protocol level solution a practical approach to solving the problem would be configuring a split view into the DNS server. In order for this to work, IPv4-only and dual-stack capable end hosts must configure their own prefixes from a different prefix than IPv6-only end hosts. Then a simple Access Control List (ACL) matching against source prefixes can be used to select appropriate DNS server internal behavior. There are still cases where views do not work as hoped. For instance, in 3GPP network deployments the same APN can be provisioned for both IPv4-only UEs, IPv6-only UEs, and dual-stack UEs. If the APN uses exactly the same IPv6 prefix pool for all IPv6 subscribers, then the DNS view solution will also fail. However, implementing proper IP pool management in the Authentication, Authorization and Accounting (AAA) backend and using PDP/PDN Type as a selector for a correct IPv6 prefix pool, then DNS views can also be made to work. Internet Systems Consortium (ISC) `Bind` DNS server open source software starting from version 9.8.0 would support the required DNS64 and view configurations.

Early trial deployments of NAT64 technology have already provided initial operational experience [36]. High availability is a real concern in large scale deployments. However, as in any stateful network nodes, it may not be trivial to move the state to the secondary nodes when the primary dies, unless hot standby solutions are used. The same also applies to NAT64. Another operational concern is traceability that is increasingly demanded by authorities for various reasons. As in any NAT solution, the connection traceability solutions might produce unbearable volumes of log data. Solutions such as predefined ports ranges and such may help but are not necessarily the final solutions.

NAT64 Discovery

While NAT64 and DNS64 work well for applications that are able to utilize IPv6, some issues remain. Specifically, if an IPv6-enabled application in an IPv6-only network receives an IPv4 address literal as referral (e.g. with HyperText Transfer Protocol (HTTP) or Session Initiation Protocol (SIP)), it cannot directly use the IPv4 address for communications. An application, or in some cases a DNS stub resolver on a host, could synthesize an IPv6 address out of an IPv4 address in the same way as DNS64 performs, if only the `Pref64::/n` were known. IETF is in the late stages of defining a solution that uses a heuristic algorithm to find the `Pref64::/n` used in an access network [37]. The heuristic approach is based on the idea that if a node asks for an IPv6 address for a name that has only an IPv4 address in the DNS, the possible DNS64 in an access network will synthesize the IPv6 address using the `Pref64::/n` in use in the access network. After the node receives the synthesized IPv6 address, it can dig out the `Pref64::/n` as the node knows the IPv4 address used in the synthesis and also the possible synthetic address formats illustrated in Figure 5.3.

The heuristic-based solution Internet-Draft, currently includes definitions for a well-known name *ipv4only.arpa* and well-known IPv4 addresses 192.0.0.170 and 192.0.0.171. These are the pieces of information on which nodes performing heuristic discovery will base their algorithm. Figure 5.8 shows an example message flow for the NAT64 discovery and the Pref64::/n learning.

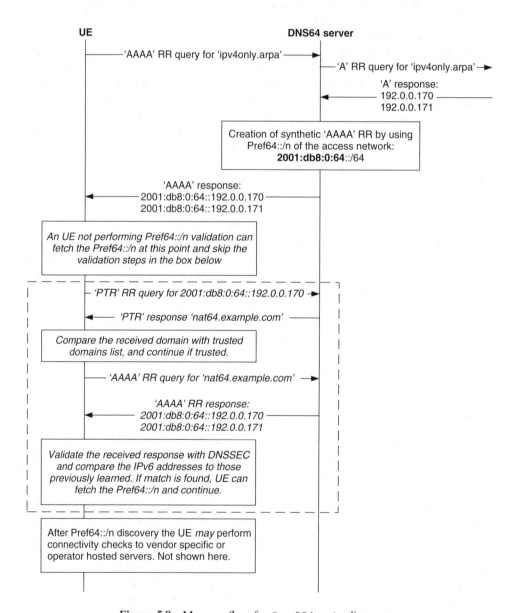

Figure 5.8   Message flow for Pref64::/n discovery.

While one reason for the `Pref64::/n` discovery is the local synthesis of IPv6 addresses based on IPv4 address literals, there are also other uses. If a node implements a validating DNSSEC [11, 38] resolver, all synthetic AAAA responses would be rejected, as by definition they are modified by a middlebox. This would yield in undesirable user experience. Hence, a UE having a validating DNSSEC resolver would want to find out the `Pref64::/n`, and then use the prefix to perform local IPv6 address synthesis based on DNSSEC validated IPv4 addresses. A third reason is to use the `Pref64::/n` in double protocol translation schemes as described in Section 5.3.4.

5.3.4 464XLAT

Sections 5.1 and 5.4.2 stated that dual-stack is the 3GPP transition mechanism of choice. As we know from Sections 4.2.2 and 4.5 dual-stack is currently not readily available, especially for the 3rd Generation (3G) General Packet Radio Service (GPRS) systems. The 3GPP standardized fallback from the PDN Type IPv4v6 (i.e. dual-stack bearer) would be using *two bearers* to mimic a dual-stack behavior. However, two bearers come at a cost. Each 3GPP bearer and connection consumes both radio network and core network resources that are typically subject to licensing, while the equivalent functionality could be accomplished just by allowing multiple addresses on a single dual-stack bearer. There are economic and deployment timing reasons for trying to come up with transition solutions that would allow a dual-stack UE to work over an IPv6-only access such as PDN Type IPv6.

Tunneling IPv4 over IPv6 is one approach, but it also comes with a cost of extra header overhead. Translating IPv4 to IPv6 does have less overhead, but it then requires a second pass of translation back from IPv6 to IPv4; hence it is often referred to as the double translation (see Section 5.4.4 for further discussion on this topic regarding 3GPP networks specifically). The first serious attempt to standardize a double translation solution in IETF was Prefix NAT (PNAT) [39]. However, at that time it faced a lot of opposition from both 3GPP and IETF communities. Eventually the authors abandoned the original idea and the solution evolved into BIH [34] (see Section 5.3.5 for a detailed discussion of BIH). The later attempt in IETF to enable dual-stack connectivity over an IPv6-only access network is called 464XLAT [19]. This time the overall IPv6 transition climate was more mature to accept a double translation based solution. Funnily enough, at both times the double translation solution was driven by a reputable mobile operator deploying 3GPP-based networks.

464XLAT can be deployed in fixed networks as well as in mobile networks. Our emphasis and interest is on the deployment on 3GPP networks. Figure 5.9 shows the basic 464XLAT architecture when deployed in 3GPP networks. 464XLAT heavily reuses existing IETF standards developed for IPv4/IPv6 translation and NAT64 [12, 14], DNS64 [26], and the companion IPv6 addressing [28]. 464XLAT supports three different connectivity scenarios for a dual-stack UE over an IPv6-only network access: 1) native IPv6 connectivity for IPv6 applications, 2) translated IPv6 access to IPv4 services for IPv6 applications, and 3) double translated access for IPv4 applications to IPv4 services. Table 5.4 shows all three scenarios and the nodes that are involved on each of those.

The Provider Side Translator (PLAT) is basically a stateful NAT64 [26] and does the translation from IPv6 to IPv4 when IPv6 translated IPv4 traffic is sent over the

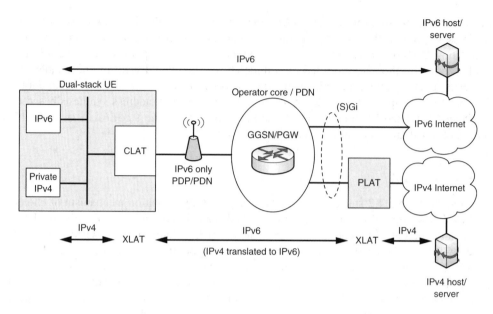

Figure 5.9 464XLAT used in 3GPP network architecture.

Table 5.4 464XLAT supported traffic treatment and translation scenarios

Application	Destination	Traffic treatment	Location of translation
IPv6	IPv6	IPv6 end-to-end	–
IPv6	IPv4	Stateful NAT64	PLAT in network
IPv4	IPv4	464XLAT	CLAT in UE and PLAT in network
IPv4	IPv6	Not supported (requires BIH)	– (BIH in UE)

IPv6-only connection from the UE. The Client Side Translator (CLAT) is, in the 3GPP case, located within the UE and does a stateless algorithmic translation from a private RFC 1918 IPv4 address to a global IPv6 address. The beauty of the 464XLAT solution is that it can be made work using the single /64 IPv6 prefix assigned to the UE during the PDN Connection activation, and it also works on pre-Release-8 networks. There is no additional provisioning specific to 464XLAT technology needed. This is because the private IPv4 address used by the UE can be picked up by the CLAT and there is no need to provision it, since every IPv4-embedded IPv6 address the UE may initiate is made unique by the /64 IPv6 prefix assigned to the UE. For IPv4-only DNS resolvers (in applications or in the case of tethering in separate devices, the CLAT needs to implement a DNS proxy [40], because the IPv4-only DNS resolvers obviously have no means of sending DNS queries over the IPv6 transport and the UE does not know anything other than IPv6 reachable DNS servers provisioned to the IPv6-only APN.

As said earlier, the 464XLAT works with a single /64 IPv6 prefix for both translated and native IPv6 traffic. In this case the CLAT has to be bound to a specific IPv6 address

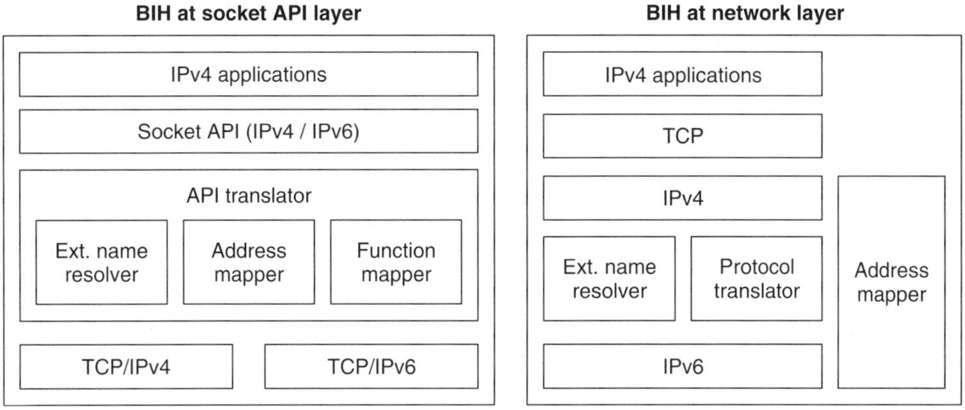

BIH at socket API layer

IPv4 applications
Socket API (IPv4 / IPv6)

API translator		
Ext. name resolver	Address mapper	Function mapper

TCP/IPv4	TCP/IPv6

BIH at network layer

IPv4 applications
TCP
IPv4

Ext. name resolver	Protocol translator	Address mapper

IPv6	

Figure 5.10 Bump-in-the-host architecture options [34].

(derived from the PDN Connection assigned /64 IPv6 prefix) for translated IPv6 traffic. Native IPv6 can then use any other address from the same prefix, thus there is the theoretical possibility of an address collision within the UE (or with other devices in LAN in case of tethering). Therefore, the use of DHCPv6 Prefix Delegation (PD) is recommended, and the UE can the dedicate a specific prefix for translation purposes. However, DHCPv6 PD was not available until 3GPP Release-10 and at that time the network should also have been fully dual-stack capable as well.

5.3.5 Bump-In-the-Host

A BIH describes a tool where IPv4-only application inside a UE can talk to an IPv6-only destination with the help of UE-internal IPv4 to IPv6 protocol translation [34]. With this feature the BIH complements 464XLAT, as show in Table 5.4.

The BIH supports two architectures for protocol translation. The first is socket Application Programming Interface (API) layer translation and the second is network layer translation. The architectural options are illustrated in Figure 5.10.

In the socket API layer implementation option, the `Function mapper` translates IPv4 socket API calls made by an application to IPv6 API calls. In the network layer implementation option, the host implements `Protocol translator` at the network layer by using technologies more commonly used on routers: IP/ICMP translation defined in RFC 6145 [12], NAT64 defined in RFC 6146 [14], and DNS64 defined in RFC 6146 [26].

The protocol translation is supported by modules called `Extension name resolver` and `Address mapper`. The extension name resolver performs functions required to fetch IPv6 address from DNS even if the application requested only IPv4. This can include intercepting DNS queries as in the network layer model, or simply by intercepting socket API calls as in API-layer mode. The address mapper entity is responsible for creating local IPv4 address representations of destinations' IPv6 addresses, and maintaining mapping tables. The local addresses used by the address mapper are chosen from the private address space defined by RFC 1918 [25], as that is the only address space that applications are certain to understand as being address translated, and hence triggers applications to use NAT traversal techniques.

The BIH comes with an inbuilt limitation to perform protocol translation only for IPv6-only destinations. This means that even if a node implementing BIH is in an IPv6-only network, it should not perform protocol translation if the destination has an IPv4 address. This design decision was based on the dominant view at the IETF during the standard writing, which was against definition of double protocol translation schemes. However, as time has passed the double protocol translation has become acceptable (see Section 5.4.4) and the current definition of BIH actually complements 464XLAT. In any case, with BIH, an implementation can be made to do protocol translation even if a destination has IPv4 address. This is done by defining special exclusion sets (RFC 6535, Section 2.3.1) to exclude all received IPv4 addresses, hence making the destination look like IPv6-only for the rest of the BIH implementation.

5.3.6 Mapping Address and Port Number

Among certain IPv6 transition/migration solutions it has become common to embed IPv4 address and port information into the IPv6 prefix assigned to the end host. Mapping of IPv4 addresses inside IPv6 addresses is an old and known technology dating back to the days of automatic tunneling [41]. These specifically concern transition solutions where IPv4 access is provided for leaf networks where parts (or the entire) ISP backbone is IPv6 only. The basic idea is to embed enough IPv4 related information into the IPv6 prefix, and in some cases also into the IID, so that the ISP network can do a stateless algorithmic mapping/translation from an IPv6 address to an IPv4 address/prefix and vice versa [42]. Often, IPv4 address sharing using *address plus port* (A+P) [20] techniques go hand in hand with mapped IPv6 addresses.

The mapping scheme described in this section supports transportation of IPv4 packets in IPv6 in both mesh and hub-and-spoke topologies (see Section 5.2.3), including address mappings with full independence between IPv6 and IPv4 addresses. Within the MAP context we mean the address family translation similar to NAT64 as the *translation mode* and tunneling IPv4 over IPv6 as the *encapsulation mode*. A generic network topology for mapped address and port number deployments is shown in Figure 5.11.

MAP specifies an algorithm to construct an IPv6 address from an IPv4 address and a port range. Figure 5.12 shows the construction of the MAP IPv6 address [21]. The figure contains the Basic Mapping Rule (BMR), a general presentation of the mapping scheme, and the Forwarding Mapping Rule (FMR) for both translation and encapsulation modes. It also has the Default Mapping Rule (DMR) for the translation mode. In the case of encapsulation mode DMR the IPv6 address would just be a normal (anycast) address of the BR. In the case of the FMR the IID contains the (public) IPv4 address of the CPE or the intended destination and the used Port-set Identifier (PSID).

For the actual algorithm of how to derive the Embedded Address (EA) bits and the PSID, please refer to Section 5 of [21]. The port set calculation algorithm allows automatic exclusion of the low range ports, for example, ports from 0 to 1023. Depending on the provisioning each shared address could, for example, be allocated 1024 ports, which would end up to 1/63 address sharing ratio (when the low ports are excluded). As a reminder,

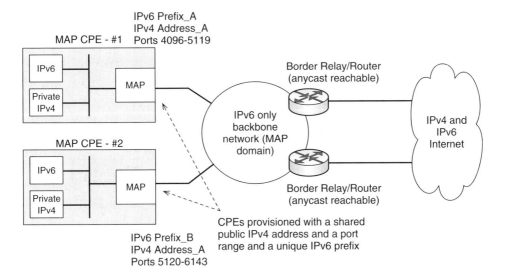

Figure 5.11   Network topology for a generic *mapped address and port* deployments.

there are three different possibilities how to construct the EA bits (we shortcut a bit and discard the operator set prefix part for the EA bits):

1. It contains a complete IPv4 address, that is no address sharing is in use.
2. It contains a suffix of a shared IPv4 address and PSID for identifying the set of ports out of total 65536 unique to a CPE.
3. It contains an IPv4 prefix.

Bootstrapping

In order for the MAP enabled devices to be able to do the algorithmic processing of the prefixes, a set of information has to be provisioned to these devices. Manual configuration does not scale and is highly error prone. In the case of CPEs, DHCPv6 is a natural method for MAP specific information distribution and provisioning [43]. A CPE would use DHCPv6 in the majority of cases anyway to bootstrap itself. The information delivered includes: transport mode (encapsulation or translation), used IPv6 and IPv4 prefixes, PSID related information, relay addresses, and so on.

From the 3GPP networks and deployments point of view, the dependency on DHCPv6 could be an issue. As we know, DHCPv6 is not a mandatory part of the 3GPP system architecture. Furthermore, since MAP is not even part of the recommended transition tool set for 3GPP networks, there is little hope, for example, for Non-Access Stratum (NAS) protocol support for bootstrapping the required provisioning information.

Examples

As we saw from Figure 5.12 the algorithmic generation of the IPv6 address used in MAP is not the most straightforward. Below we show an example how to build the address to

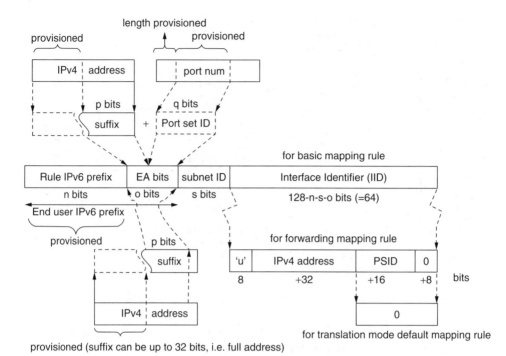

Figure 5.12 Basic mapping rule and forwarding mapping rule for both translation and encapsulation mode IPv6 address derivation. For the default mapping rule in translation mode the interface identifier needs to be modified.

and from the Internet to a host behind a CPE. These examples are almost *directly taken* from the IETF draft [21]. A MAP enabled border relay receives a packet from the Internet destined to a host behind a MAP enabled CPE. The border relay has the following FMR and BMR configuration:

- Rule IPv6 prefix `2001:db8:0000::/40`.
- Rule IPv4 prefix `192.0.2.0/24`.
- PSID offset 4 and length 8.
- Rule EA-bits length 16 bits, which means 8 bits suffix from the rule IPv4 prefix is used.
- Border relay prefix `2001:db8:ffff::/64`.
- The subscriber is assigned with a /56 prefix.

When a packet is received from the Internet with a source address `12.13.14.15` (0x0c0d0e0f) and port `80` destined to an address `192.0.2.18` (0xc0000212) and port `9030` (0x2346), then the FMR destined to the CPE becomes:

- EA-bits will be `0x1234`; the 8-bit suffix of `192.0.2.18/24` is `0x12` and the PSID of `9030` is `0x0034`.
- IPv6 prefix is `2001:db8:0000::/40`.

When put together the FMR is: `2001:db8:00 + 12 + 34 + 00 + IID` and remembering the specific formulation of the IID the final FMR IPv6 address is: `2001:db8:0012:3400:00c0:0002:1200:3400/128`. Similarly, the source address of the border relay becomes: `2001:db8:ffff:0000 + IID` and based on the specific formulation of the IID in the case of DMR the final IPv6 address is: `2001:db8:ffff:0000:000c:0d0e:0f00:0000/128`. Summarizing:

- IPv6 source address: `2001:db8:ffff:0:c:d0e:f00:0000`.
- IPv6 destination address: `2001:db8:12:3400:c0:2:1200:3400`.
- IPv6 source port: `80`.
- IPv6 destination port: `9030`.

5.3.7 Other Tunneling or Translation Based Transition Mechanisms

Application Proxies

As described previously, NAT64 provides a generic mechanism to translate between IPv4 and IPv6. However, sometimes a generic solution is not desirable for certain reasons. These reasons include performance, manageability, and protocol incompatibility with middle-boxes like NAT64. An alternative can be provided by application specific proxies. An application specific proxy can be an explicit proxy where the client is configured to support a proxy. This category includes web (HTTP) proxies [44] that are used in many enterprises anyway. Some protocol architectures are also in general adapted well for proxying – for instance Internet mail [45].

A proxy is, simplifying heavily, an entity that acts like a server to the client, and a client towards the upstream server. For instance, a web proxy takes HTTP requests from the client side and sends the request to the server, which is supposed to provide the actual service. As there are two separate connections, the connection between the client and the proxy, and the proxy and the server, these can also use different IP versions. For instance, a client can be in an IPv6 network and use a proxy to access a service in an IPv4 network as long as the proxy has access to both networks.

Internet mail was used as an example earlier for a protocol that is easily proxied. This is due to the email protocol architecture. The email clients connect to servers for receiving and sending email. The servers are then connected to other servers to deliver the mail to the server, which serves the email recipient. Thus, as in web proxies, the connection between the client and the server is completely separate from the connections between the servers. Again, the email servers can provide the service to their clients on an IP version independent of the connections between the servers themselves.

Address and Port Sharing

Address and port sharing, often referred to as *A+P* or *port-restricted IPv4 addressing*, was originally described in RFC 6346 [20]. Section 5.3.6 has already described the basics of address sharing using port ranges, and a MAP based solution approach. Concatenating an IPv4 address and a transport protocol (e.g. Transport Control Protocol (TCP) or User Datagram Protocol (UDP)) port number effectively gives 48 bits of total 'IPv4' address space. Of course specialized A+P aware routing functions have to be deployed in the access network to make address sharing happen.

This section will not repeat the content of Section 5.3.6, but there have been multiple other proposals that are similar in spirit to MAP such as the DS-Lite variant described in [46, 47]. The basic principle is always the same, though, when it comes to address sharing: a number of end hosts that share the exactly same IPv4 address are differentiated by the port range (e.g. 1024 ports each) assigned to them. The port range does not need to be contiguous but distributed into the 65536 ports space using some method [48]. There are obvious similarities to NAT44 at the conceptual level.

Address and port sharing may have additional concerns and issues beyond the normal known NATting issues. For instance, some protocols check for IP addresses and limit a number of connections seemingly originating from or terminating at the same host. When address and port sharing is used, this can obviously lead to false positives, since the same shared address can belong to multiple end hosts [49].

DS-Lite with B4 in UE

Dual-stack Lite (DS-Lite) [8] is yet another IPv6 transition solution, which tunnels IPv4 packets over IPv6 tunnels. DS-Lite was aimed at deployments where the ISP core network is IPv6-only and still customer networks are dual-stack or IPv4-only. A DS-Lite solution has two functional entities: a Basic Bridging BroadBand (B4) and an Address Family Transition Router (AFTR). The B4 resides at the customer premises or in the end host, and terminates the IPv6 tunnels that carry IPv4 traffic. The AFTR is the ISP side IPv6 tunnel concentrator, which also contains a centralized Carrier Grade NAT (CGN). IPv6 traffic is delivered natively over the ISP core network between the customer sites and the Internet. Customers, or rather their CPEs, are identified and numbered using the unique IPv6 address of the tunnel endpoint used for tunneling the IPv4 traffic. This identifier is often referred to as the *Softwire-ID*.

In a typical DS-Lite deployment the B4 function resides inside the CPE device. However, RFC 6333 also allows a deployment model where the B4 is located inside the end host, which in our 3GPP architecture case would be the UE. In the case of 3GPP architecture and DS-Lite deployment, the Software-ID would be the IPv6 address that the UE configured using the PDN Connection assigned IPv6 prefix. Figure 5.13 illustrates how a DS-Lite deployment could be overlaid on 3GPP architecture. The AFTR most probably

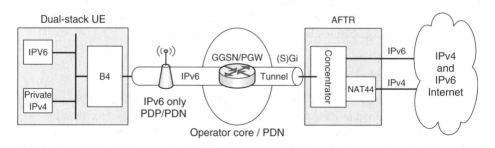

Figure 5.13 DS-Lite with a B4 in the UE deployed in 3GPP network.

is a standalone node but could also be integrated into a PGW. Since DS-Lite relies on the NAT44 in the network, and the customer networks are identified using the IPv6 address of the tunnel endpoint, the private IPv4 addressing has no real function than to make the UE side IP stack work. In that sense DS-Lite resembles GTP whose user plane IP addressing is merely to please the UE side IP stack, and Softwire-ID would be used in a similar manner to Tunnel Endpoint Identifiers (TEIDs).

Since a dual-stack UE may want to use Dynamic Host Configuration Protocol version 4 (DHCPv4) for configuring itself (e.g. DNS server addresses), the B4 typically has to implement a DHCPv4-proxy functionality. The DHCPv4-proxy would essentially do an address family translation to DHCPv6 and use that over the IPv6-only access. DS-Lite has been criticized for a number of reasons (and not only in 3GPP network architectures). First, it would use tunneling over the wireless leg, which is highly undesirable in 3GPP networks. Second, DS-Lite relies on a powerful NAT44 function (i.e. CGN) in the core network side. Third, DS-Lite with B4 inside the UE requires end host modifications.

For any larger deployment the bootstrapping of the DS-Lite service has to be automated. RFC 6334 [50] defines the required DHCPv6 options that the B4 can use for bootstrapping the AFTR address. From the 3GPP networks and deployments point of view the dependency on DHCPv6 could be an issue. As we know, DHCPv6 is not a mandatory part of the 3GPP system architecture. Furthermore, since DS-Lite is not even part of the recommended transition tool set for 3GPP networks, there is little hope, for example, for NAS protocol support for bootstrapping the required bootstrapping information.

6RD

6RD [5] is an automatic tunneling protocol based on 6to4. The difference from 6to4 is that instead of the 6to4 IPv6 prefix, 6RD uses the operator's own prefix. This solves many of the problems manifested in 6to4 the IPv4 address does not have to be globally unique as long it is unique in the operator's network, and the asymmetric routing issues are not present in 6RD as the routing is based on the operator's own address prefix.

6RD has been designed to be run from the customer premises equipment, such as a residential gateway or a home router. The 6RD address is created by combining the operator's IPv6 prefix with a part of the home router's IPv4 address. This creates a prefix that is equal or shorter than /64. As the used IPv4 address is configured by the operator, a similar stateless tunneling mechanism to 6to4 can be used. 6RD is an example of a transition solution that can implement a mesh topology.

Interestingly, 6RD has been one of the most successful transition technologies with many large live network deployments. Its success has been on the fixed network access where CPE devices could be upgraded (or switched) to 6RD supporting firmware. There has not been a similar level of success or even interest in using 6RD on the wireless networks, because of obvious reasons discussed earlier in this chapter.

5.4 Transition Scenarios for 3GPP

3GPP started the specification work on IPv6 and on the IPv6 transition at the end of 1990s. When looking at the IPv4 address exhaustion problem timeline, and the IPv6 technical maturity, 3GPP started the specification work relatively early. While starting the specification early is admirable, the implementation and especially deployment has lagged behind significantly. In practice, this means that the technology, network environment, and even the usage of the 3GPP networks has changed significantly during that time. Especially drastic changes in the environment have been the recent exhaustion of the IPv4 address space on the other side, and the exponentially increased data use in mobile broadband networks on the other. Hence, the original transition scenarios, which relied on IPv4 addresses to be relatively easily available, or that took shortcuts in performance due to assuming transition mechanisms to be relatively short lived workarounds, did not serve the changed environment at all. In addition, the availability of IPv6 has increased significantly in infrastructure equipment, and in the infrastructure on which the operators build their mobile networks. Whereas, at the beginning of the transition specification work a real concern for the mobile operators was obtaining IPv6 access, this is hardly a concern anymore.

Therefore, the 3GPP community has had to go through multiple transition scenario iterations for different environmental situations. In addition to the transition scenarios, the 3GPP community has had multiple opportunities to change its mind on the role of IPv6 in general. For instance, the IP Multimedia Subsystem (IMS) [51] IP version has swung between being exclusively IPv6, dual-stack, and mostly IPv4 over the years. The first transition scenarios were made only for 2nd Generation (2G) and 3G-GPRS because the specification for Evolved Packet System (EPS) has not even started yet. In addition, the support for dual-stack PDP Context was not originally in the 3GPP specifications. Later on, scenario work did take these changes and enhancements into account. In this section, we will first look at the evolution of the transition scenarios in 3GPP networks, and then look at the scenarios that have survived over time.

5.4.1 Transition Scenario Evolution

The first attempt to agree on the 3GPP transition scenarios was done at the IETF in the early 2000s. The result was two documents discussing IPv6 transition in GPRS, RFC 3574 [52], and RFC 4215 [53]. The former document describes a series of transition scenarios providing connectivity for the UE. GPRS network internal transition scenarios were not discussed. The document describes the transition scenarios for GPRS and IMS use cases. However, we will focus here on the GPRS use cases. Figure 5.14 describes the GPRS scenarios. The idea of listing scenarios in RFC 3574 was to document the likely use cases, but not to impose any particular solutions.

RFC 4215 documented an analysis of the scenarios described in RFC 3574. The analysis strongly favored the use of dual-stack in the 3GPP environment, and discouraged the use of translators between IPv4 and IPv6. This was logical at the time as adequate number of IPv4 addresses were still available. The analysis rather suggested turning on IPv6 in addition to IPv4. While work was ongoing on RFC 3574 and RFC 4215, obtaining IPv6 upstream, was still a real issue. In addition, the operators' own network equipment infrastructure, such as routers and switches, did not necessarily support IPv6 at all. This is clearly visible in the scenarios in Figure 5.14, where transporting IPv6 over IPv4 is prominently visible. Today, the scenarios are not really relevant anymore. However, the scenario for tunneling IPv4 over IPv6, which at that time did not seem important, may become relevant in the future as discussed earlier in this Chapter.

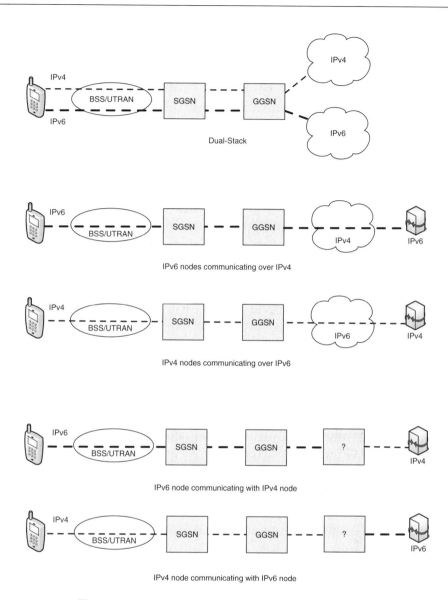

Figure 5.14   RFC 3574 and RFC 4215 transition scenarios.

Although, IMS is not really within the scope of this book, the discussion about the IP version used in IMS has had implications for the transition scenarios of the packet service as well. 3GPP documented IMS transition scenarios in the technical report 23.981 'Interworking aspects and migration scenarios for IPv4-based IP Multimedia Subsystem (IMS) implementations' [54]. Although most of the document concentrates on IMS scenarios, it considers the applicable GPRS network scenarios as well. In early 2000, when IMS was considered, the support for PDP Type IPv6 was not considered to be widespread in the networks deployed at the time. Hence, considerable effort had to go into discussing what

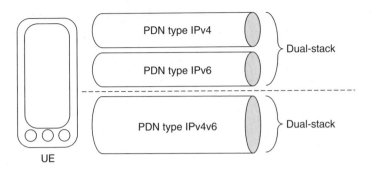

Figure 5.15   Dual-stack with two PDN Connections (upper) or single PDN Connection (lower).

should happen in situations where some of the elements in a GPRS network did support IPv6 and others did not. 3GPP decided that it would not support any translation of PDP Type during the network transit, for example between Serving Gateway Support Node (SGSN) and GGSN. Practically, if one of the elements did not support PDP Type IPv6, the UE would need to use IPv4. This simplified the transition scenarios significantly. The conclusion here was also that supporting dual-stack in the UE is key to transition, and the networks would also need to be upgraded to support PDP Type IPv6 over time.

More recently, 3GPP has agreed on a new set of IPv6 transition scenarios. These scenarios are listed in the 3GPP 23.975 'IPv6 migration guidelines' document [55] and include *dual-stack*, *IPv6-only*, and *Gateway-Initiated Dual-Stack Lite*. The last of which does not seem to have traction in the industry, and hence we do not consider it in any in depth this section.

Appendix B of 23.975 also includes an set of other technologies that were topical when the core of 23.975 was written. These are *Dual-Stack Lite*, *Address+Port*, *Protocol Translation*, *Per-interface NAT44* [56], and *BIH/NAT64*. From this set of additional technologies, the Protocol Translation is the one currently seeing market adoption, and hence we have included it in this section.

5.4.2   Dual-stack

As described above, the recommended transition model is the dual-stack deployment model. This is true for the more generic case recommended by the IETF, and for 3GPP networks in particular. In the 3GPP networks there are two ways to provide UEs with dual-stack connectivity: two PDN Connections (PDP Contexts or EPS Bearers) – one for IPv4 and one for IPv6, or a single dual-stack IPv4v6 type of PDN Connection. Figure 5.15 illustrates both these setups, with parallel IPv4 and IPv6 PDN Connections above and single IPv4v6 PDN Connection below.

Prior to Release-8, 3GPP supported only separate PDN Connections for IPv4 and IPv6. Dual-stack bearer type was introduced into the EPS architecture in Release-8, but for GPRS the support only came in Release-9. PDN Connections, be they PDP Contexts or EPS Bearers, require state in the networks, so using a single connection burdens the network much less than two connections. Therefore, it is advisable to use the dual-stack PDN Connection. However, due to possible limitations of continuing to use the

IPv4v6 PDN Connection type in the networks, two parallel connections are needed in fallback cases.

These fallback cases are needed because there is uncertainty over the support of the dual-stack PDN Connection in the networks. As described in depth in Section 4.5, this causes a set of issues and complications in PDN Connection setup procedures.

The dual-stack approach obviously does nothing for the IPv4 address exhaustion problem, and hence the dual-stack cannot really be a long-term solution. In the short term, IPv4 address shortage is managed by implementing IPv4 NAT functions and providing UEs with, possibly overlapping, private IPv4 addresses from the space allocated by RFC 1918 [25].

5.4.3   IPv6-only

Until most, if not all, Internet services have transitioned to IPv6, the strictly IPv6-only scenarios with no access to the IPv4 portion of the Internet are only applicable in limited deployments. The possible use cases include applications where the usage is restricted to a limited set of well-managed applications. Such use cases include, Machine-to-Machine (M2M) applications where the application set is very limited (for instance temperature or electricity meter reading), or the application is set and operated by the same party. The latter case includes operator services like Internet Protocol Television (IPTV), Voice over IP (VoIP), services provided by IMS or other walled garden services.

Restricting the generic Internet access use case to IPv6-only services would, as of now, be too limiting because a major portion of Internet services are still accessible only via IPv4. Although IPv6 transition is moving forward at a steady pace, it will most probably still take a significant time before IPv4 Internet access can be completely removed from a normal user.

That said, it is possible to use IPv6-only PDN Connection for a UE as long as the connection to the IPv4 portion of the Internet can be guaranteed. Practically, this means that the UE, including all applications used in it, is IPv6 capable and the network provides translation services between IPv6 and IPv4. The IPv6 to IPv4 translation enables the connectivity to the IPv4 services in the Internet regardless of the IP version that the UE is using. We described a suitable tool for protocol translation, NAT64, in Section 5.3.3. This use case is not very different from the Internet access much of the users get, where the UE gets a private RFC 1918 address, and a NAT44 is used.

In an IPv6-only use case, IPv4 applications could also be supported by providing Network Address Translation from IPv4 to IPv6 (NAT46) within the UE. In this case, the UE internal NAT46 would provide an IPv4 address to the applications, and thus create an illusion of IPv4 access. However, as discussed, access to IPv4 services will still be needed for the foreseeable future. Supporting the access to IPv4 services for the IPv4 applications would require double translation. We discuss double translation in the following section.

5.4.4   Double Translation

The requirement of the IPv6-only approach to have all applications IPv6 enabled is a tough request. Some operating systems provide IP version agnostic APIs, which considerably simplifies transition for applications that correctly use the provided APIs (see more in

Section 3.9.2). However, there are many cases where either a significant portion of the applications have not correctly used the provided APIs, or the operating system has different APIs for IPv4 and IPv6. Until recently, some mobile operating systems did not support IPv6 at all. Therefore, there are still a significant number of (legacy) applications that work only with IPv4. The lack of IPv6 deployment in 3GPP networks has not helped either, as application developers have not had good opportunities for verifying IPv6 support of their applications.

A solution for allowing a UE to have IPv6-only PDN Connection while still serving IPv4-only applications is to implement a double translation scheme. Double translation works by combining BIH-like functionality on a UE with NAT64 on a network. The IPv4-only applications send IPv4 packets to IPv4 addresses as usual, but before sending the packet out from the UE it is translated into IPv6. The resulting IPv6 packet is sent over 3GPP access and then translated back to IPv4 at the NAT64.

While the double translation approach is not part of the official transition scenarios listed in 23.975 [55], it is used in some pre-commercial deployments and trials. It is well possible that it will be used as part of commercial transition from IPv4-only 3GPP access to IPv6-only access. The benefits of double translation when compared to dual-stack is the lack of need for parallel PDN Connections or support for the new IPv4v6 type of PDN Connection.

464XLAT [19], that was discussed in Section 5.3.4, is another example of a double translation solution. It has been trialed in some IPv6-only mobile networks as a possible consumer-grade solution. Like many other transition solutions, 464XLAT has its deployment reality issues. It is not part of the official 3GPP architecture and UE requirements, which makes network and device vendors reluctant to implement its required components as part of their default feature sets.

5.5   Transition Impacts on 3GPP Architecture

We have now discussed the different transition technologies, and the 3GPP specific scenarios where they apply. Now it is time to look at how the transition affects the 3GPP architecture itself, and its support functions. The 3GPP networks include quite a lot of elements and functionality that is not visible in the 3GPP architecture, but is still vital for the system. These include DHCP, AAA, and DNS servers. Sometimes the operators also have additional infrastructure that is transparent to the user, but very much part of the operator's infrastructure.

In this section, we will look at the impacts that the transition technologies have on the 3GPP architecture – including new elements, and what the impact transition to IPv6 has on all the functionality that 3GPP networks rely on.

5.5.1   Transition Impact on the Supporting Infrastructure

The 3GPP network architecture pictures we saw in Chapter 2 show the 3GPP network elements, and present almost a sterile picture of the network architectures. However, in reality the network is much bigger than what is visible from those pictures. The little lines between the little boxes are actually more than just point-to-point connections (except for the radio interface) – they are complete networks with cables, wireless microwave links, switches, and routers. In addition to these networks, there are also other networks that are not visible. These networks include the operations and management network, which is used to configure, monitor, and maintain the network elements themselves.

And it is not only packet network infrastructure that is needed. A complete infrastructure of support functions is needed to drive a mobile network. The infrastructure includes DNS servers, AAA servers as described in Chapter 4, network management infrastructure, and the complete billing and charging infrastructure. In addition, operators have additional infrastructure for their own services, and also infrastructure, for instance, for web acceleration.

All these are also affected by the transition to IPv6. Sometimes the trickiest to transition are the billing and charging, operations and management, and the operators, own other infrastructure. The reason is that these infrastructures often include a lot of best practices proprietary to the operator – their own code that was written a long time ago, but no one knows by whom or when. If the software includes, for instance, literal IPv4 addresses or data structures only suited to IPv4, the code has to be modified, or even completely replaced. In this section, we will look at some of the parts of the system that may be affected by the transition to IPv6 and how to mitigate those impacts. It has been jokingly said that IPv6 transition is a year 2000 problem without a firm deadline.

5.5.2 IP Network Support Systems

As stated previously, the 3GPP network needs a set of support systems that are vital for running an IP network. In a way, there are two completely disjoint IP network infrastructures: the network that connects the network elements to each other, and the network elements to the management infrastructure which is its own network with its own systems. The second network is the network that provides the services for the UE. This network may include operator services, service enhancement functions, and the connection to the Internet. These two networks are completely isolated from each other for security purposes. We will examine those networks separately below.

Operator's Internal Infrastructure Network

The operator's internal network may include one network or different network segments for operations and management for instance. This is the network that connects the different 3GPP elements to each other. Hence, it is the 3GPP transport plane that creates the IP layer in the protocol stack pictures in Chapter 2, and the lines between the elements in the architecture pictures. As described in Chapter 2, the transport plane is completely separate from the user plane. Thus, the transition to IPv6 in the transport plane does not have to happen at the same time as the user plane. In addition, as the operator's internal network is only for the operator network elements, there are extremely few network nodes compared to the network where the UEs are. Practically, this means that the operators have little incentive to transition their internal network infrastructure anytime soon. However, eventually the operator internal infrastructure will be transitioned to IPv6 as well.

As mentioned above, the operator internal network needs its own switches and routers, which obviously have to be upgraded to support IPv6 when transition becomes relevant. In addition, the operator's internal network also has its own DNS, and operations and management infrastructure. The operations and management infrastructure has to be updated to support the configuration of the IPv6 addresses to the switches and routers, enabled to support configuration and monitoring over IPv6, and be able to receive monitoring data with IPv6 addresses in it. Though, on paper, this all seems relatively straightforward, the operators may have equipment and software in their operations and management infrastructure which is relatively old. The infrastructure has been created at the same time as the original IP network, and perhaps new features have been added to it over the years. If it has worked well for the operator, there has been little incentive to upgrade it. Making a fundamental change like IPv6 transition may not be straightforward.

As in all normal IP networks, DNS infrastructure is used within the operator network infrastructure. The network elements will have their own DNS names, and as described in Chapter 2, the APNs are resolved using the DNS. Hence, the DNS has an important role in a 3GPP network. When considering IPv6 transition, DNS has two parts that have to be updated – the access to the DNS server, and the DNS records. Obviously, if the network is transitioned to IPv6 the network elements will contact the DNS server on IPv6 as well. Hence, the DNS server needs to be IPv6 capable. In addition, when network elements are configured with IPv6 addresses, the DNS records pointing to those nodes have to have the related AAAA records added.

Operator Service Network

Here we call the network that connects the UE to the operator services, and to the Internet, the operator service network. Sometimes there may actually be different networks, and hence different APNs, for the operator's own services and for the Internet connectivity. The transition process does not really change even if there are one or multiple networks, or network segments. The only difference is that if some services are in their own networks behind a separate APN, the networks can be transitioned at different times. This may be useful if the operator's own services are difficult to transition when they have been built perhaps on platforms that are not upgradable. However, this also means that multiple networks have to be upgraded instead of just one.

The upgrading of the network itself, including the switches and routers, the DNS, and the operations and management infrastructure, follows the same lines as the transition of the operator's infrastructure network, or any other IP network. The switches and routers have to be IPv6 capable to transport IPv6 packets. The IPv6 access to the DNS servers has to be added – and it has to be ensured that the DNS servers do support queries over IPv6, configuring the infrastructure providing the protocol configuration options to the UE to include the IPv6 address of the DNS servers, and obviously connecting the DNS server to the operator's IPv6 network. As in the operator's internal network, the operation and management infrastructure has to be upgraded to support configuring and monitoring of the IPv6 network and its nodes.

In Section 4.4, we described the address allocation support infrastructure including AAA, and DHCP servers. If the operator is using this kind of infrastructure to manage IP address allocation, the infrastructure has to be upgraded to support IPv6 as well. In addition, the operator must obtain its own IPv6 address block from the local RIR.

In addition, the services themselves have to be transitioned. For Internet access service, it should be obvious that the operator must make sure it has IPv6 capable upstream Internet access. This is a relatively straightforward nowadays in most countries. Upgrading the operator's own services may, however, be more difficult, depending on the platform used to create the services, and the age of the platform. However, sometimes it may be possible to put a protocol translator (i.e. a NAT64) in front of an old service, and that way prolong the life of an older service platform.

Some operators also run an infrastructure to enhance Internet or web access, and to monitor the traffic usage. This service infrastructure has sometimes been created over the years of operation, and may prove difficult to upgrade. However, if the operator wishes to use the same or similar infrastructure, the infrastructure has to be upgraded or replaced, or a new infrastructure for IPv6 access has to be created.

5.5.3 Tools to Divide Subscribers Per IP Capability

Moving a large number of subscribers to IPv6 might not be something that happens overnight or is even close to trivial to provision and roll out. There are some known (not necessarily best) practices on how to start moving subscribers in reasonable blocks into IPv6 enabled service offering. We will present principles for some approaches. They all, more or less, circulate around provisioning tricks that are doable using APNs. We have left out various provisioning 'tricks' that concentrate on overwriting APN names based on various vendor-specific means, like charging characteristics (configured in the subscription profile), International Mobile Subscriber Identity (IMSI) and Mobile Station International Subscriber Directory Number (MSISDN) analysis, or transparently switching the PDN Connection to a different APN (using triggers from Policy and Charging Control (PCC) or AAA infrastructure).

Dedicated IPv6 APN and a New Subscription

A completely new non-overlapping IPv6 or dual-stack enabled APN is created and provisioned into the network. Subscribers and their UEs can be provisioned to the new APN on an *opt-in* basis. Since activating the new APN requires active participation from the subscriber side, there is a strong chance that the subscriber actually knows what he or she is doing. This approach works specifically for trialing purposes, but does not scale to large subscriber populations.

Dedicated APNs allow IPv6 rollouts where the existing IPv4-only deployment remains completely untouched and new IPv6 features are rolled out possibly using more recent platforms. This can also work out well with new subscribers.

IPv6 Enabled APN and Re-provisioning an Old Subscription

In situations where an operator has an existing IPv4-only APN, but gradually wants to move subscribers to IPv6, an overlapping (name wise) IPv6-only or dual-stack APN is created and provisioned into the network. For simplicity, it makes sense to configure it into the same Gateway GPRS Support Nodes (GGSNs/PGWs) where the legacy IPv4-only APN is configured and hence save trouble on DNS configurations.

If the network supports dual-stack – that is Release-9 or better for GPRS or Release-8 or better for EPS – then existing subscription profiles in HLRs or HSSs are updated to

PDP or PDN Type IPv4v6. Otherwise, if the network does not support PDP or PDN Type IPv4v6, then an IPv6-only APN with the same name as the IPv4-only APN is provisioned into the subscription profiles. The provisioning may need to be done on an *opt-in* basis, when subscribers switch to a new mobile device or even as a mass re-provisioning in HLRs/HSSs. Mass provisioning always has the danger that some old UEs might get confused.

However, there is one small issue in this approach. Prior Release-8, 3GPP specification did not explicitly spell out that APNs may overlap name-wise when the PDP types differ. There are old GGSNs that cannot support two APNs with the same name, but different PDP type, in the same box.

All-in-one APN

An existing APN is re-configured into the networks as an *all-in-one* APN. This means the same APN serves IPv4-only subscribers, IPv6-only subscribers, and dual-stack subscribers. The APN information in the HLR/HSS subscription profiles may then be re-provisioned on an opt-in basis or as a mass rollout. Then the selection of the used PDP or PDN type is a combination what the UE asks for and what is provisioned in the subscription profile. This approach works nicely when the operator knows that the rest of the network is dual-stack capable throughout and UEs are also up to date.

The selection of the PDP or PDN type can also be controlled per subscriber by using the backend AAA infrastructure. There are many possibilities to override what the UE asked for based on various pieces of information available in the (S)Gi AAA interface profile. See Section 4.4 for more details.

Dedicated IPv6-enabled GGSN or PGW

Specifically, in roaming cases it might make sense to dedicate a GGSN or a PGW only for inbound roamers. The GGSN, or the PGW, would have a plain configurations with for example, IPv4-only APNs. The same APNs could be IPv6 enabled when the UE is not roaming, and gets anchored to a different GGSN or PGW. In this way, roaming users can be turned IPv6-less without any kind of provisioning in the subscription profiles or UEs.

5.5.4 Translation Implications

Today, still relatively in the beginnings of the transition, there are many services in the Internet that are only accessible over IPv4. If an operator would like to move many of its customers to IPv6, the operator may choose to provide IPv6-only access to the subscribers, and translate between IPv6 and IPv4. Translating between IPv6, and IPv4 has its implications, though.

The most obvious implication is the additional equipment that is needed. When translation such as NAT64 is used, the NAT64 equipment has to be installed. The dimensioning of the NAT64 infrastructure has to be done carefully. Usually, when a new service or technology is introduced, the amount of data traffic is small at first, and grows over time. With NAT64, at the beginning of the transition, the amount of IPv4 content is high. The data amounts should diminish over time as the transition moves forward. Therefore, the peak capacity of the translation equipment is needed at the beginning, and then the need for the capacity falls gradually over time. Obviously, whether the real deployment actually follows this trend depends strongly on the operator. In any case, as the IPv6 transition moves forward and content becomes increasingly available over IPv6, the use of translation will decrease.

New equipment has also to be configured and maintained. Therefore, there is a clear new overhead for the operator's operational people. In addition, as we described in Section 5.2.1 how, with a NAT64, a DNS64 has to be set up as well. Where NAT64 can be new a equipment and a new system, DNS64 has to be added to the operator's DNS infrastructure.

We described earlier in Section 5.2.1, how modifications made by DNS64 to DNS responses have implications both for protocols like DNSSEC, and also for the UE itself. These are no real implications for the network, but there are for the end user equipment. The problem with these implications is that different equipment, and even different software versions, can be differently affected. Therefore, the user experiences can vary.

5.5.5   Transition Support in the Transport Plane

We described some of the implications to the transport plane, and thus, to the operator's internal infrastructure network in Section 5.5.2. However, there are also implications for the higher layers – the GTP. For instance, there can be a handover situation where the serving SGSN changes from a dual-stack capable SGSN to a single-stack IPv4 capable SGSN. If the dual-stack SGSN has had the GTP tunnel to the GGSN over IPv6, and the dual-stack SGSN would only pass the IPv6 address of the GGSN to the IPv4-only capable SGSN, the new serving SGSN would have to drop the connection – even if the GGSN were dual-stack capable.

Therefore, 3GPP has included a mechanism in the GTP protocol to pass the network elements' IPv4 and IPv6 addresses. This ensures correct functionality even in an environment where there are mixed IPv4-only capable and dual-stack capable network elements. This functionality would not help in an environment where there were single-stack IPv4 and single-stack IPv6 elements. However, the 3GPP standards do not support this kind of setup, and it has not defined how GTP should be run through translators. It is, however, unlikely that anybody would configure this kind of environment – at least not intentionally.

5.5.6   Roaming

We have now learned that transitioning to IPv6 is a relatively difficult task that needs focus and careful planning. Even inside one operator's network, the task can be daunting, and if it is not planned carefully and executed well, multiple problems on many levels can happen. The reassuring thing about transitioning an operator's network is the fact that everything can be controlled by that operator, and that the first operators are already running commercial IPv6 networks. Unfortunately roaming removes the luxury of full control. Basically, the operator does not have to only work with its own network, but has to make sure that its service and network will be able to cooperate with all of the roaming partners' networks and services. Section 4.2.5 discussed a number of techniques of how to turn on roaming restrictions that are specifically targeted to selectively disable

IPv6 roaming. In certain cases that can be done in the granularity of an APN per visited operator (or rather per Public Land Mobile Network (PLMN)).

When an operator turns IPv6 on as an end user service, it does not only expose its own network to IPv6, but also possibly all its roaming partners' networks. The subscribers expect that the services they are getting in their home network will also be available when they are roaming in different networks. If nothing else, they will try to use the same service also when roaming. This can cause surprises both for the user and for the network in which the user is roaming. If the service is not available, the user experience may be disappointing, which by itself is already undesirable. However, the user may also call the operator's support line causing additional costs or, depending the level of disappointment, may even cause them to change operators. On the other hand, the visiting network might be a little surprised at the IPv6 traffic running on their network. This causes IPv6 addresses to be present in the visiting operator's logs, billing records, and other records that the operator might collect for different purposes. This traffic can either cause warning bells to ring as the IPv6 traffic may be considered to be an anomaly, or make the unsuspecting backend software such as billing and charging software behave in an undefined way, or even fail completely. Obviously, it is every operator's own responsibility to make sure the network is well prepared and protected when IPv6 is turned on by the operator itself or by a roaming partner, even if the operator itself has not yet turned IPv6 services on.

With roaming, the situation is similar on the transport plane. When an operator turns on IPv6 in its transport plane network, the change can be visible to the roaming partners as well. The DNS queries for the network elements resolve to both A and AAAA records, and the GTP signaling may start to contain IPv6 addresses. The operators will have to be ready for this, and make sure that at least a roaming partner's transition does not cause adverse effects on the operator's own network, or vice versa. Surprising issues might show up, such as Border Gateways (BGs) or GTP aware firewalls not being IPv6 Information Element (IE) friendly.

Until recently, the operator community has neglected IPv6 roaming details. Around mid 2010 the GSM Association (GSMA) started working on a white paper that covers IPv6 impacts on roaming [57]. Unfortunately, the white paper was not publicly available for non-GSMA members, although those who care about roaming in 3GPP networks most probably have some level of membership in GSMA. Based on the white paper discoveries, a number of GSMA documents were identified that are impacted: GPRS roaming guidelines IR.33 [58], GPRS Roaming eXchange (GRX)/IP Packet eXchange – evolved GRX (IPX) backbone specification IR.34 [59], IP addressing and Autonomous System (AS) numbering guidelines IR.40 [60], GRX/IPX DNS guidelines IR.67 [61] and LTE roaming guidelines IR.88 [62]. At the time of writing, at least the IR.33 includes a number of IPv6 transition scenarios in a roaming environment. The guidelines only concern the user plane, since the GRX/IPX that tunnels the roaming GTP traffic is not going to be migrated to IPv6 transport in the foreseeable future.

5.5.7 Impact of Delayed Transition to IPv6

We have already described problems and difficulties that may occur while executing the transition to IPv6. It is important to also understand that it is not an easy task. However, not preparing for the transition to IPv6 has its impacts and costs.

Today, the easily available IPv4 addresses have been exhausted in the Asia-Pacific region under Asia-Pacific Network Information Center (APNIC) responsibility, and the European and Middle-East Region under Réseaux IP Européens Network Coordination Centre (RIPENCC) responsibility. Practically, this means that the traditional way of getting IPv4 addresses by asking for them from the RIR is no longer possible in those regions. The North American, American Registry for Internet Numbers (ARIN), region is expected to follow suit during 2013. The only way for operators to get additional IPv4 address space is to purchase it from somebody that is willing to sell it. The operators depend on IP addresses to be able to grow the number of subscribers they serve. A new risk for an operator's business is that the is growth becomes, at least partially, dependent on somebody else being willing to sell their IPv4 addresses. In addition, it is expected that the purchased address blocks will be smaller than the address blocks provided by the Regional Internet Registrys (RIRs). This adds cost to the operator's address management.

One strategy for reducing the dependency on the globally routable IPv4 addresses is NAT. Most operators, if not all, already provide private IPv4 addresses to their subscribers, and thus 'NAT' the Internet access. The ratio of connections NATted to a single public, globally routable IPv4 address will only grow in the near future. Using NAT in a big operator's network means that the NAT equipment must be highly scalable and have high capacity. This kind of equipment is not cheap, and thus it adds cost to the operator – both the capital cost of the equipment, and the operational cost of the management of the additional equipment. And all this at the time when the data volumes are increasing. In addition, the increasing ratio of multiplexing of connections will eventually cause some of the connections to fail, and thus may add random failures to the subscribers' service.

Many of the major Internet content providers have turned IPv6 on in their services, and most, if not all, modern computing platforms can now use IPv6. This means that when operators turn on IPv6 services, they have to be of at least the same quality as the IPv4 service. Thus, there is little space to train or trial IPv6 anymore. As time goes by, even more traffic will be moving to IPv6 and its importance will only grow. We have explained that the transition to IPv6 is difficult, and needs quite a bit of work. Hence, there is also a lot that can go wrong. Trying with limited traffic amounts, and limited user groups, to mature the service is important. Therefore, it is important to enable IPv6 while there is still time to fix errors. Leaving the transition to a time when the operator has to transition, will just make the transition more painful.

As we described in Section 5.5.6, the operator is both impacted by its own transition, and also by the transition of other operators. The impact of not being prepared can have unexpected consequences for the operator, for instance when the roaming partners transition.

We hope that we have made it clear that while the transition to IPv6 is difficult, and potentially expensive, the cost of not transitioning is even higher.

5.6 Transitioning to IPv6

The previous sections explained the different transition mechanisms, the transition scenarios in 3GPP environment, and finally the impacts of IPv6 transition. This section tries to give an overview of how to prepare, and execute transition to IPv6 in the 3GPP environment. We are fully aware that every network is different, and therefore it is neither possible nor even desirable to try to provide a detailed, universal transition plan.

5.6.1 Application Developer's Transition Plan

As the transition to IPv6 is pervasive at all layers of network protocols, the application developers also need to be sufficiently aware of IPv6. We explained in Section 5.4.4 that many operating systems provide IP version agnostic APIs. At the minimum, the developers have to use these. In addition, application developers have to make sure that the data structures that handle IP addresses are capable of storing both IPv4 and IPv6 addresses. For a small class of applications this is a major task.

Applications may handle IP addresses at different levels. Clearly, an application may encounter IPv6 addresses when opening a connection, for instance, as a result of a DNS query. However, there are also other circumstances where an application may encounter IPv6 addresses. For instance, if the application implements or uses a protocol over the Internet, which exchanges literal IP addresses. In addition to internal data structures, the applications that represent or expect as input literal IP addresses in the User Interface (UI), have to make sure it is actually possible to show and to take as input also literal IPv6 addresses.

5.6.2 Phone Vendor's Transition Plan

The IPv6 capability of a mobile phone or similar device is determined by a multitude of factors. We have already described the application layer. In addition, the operating system and the middleware, which create the software platform for the device, have to support IPv6. Almost all modern operating systems used in the modern smartphones support IPv6. However, theoretical IPv6 support may not be adequate in practice. The IPv6 has to be compiled in and turned on as well. In addition, the middleware in the software platform has to be able to support IPv6. The operating system support is not very useful, if the applications cannot actually use IPv6.

In addition to software, the hardware also has to be capable of supporting IPv6. The main component is the wireless modem – the part of the hardware and firmware that implements the 3GPP cellular technology, and actually connect a device to the network. As a sign of good progress, the support for IPv6 in the cellular chipsets has improved greatly in the last couple of years.

Thus, in the phone the whole stack of software and hardware has to be prepared for IPv6. A couple of years ago, practically only one phone manufacturer supported IPv6 in their phones. At that time the whole stack from the software platform to the hardware of the wireless modem was under the control of that company, and hence it was easier to have IPv6 support early on. Today, the list of equipment and vendors supporting IPv6 has grown greatly. However, IPv6 is not necessarily a feature that phone vendors advertise, and hence it may be difficult to see which models support IPv6, and which do not.

5.6.3 Network Operator's Transition Checklist

As stated earlier, every network operator, and thus every network, is different. There are multiple variables that create the unique characteristics of an operator. For instance, geography, population, and demography of the country could have influenced the operator's network greatly. In addition, the time when the network has been taken into use, and how it has been set up, influence the state of the current network. Therefore, it is difficult to give a planning template for IPv6 transition that would fit all or even most operators. However, there are certain steps that are common for all operators regardless of the characteristics of the network. Here we try to provide a checklist of steps that an operator should consider when planning and executing the transition to IPv6.

Prioritize end-user service over operator internal infrastructure. The biggest pressure on addressing is on the end-user services, as the UEs use the most addresses. As the operator's internal network is more static, and there are fewer nodes in the network, the use of addresses is much less, and the address planning even with few addresses is easier.

Survey the network carefully. It is especially important to check all the support functions such as billing and charging, operations and management, and subscriber management functions, as these are the systems that are most likely to have proprietary software, created over the years. The proprietary software is also the most difficult to upgrade to support IPv6, as company-internal software tends to be less well documented than commercial software. In addition, maintaining software needs fewer people than upgrading the software. It is important to understand not only which systems will be directly connected to the IPv6 network, but also which systems will have to deal with IPv6 addresses. For instance, the operations and management system may be connected to the network elements with IPv4, but it has to be able to configure an element with IPv6 interfaces. Therefore, it is important to understand, early on, which systems in the network may be potentially problematic. Furthermore, quite often old hardware equipment has to replaced since the equipment vendor plain refuses to support IPv6 or any new features (excluding critical patches for software errors) on the old hardware.

Plan carefully the transition. Based on the survey mentioned above, the system upgrades have to prioritized, scheduled, and the dependency of the systems understood.

Start small in the beginning. Practically, the simplest service should be handled first – for example, Internet access with no web acceleration. In addition, it is important to start with a closed user group, or with a trial. When subscribers are participating in a trial, they understand better that the service may not always be of commercial quality, and even may have glitches now and then. In addition, a trial directed towards a technically inclined user group usually also provides more useful feedback than a commercial service would. Small user groups can also be more easily provisioned than a commercial, widely used service, especially if the subscriber provisioning system is not able to support IPv6 at the beginning.

Start training early on in the transition project. IPv4 and IPv6 are very similar in many ways, but there are also many differences. Even technically capable engineers do benefit from training, as the knowledge accumulated with IPv4 may prove to be inadequate with IPv6. Training is especially important for customer support people, when the IPv6 service goes mainstream. There are guaranteed to be unexpected problems when the service is commercialized. It is important that customer support agents do know how to solve the problems, or are able to suspect IP-address family issues and know to where they can escalate them.

Grow smartly as the service matures. A closed user group trial can mature the service nicely, but may not be adequate to hash out all the problems. Only adequate traffic volumes can guarantee that scalability issues are found. On the other hand, opening the service to too large a group too fast may be problematic. It may be difficult to roll back once millions of phones have been provisioned, and the customer support hotlines are congested.

Participate in network operator groups and other conferences where operators share their experiences. This may help avoiding some of the potholes other people have already found. In addition, sharing one's own experiences also helps finding solutions to common problems, and also helps the vendors to enhance their products. The transition to IPv6 is an industry-wide project. It makes good business sense to participate in that.

Try it out! There is no substitute for hands-on experience. Only trying it out will provide the experience and knowledge on what can go wrong and what works well. It is best to find the problems as early as possible rather than after the commercial launch.

Start now! Unless the planning and transitioning has not even been started yet, it is time to start now. The only thing that can be assured for the transition journey is that it is going to be difficult, and issues about the network will be discovered that will need some problem solving. The earlier they can be found, the quicker solving of the problems can be started, and therefore, there is more time to solve them. It is difficult to fix problems, but it is definitely not nice to have to fix them in a hurry!

As stated earlier, this is an overview at best. However, we hope that we can import some structure to the transition process, and perhaps help the transition process. The most important part is to start the enabling of IPv6 as a service. The other steps will follow naturally.

5.7 Chapter Summary

In this chapter we looked into IPv6 transition in 3GPP networks and into a selected set of IPv6 transition mechanisms. We boldly chose the IPv6 transition mechanism for a closer inspection that we find fitting into wireless 3GPP architecture environment, where the majority of the wireless end hosts are assumed to be handsets. This obviously leaves a large group of ingenious variations of different IPv6 transition tools out of the detailed discussion.

3GPP has advocated dual-stack as the preferred transition mechanism since the beginning of the whole IPv6 transition discussion. As we realized and illustrated, dual-stack is not always a commercial or even a realistic technical option during the phased rollout of IPv6. Therefore, considerable effort was also directed to the development of transition mechanisms that are tailored for single IP version access networks. Within the 3GPP mindset the transition mechanisms that rely on tunneling over the cellular link have systematically been played down, which have led to moderate success of address family translation solutions, NAT64 being a textbook example. In order to alleviate known application level issues of a single IP version in an end host, so-called double translation schemes came into being, which eventually deliver dual-stack to the applications on an end host over single IP version access networks.

We also went through a number of known operational concerns, scalability issues, and identified impacts to the 3GPP system architecture. We even touched the surface of roaming challenges from the IPv6 point of view. At the end, we specifically laid out and described few rules of thumb on how to get started with a phased IPv6 rollout in a 3GPP network, and pictured the favored transition scenarios. Although the guidance may sound simple and straightforward, the operational reality has proven it to be far from a quick exercise with a timely rollout.

References

1. Gilligan, R. and Nordmark, E. *Transition Mechanisms for IPv6 Hosts and Routers*. RFC 1933, Internet Engineering Task Force, April 1996.
2. Nordmark, E. and Gilligan, R. *Basic Transition Mechanisms for IPv6 Hosts and Routers*. RFC 4213, Internet Engineering Task Force, October 2005.
3. Carpenter, B. and Moore, K. *Connection of IPv6 Domains via IPv4 Clouds*. RFC 3056, Internet Engineering Task Force, February 2001.
4. Carpenter, B. and Jung, C. *Transmission of IPv6 over IPv4 Domains without Explicit Tunnels*. RFC 2529, Internet Engineering Task Force, March 1999.
5. Townsley, W. and Troan, O. *IPv6 Rapid Deployment on IPv4 Infrastructures (6rd) – Protocol Specification*. RFC 5969, Internet Engineering Task Force, August 2010.
6. Templin, F., Gleeson, T., and Thaler, D. *Intra-Site Automatic Tunnel Addressing Protocol (ISATAP)*. RFC 5214, Internet Engineering Task Force, March 2008.
7. Huitema, C. *Teredo: Tunneling IPv6 over UDP through Network Address Translations (NATs)*. RFC 4380, Internet Engineering Task Force, February 2006.
8. Durand, A., Droms, R., Woodyatt, J., and Lee, Y. *Dual-Stack Lite Broadband Deployments Following IPv4 Exhaustion*. RFC 6333, Internet Engineering Task Force, August 2011.
9. Tsirtsis, G. and Srisuresh, P. *Network Address Translation - Protocol Translation (NAT-PT)*. RFC 2766, Internet Engineering Task Force, February 2000.
10. Aoun, C. and Davies, E. *Reasons to Move the Network Address Translator – Protocol Translator (NAT-PT) to Historic Status*. RFC 4966, Internet Engineering Task Force, July 2007.
11. Arends, R., Austein, R., Larson, M., Massey, D., and Rose, S. *DNS Security Introduction and Requirements*. RFC 4033, Internet Engineering Task Force, March 2005.
12. Li, X., Bao, C., and Baker, F. *IP/ICMP Translation Algorithm*. RFC 6145, Internet Engineering Task Force, April 2011.
13. Weil, J., Kuarsingh, V., Donley, C., Liljenstolpe, C., and Azinger, M. *IANA-Reserved IPv4 Prefix for Shared Address Space*. RFC 6598, Internet Engineering Task Force, April 2012.
14. Bagnulo, M., Matthews, P., and vanBeijnum, I. *Stateful NAT64: Network Address and Protocol Translation from IPv6 Clients to IPv4 Servers*. RFC 6146, Internet Engineering Task Force, April 2011.
15. Perreault, S., Yamagata, I., Miyakawa, S., Nakagawa, A., and Ashida, H. *Common requirements for Carrier Grade NATs (CGNs)*. Internet-Draft draft-ietf-behave-lsn-requirements-10, Internet Engineering Task Force, December 2012. Work in progress.
16. 3GPP. *General Packet Radio Service (GPRS); GPRS Tunnelling Protocol (GTP) across the Gn and Gp interface*. TS 29.060, 3rd Generation Partnership Project (3GPP), March 2012.
17. Li, X., Dawkins, S., Ward, D., and Durand, A. *Softwire Problem Statement*. RFC 4925, Internet Engineering Task Force, July 2007.
18. RIPE. *IPv6 Address Allocation and Assignment Policy*. PDF 552, RIPE, May 2012.
19. Mawatari, M., Kawashima, M., and Byrne, C. *464XLAT: Combination of Stateful and Stateless Translation*. Internet-Draft draft-ietf-v6ops-464xlat-09.txt, Internet Engineering Task Force, January 2013. Work in progress.
20. Bush, R. *The Address plus Port (A+P) Approach to the IPv4 Address Shortage*. RFC 6346, Internet Engineering Task Force, August 2011.
21. Troan, O., Dec, W., Li, X., Bao, C., Matsushima, S., and Murakami, T. *Mapping of Address and Port with Encapsulation (MAP)*. Internet-Draft draft-ietf-softwire-map-02, Internet Engineering Task Force, September 2012. Work in progress.
22. Soliman, H. *Mobile IPv6 Support for Dual Stack Hosts and Routers*. RFC 5555, Internet Engineering Task Force, June 2009.

23. 3GPP. *Mobility management based on Dual-Stack Mobile IPv6; Stage*. TS 24.303, 3rd Generation Partnership Project (3GPP), September 2011.

24. Carpenter, B. *Advisory Guidelines for 6to4 Deployment*. RFC 6343, Internet Engineering Task Force, August 2011.

25. Rekhter, Y., Moskowitz, B., Karrenberg, D., deGroot, G. J., and Lear, E. *Address Allocation for Private Internets*. RFC 1918, Internet Engineering Task Force, February 1996.

26. Bagnulo, M., Sullivan, A., Matthews, P., and vanBeijnum, I. *DNS64: DNS Extensions for Network Address Translation from IPv6 Clients to IPv4 Servers*. RFC 6147, Internet Engineering Task Force, April 2011.

27. Baker, F., Li, X., Bao, C., and Yin, K. *Framework for IPv4/IPv6 Translation*. RFC 6144, Internet Engineering Task Force, April 2011.

28. Bao, C., Huitema, C., Bagnulo, M., Boucadair, M., and Li, X. *IPv6 Addressing of IPv4/IPv6 Translators*. RFC 6052, Internet Engineering Task Force, October 2010.

29. Nordmark, E. *Stateless IP/ICMP Translation Algorithm (SIIT)*. RFC 2765, Internet Engineering Task Force, February 2000.

30. Kawamura, S. and Kawashima, M. *A Recommendation for IPv6 Address Text Representation*. RFC 5952, Internet Engineering Task Force, August 2010.

31. Narten, T., Draves, R., and Krishnan, S. *Privacy Extensions for Stateless Address Autoconfiguration in IPv6*. RFC 4941, Internet Engineering Task Force, September 2007.

32. Arkko, J. and Keranen, A. *Experiences from an IPv6-Only Network*. RFC 6586, Internet Engineering Task Force, April 2012.

33. vanBeijnum, I. *An FTP Application Layer Gateway (ALG) for IPv6-to-IPv4 Translation*. RFC 6384, Internet Engineering Task Force, October 2011.

34. Huang, B., Deng, H., and Savolainen, T. *Dual-Stack Hosts Using Bump-in-the-Host (BIH)*. RFC 6535, Internet Engineering Task Force, February 2012.

35. Korhonen, J. and Savolainen, T. *EDNS0 Option for Indicating AAAA Record Synthesis and Format*. Internet-Draft draft-korhonen-edns0-synthesis-flag-02, Internet Engineering Task Force, February 2011. Work in progress.

36. Chen, G., Cao, Z., Byrne, C., Xie, C., and Binet, D. *NAT64 Operational Experiences*. Internet-Draft draft-ietf-v6ops-nat64-experience-01, Internet Engineering Task Force, January 2013. Work in progress.

37. Savolainen, T., Korhonen, J., and Wing, D. *Discovery of IPv6 Prefix Used for IPv6 Address Synthesis*. Internet-Draft draft-ietf-behave-nat64-discovery-heuristic-13, Internet Engineering Task Force, November 2012. Work in progress.

38. Arends, R., Austein, R., Larson, M., Massey, D., and Rose, S. *Protocol Modifications for the DNS Security Extensions*. RFC 4035, Internet Engineering Task Force, March 2005.

39. Huang, B. and Deng, H. *Prefix NAT: Host based IPv6 translation*. Internet-Draft draft-huang-pnat-host-ipv6-01, Internet Engineering Task Force, July 2009. Work in progress.

40. Bellis, R. *DNS Proxy Implementation Guidelines*. RFC 5625, Internet Engineering Task Force, August 2009.

41. Gilligan, R. and Nordmark, E. *Transition Mechanisms for IPv6 Hosts and Routers*. RFC 1933, Internet Engineering Task Force, April 1996.

42. Boucadair, M., Matsushima, S., Lee, Y., Bonness, O., Borges, I., and Chen, G. *Motivations for Carrier-side Stateless IPv4 over IPv6 Migration Solutions*. Internet-Draft draft-ietf-softwire-stateless-4v6-motivation-05, Internet Engineering Task Force, November 2012. Work in progress.

43. Mrugalski, T., Troan, O., Bao, C., Dec, W., and Yeh, L. *DHCPv6 Options for Mapping of Address and Port*. Internet-Draft draft-ietf-softwire-map-dhcp-01, Internet Engineering Task Force, August 2012. Work in progress.

44. Fielding, R., Gettys, J., Mogul, J., Frystyk, H., and Lee, T. B.. *Hypertext Transfer Protocol – HTTP/1.1*. RFC 2068, Internet Engineering Task Force, January 1997.

45. Klensin, J. *Simple Mail Transfer Protocol*. RFC 5321, Internet Engineering Task Force, October 2008.

46. Cui, Y., Sun, Q., Boucadair, M., Tsou, T., Lee, Y., and Farrer, I. *Lightweight 4over6: An Extension to the DS-Lite Architecture*. Internet-Draft draft-cui-softwire-b4-translated-ds-lite-09, Internet Engineering Task Force, October 2012. Work in progress.

47. Farrer, I. and Durand, A. *lw4over6 Deterministic Architecture*. Internet-Draft draft-farrer-softwire-lw4o6-deterministic-arch-01, Internet Engineering Task Force, October 2012. Work in progress.

48. Boucadair, M., Levis, P., Bajko, G., Savolainen, T., and Tsou, T. *Huawei Port Range Configuration Options for PPP IP Control Protocol (IPCP)*. RFC 6431, Internet Engineering Task Force, November 2011.

49. Boucadair, M., Zheng, T., Deng, X., and Queiroz, J. *Behavior of BitTorrent service in PCP-enabled networks with AddressSharing*. Internet-Draft draft-boucadair-pcp-bittorrent-01, Internet Engineering Task Force, May 2012. Work in progress.

50. Hankins, D. and Mrugalski, T. *Dynamic Host Configuration Protocol for IPv6 (DHCPv6) Option for Dual-Stack Lite*. RFC 6334, Internet Engineering Task Force, August 2011.

51. 3GPP, *IP Multimedia Subsystem (IMS); Stage 2*. TS 23.228, 3rd Generation Partnership Project (3GPP), September 2010.

52. Soininen, J. *Transition Scenarios for 3GPP Networks*. RFC 3574, Internet Engineering Task Force, August 2003.

53. Wiljakka, J. *Analysis on IPv6 Transition in Third Generation Partnership Project (3GPP) Networks*. RFC 4215, Internet Engineering Task Force, October 2005.

54. 3GPP, *Interworking aspects and migration scenarios for IPv4-based IP Multimedia Subsystem (IMS) implementations*. TR 23.981, 3rd Generation Partnership Project (3GPP), December 2009.

55. 3GPP. *IPv6 migration guidelines*. TR 23.975, 3rd Generation Partnership Project (3GPP), June 2011.

56. Arkko, J., Eggert, L., and Townsley, M. *Scalable Operation of Address Translators with Per-Interface Bindings*. RFC 6619, Internet Engineering Task Force, June 2012.

57. GSMA. *IPv6 EMC Task Force; Roaming and Interoperability Impacts of IPv6 Transition Whitepaper*. White Paper 1.2, GSM Association (GSMA), March 2011.

58. GSMA. *GSMA PRD IR.33 GPRS Roaming Guidelines*. PRD IR.33 6.0, GSM Association (GSMA), May 2011.

59. GSMA. *Inter-Service Provider IP Backbone Guidelines*. PRD IR.34 5.0, GSM Association (GSMA), December 2010.

60. GSMA. *GSMA PRD IR.40 'Guidelines for IP Addressing and AS Numbering for GRX/IPX Network Infrastructure and User Terminals'*. PRD IR.40 5.0, GSM Association (GSMA), December 2010.

61. GSMA. *GSMA PRD IR.40 DNS/ENUM Guidelines for Service Providers and GRX/IPX Providers*. PRD IR.40 7.0, GSM Association (GSMA), May 2012.

62. GSMA. *GSMA PRD IR.88 LTE Roaming Guidelines*. PRD IR.88 6.0, GSM Association (GSMA), August 2011.

6

Future of IPv6 in 3GPP Networks

The evolution of Internet Protocol version 6 (IPv6) in the 3rd Generation Partnership Project (3GPP) networks is nowhere near complete. In fact, the commercial deployments of the features we have described in the first five chapters have hardly even begun yet.

The future is likely to bring fine tuning of the basic IPv6 features, as the industry gains more operational experience from the actual deployments and as academia discovers improvement needs via research. In 3GPP networks the Stateful Dynamic Host Configuration Protocol version 6 (DHCPv6) (see Section 3.5.2) may gain traction, if more controlled enterprise-style address allocation means are going to be required, or maybe multicasting of content directly between end hosts will find commercially viable use cases. The experience that will be gained with widespread IPv6 deployment will undoubtedly also mean that inefficiencies, security issues, and vulnerabilities will be found and addressed in due course.

Besides those particular improvements that are perhaps best categorized as *IPv6 maintenance*, there will be very interesting new applications that will use IPv6 and bring both growth and significant technical improvements. In this chapter we will talk about five IPv6 related future topics that the authors are finding to be of particular interest, but these five are by no means the only IPv6 related improvements that we will (or we hope to) see in 3GPP networks during the coming exiting years! It should be noted that some of the material presented in this section is rather futuristic and just presents ideas with which the authors have been experimenting.

6.1 IPv6-based Traffic Offloading Solutions

The significant improvements in the 3GPP networks' throughput and latency since the introduction of 3rd Generation (3G), and rapid developments in smart phones, has changed usage patterns of 3GPP access. Nowadays users are increasingly using both always-on and data-intensive applications. These applications include social media, instant messaging, Voice over IP (VoIP), and web browsers, which are often extremely data intensive because of transmitting video, audio, and photographs. As a result, the traffic and signaling volume increase that has occurred in the core network side has been phenomenal. Operators are facing tough competition, and profit margin growth is slower than traffic volume growth. At the same time operators have investment needs for cellular core infrastructure, radio

Deploying IPv6 in 3GPP Networks: Evolving Mobile Broadband from 2G to LTE and Beyond, First Edition.
Jouni Korhonen, Teemu Savolainen and Jonne Soininen.
© 2013 John Wiley & Sons, Ltd. Published 2013 by John Wiley & Sons, Ltd.

access, and Internet Protocol (IP) transmission capabilities. The fast growth of traffic and signaling volumes combined with relatively slower profit growth have forced both operators and equipment vendors to seek for IP traffic offloading solutions.

It has been identified that offloading of the bulk Internet traffic (bulk from the 3GPP operators' point of view, not necessarily from the end users' point of view) to alternative Internet access technologies could be a viable solution to temporarily get away from the investment pressure for more powerful network infrastructure. Quite unsurprisingly, due to almost ubiquitous support in smart phones and the huge numbers of both private and public access points, the Wireless Local Area Network (WLAN) is visioned to be the technology to which traffic would be offloaded. In the grand scheme of things, the offloading needs to happen from the wide-area radio technologies into local-area radio technologies – hence enabling better spectrum usage and efficiency. It just happens to be that the 3GPP is effectively the main wide-area radio technology, and WLAN is the local-area radio technology, even though the suite of radio technologies is large.

To make offloading happen, 3GPP has placed significant effort on standardizing IP traffic offloading solutions for the Evolved Packet Core (EPC). The solutions under standardization include Local IP Access (LIPA) (see Section 4.3.16 of [1]), Selective IP Traffic Offload (SIPTO) enabled bearer (see Section 4.3.15 of [1]), IP Flow Mobility and Seamless WLAN Offload (IFOM) [2], S2a Mobility based on GTP and WLAN access to EPC (SaMOG) [3], and non-seamless WLAN offload. As is traditional in 3GPP, the described solution approaches rely on tight cellular operator control and integration to the 3GPP network architecture. The tight integration comes with typical cons, such as the difficulties of easily interworking with parallel non-3GPP accesses, but also with typical pros such as tight integration which can bring security and ease-of-use benefits. A thing common to all 3GPP solutions is the utilization of a new Access Network Discovery and Selection Function (ANDSF) [4], which is needed for providing operator policies to the User Equipments (UEs) [5].

Only time will tell how widespread the adoption of 3GPP offloading solutions will be among the various general purpose Operating System (OS) vendors, especially if the 3GPP's solution, namely the ANDSF, is not needed for any other purposes. In fact, it is not yet well understood what level of management burden the 3GPP operators themselves are willing to bear for micromanaging IP traffic offloading policies.

In this section we will look in more detail at how different Internet Engineering Task Force (IETF)-defined IPv6-based offloading solutions work. The IP version agnostic 3GPP specific solutions are left for other books and for the referenced 3GPP documents to describe. When it comes to Internet Protocol version 4 (IPv4) offloading, we do not believe that it is a justified use of time to try to specify sophisticated IPv4-based offloading solutions, as the future is in IPv6. The exhausted IPv4 address space, with widespread use of private IPv4 addresses, creates addressing conflicts and therefore makes the address and router selection process hard.

6.1.1 Motivations in Cellular Networks

Originally, the 3GPP General Packet Radio Service (GPRS) architecture adopted a point-to-point approach for network access. This is visible, for example, in how the mobile device – the UE – sees the network connection, the network interface, and how the connection between the UE and the network is realized. The point-to-point connection between the UE and the external Packet Data Network (PDN) is referred to as a *PDP context* or a *PDN connection*.

Traditional phones (see Section 2.6.1) typically have a silo view of the network connectivity for a single application or a group of applications. Launching a new application either shares an existing PDN Connection or creates a new parallel PDN Connection with its own IP address, effectively making the UE multi-interfaced. The host IP stack and the radio modem are typically tightly integrated. *Split-UEs* (see Section 2.6.2) have a clear separation of the host IP stack and the radio modem. They are sometimes even in physically separate devices. Hence, a Split-UE is just like any IP enabled host equipped with a wireless radio networking technology. All applications typically share the same single PDN Connection. The Split-UE model is increasingly becoming the dominant design.

The 3GPP Release-7 enhancements on the High Speed Packet Access (HSPA) radio technology and architectural enhancements aiming at a flatter network (e.g., direct tunneling) have significantly reduced the gap between the cellular and fixed broadband access. Moreover, the 3GPP Release-8 Long Term Evolution (LTE) for both radio and core network evolution has brought cellular broadband access to and even beyond the mass market fixed broadband access – such as Digital Subscriber Line (DSL) – in terms of network latency and throughput. The increase in cellular broadband usability and the near flat rate billing models have had two notable outcomes: 1) consumers use cellular broadband in a similar way to fixed access and 2) the cellular Internet service providers have experienced an exponential IP traffic growth.

Heavy traffic growth poses investment pressures on radio access, backhaul and packet core capacity. Today 3GPP packet core network architecture and live network deployments tend to favor heavy IP traffic aggregation at a Gateway GPRS Support Node (GGSN) or Packet Data Network Gateway (PGW) into relatively few sites. The IP traffic packet forwarding capacity of these gateway nodes might not sustain the traffic growth. In addition, the increased capacity investment requirements can be hard to meet due to the declining average revenue per user.

The issues discussed above have driven mobile operators to evaluate solutions to offload the bulk or low profit (Internet) traffic to alternative access technologies that are cheaper to deploy in dense hotspot areas and ideally would use 'someone else's' backhaul. There has been a vision of using managed WLAN deployments or subscribers' home WLANs as alternative accesses. We can identify two main motivations:

- *Compensate the cellular radio coverage and access capacity* with a cheaper radio technology in a dense hotspot area while still routing the traffic through the operator's packet core.
- *Bypass the operator's packet core, cellular access, and possibly backhaul*, completely to maximize 'savings'. This is our main target scenario in this chapter.

The development of UEs strengthens these IP offloading visions, since effectively all high-end and most mid-range UEs have a WLAN support. There are few issues left though. First, a UE may not be operating multiple radios simultaneously, which effectively prohibits selective offloading of IP traffic between access technologies. In this book, we assume that a UE can operate multiple radios in parallel. Second, how to determine which IP traffic to offload and which traffic to route through the mobile operator's core. Third, how to steer the offloading decision making from the network side. This can be challenging, especially in the case of Split-UEs. Fourth, how to minimize the impact on the operator network and especially in the UE.

6.1.2   Benefits of IPv6-based Offloading Approaches

The main benefits of IPv6 offloading approaches can be categorized as follows:

- *System-agnostic standardization*: Layer 3 based offloading solution by definition is agnostic of layers 2 and 1, and hence IPv6-based offloading solutions allow easy deployment and utilization on various access technologies.
- *Prefix-based policies*: Offloading policies using IPv6 prefixes, the fewer and the shorter prefixes the better, is reliable and doable with low management burden.

Effectively, the IPv6 offloading solutions are part of the 'normal' IPv6 routing, address selection (see Section 3.5.4), and next-hop selection (see Section 3.4.6) procedures with minimal administrative overheads. With working Internet connectivity to specific operator services, independent of the used access network, it is possible to view the IPv6 traffic offloading solutions as ordinary IPv6 suite components that virtually all IPv6 implementations have to support, and for which no access technology specific enhancements are needed.

6.1.3   IP-friendly Offloading Solutions

We have experimented with three IP-friendly approaches to achieving an IP traffic offloading solution for multiple interfaced UEs with network side control for the offloading policies. The first approach builds on the top of DHCPv6. The second approach builds on top of IPv6 Neighbor Discovery Protocol (NDP), and the third approach extends the second solution with IPv4 capabilities.

The IP-friendly solutions try to conform to a 'pure IP' view and have specifically designed Split-UEs in mind. None of the solutions aim at guaranteeing that the offloading policy provided by the network would work in all possible cases. More important is that the UE always has Internet connectivity, meaning that none of the used access networks should be a walled garden. The three solutions have several aspects in common:

- They rely on IPv6 features when possible. No 3GPP-specific extensions are required.
- They primarily target UEs with cellular 3GPP access. A GGSN or a PGW is used as the policy coordinator in the operator network.
- Cellular 3GPP radio is considered a *trusted* access and hence used to deliver offloading policies. When the 3GPP access is not available there is obviously nothing to offload either.
- They are designed to benefit from multiple interfaces.
- The offloading policies are typically in the form of *offload everything except a few selected destination networks*.

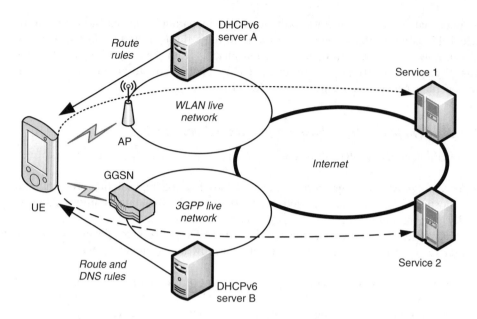

Figure 6.1   Example deployment scenario of new DHCPv6 options.

New DHCPv6 Options

During the last couple of years IETF has been working for improvements for multi-interfaced hosts suffering from problems listed in RFC 6418 [6]. This has already resulted in one new DHCPv6 option being published for improved Recursive DNS Server (RDNSS) selection in RFC 6731 [7]. The work is continuing for DHCPv6 improvements for more specific route information delivery [8].

An example scenario where these new options could be used is illustrated in Figure 6.1. In this example scenario, a host is connected to Internet via dual-stack WLAN and 3GPP accesses. DHCPv6 servers that provide the configuration information are needed in both access networks, although if all traffic from the WLAN network were tunneled to the operator's core there would be no need for a separate DHCPv6 server in the WLAN network.

UE Implications

The new DHCPv6 options, mentioned above, incur a set of changes for UEs in addition to simple DHCPv6 query changes. Namely, a UE implementing the improvements for RDNSS selection has to be able to decide to which RDNSS a Domain Name System (DNS) query should be sent. This logic is completely new for the host operating systems; however, some advanced DNS resolvers can partially support this feature with just smart and dynamic configuration. On the other hand, if the UE already supports RFC 4191 [9] described in Section 3.5.6, the DHCPv6 route option in the best case merely means having an additional source for more specific route information.

The improved RDNSS selection RFC recommends using Domain Name System Security Extensions (DNSSEC), or secure and trusted channel, to counter attackers sending malicious DHCPv6 options. By injecting malicious RDNSS selection rules attackers could cause targeted hosts to send only specific queries to attackers, hence making an ongoing attack more difficult to detect. This attack could enable eavesdropping, Denial of Service (DoS), or other kinds of man-in-the-middle attacks.

Luckily, for the 3GPP UEs, the 3GPP access is generally considered secure enough for these DHCPv6 options, and hence UEs can trust the information that they receive over the 3GPP link. Therefore, DNSSEC is not required if these new DHCPv6 options are only requested from the 3GPP access's DHCPv6 servers.

Default Router and More-specific Route Selection

The IETF standard *'Default Router Preferences and More-Specific Routes'* [9] (RFC 4191) extends IPv6 NDP with two router preference flags in the Router Advertisement message header, and a *Route Information Option (RIO)*. The former allows for a simple three-step prioritization of default routers in a host's default router list (*LOW, MEDIUM – the default – and HIGH*). The last allows for a Router Advertisement emitting router, even if not willing to be included into host's default router list, to mark up to 17 IPv6 destinations for which the router wants to serve as the first-hop. RFC 4191 can also be deployed in a multiple interfaces scenario. The only consideration is to limit the number of interfaces accepting RFC 4191 extensions to one interface that is both trusted and centrally managed by the operator. In our case, the 3GPP cellular connection fulfills these requirements. Several mainstream OSs already implement RFC 4191, including Linux, BSD variants, and Microsoft Windows, starting from XP. The interface(s) accepting an RFC 4191 extension can typically be specified just by using host side configuration.

The 3GPP architecture relies on IPv6 Stateless Address Autoconfiguration (SLAAC) for its (un)trusted 3GPP access. The PGW must always send Router Advertisements for SLAAC purposes and therefore using the 3GPP access as a command channel for NDP-based offloading purposes is a small enhancement to the PGW functionality. Furthermore, Router Advertisement can always be sent unsolicited. The NDP-based solution is both extremely lightweight and also allows an on-demand push mode of operation from the cellular operator point of view.

We have experimented with RFC 4191 and IPv6 offloading in live networks with multiple-interface UEs. Since modifying a live network PGW is typically not an option for experimentation, it is simpler to extend an external router such as L2TP Network Server (LNS) in Layer 2 Tunneling Protocol (L2TP) based external Access Point Name (APN) router deployments (see Section 4.2.4). The required Router Advertisement injecting tools can be implemented in the APN termination router. When the APN is terminated to an external router, the PGW essentially becomes a bridge and the APN terminating router is the first-hop router for UEs. The setup that the authors used for *the proof of concept* allowed sending of Router Advertisements with RFC 4191 support, among various other Neighbor Discovery messages, from the network to a specific UE.

Host OSs typically prefer WLANs over cellular access. For example, Linux implicitly prefers WLAN access over cellular access, thus prioritizing the first-hop router(s) on WLAN access over the first-hop router on a cellular access. For consistency, we should always use *LOW* default router priority on the cellular access. Since the default router priority on other interfaces is implicitly *MEDIUM*, the host IP stack will prefer any other interface for default destinations than cellular. When the cellular operator wants to route certain traffic over the cellular, it only needs to send a Router Advertisement with an RIO containing the IPv6 prefix(es) for those destinations (e.g., prefix(es) used to number the operator's own services). The default router and default address selection algorithm [10] in the host IP stack will take care of selecting an appropriate interface for the new IPv6 connections (e.g., existing TCP connections will not move). When the model of operation is *'offload everything except specific destinations'*, the number of routing rules can be kept low.

If the UE has multiple PDN Connections, the PGW on each of those cellular connections could be sending a Router Advertisement with more specific routes. That could result in a conflict of interest and offloading policies. Still, if the router preferences are set correctly on the cellular side and alternative accesses are available, the offloading to the alternative accesses would take place. And furthermore, each PGW could still deterministically mark the destinations that must go through them using more-specific routes.

Enhanced Neighbor Discovery with IPv4 Support

To overcome the limitations of the IPv6-only RFC 4191 approach, we have proposed new IPv4 traffic specific Router Advertisement options [11]. The new options enable access routers to convey an IPv4 default gateway address and more-specific IPv4 routes.

In a prototype made for IPv4 offloading, the existing Linux kernel implementation of linux/net/ipv6/ *ndisc.c* had to be extended by adding an intercepting hook function (*kernel offload hook*) and one module (*kernel offload module*) to push more-specific IPv4 route and the IPv4 default gateway address Router Advertisement options from kernel to user space via the *sysfs interface*. In the user space, an *IPv4 offload daemon* handled the main tasks for IPv4 traffic offloading by manipulating IPv4 routing tables. Figure 6.2 shows a captured Router Advertisement message, where the RIO option carries an IPv4-mapped IPv6 address (::ffff:81.90.77.0/120) to route an IPv4 subnet to the Router Advertisement originating router. The Router Advertisement also carries the default IPv4 gateway address of the dual-stacked router as the last option (10.6.6.6 i.e. 0a 06 06 06).

All this was needed because currently Linux's IP stack sends only few selected Router Advertisement options into the user space – RFC 4191 options are processed within the kernel, and IPv4-mapped IPv6 addresses in RIOs do not affect IPv4 routing, and we had no other way of changing the IPv4 default gateway on demand. Note that the proof-of-concept implementation, whose capture is shown in Figure 6.2, was based on the earlier version of [11]. The current version of [11] is independent of RFC 4191 and therefore removes the need for our own hook function and module. The existing kernel method for pushing Router Advertisement options into the user space could be reused with a few lines of code change. We recognize that our RFC 4191 extension for IPv4 more-specific routes would need a major push in standardization to reach wider acceptance.

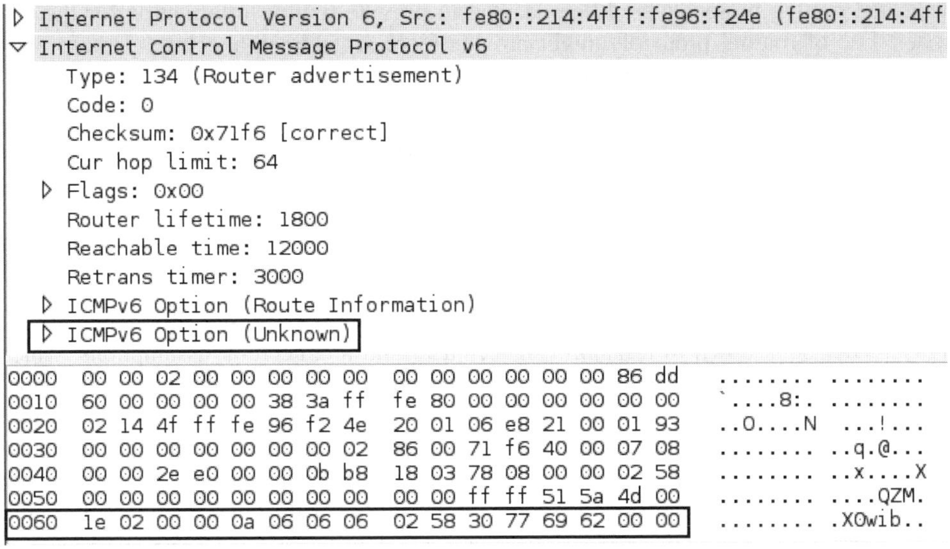

```
▷ Internet Protocol Version 6, Src: fe80::214:4fff:fe96:f24e (fe80::214:4ff
▽ Internet Control Message Protocol v6
      Type: 134 (Router advertisement)
      Code: 0
      Checksum: 0x71f6 [correct]
      Cur hop limit: 64
    ▷ Flags: 0x00
      Router lifetime: 1800
      Reachable time: 12000
      Retrans timer: 3000
    ▷ ICMPv6 Option (Route Information)
    ▷ ICMPv6 Option (Unknown)

0000  00 00 02 00 00 00 00 00  00 00 00 00 00 00 86 dd   ........ ........
0010  60 00 00 00 00 38 3a ff  fe 80 00 00 00 00 00 00   `....8:. ........
0020  02 14 4f ff fe 96 f2 4e  20 01 06 e8 21 00 01 93   ..O....N  ...!...
0030  00 00 00 00 00 00 00 02  86 00 71 f6 40 00 07 08   ........ ..q.@...
0040  00 00 2e e0 00 00 0b b8  18 03 78 08 00 00 02 58   ........ ..x....X
0050  00 00 00 00 00 00 00 00  00 00 ff ff 51 5a 4d 00   ........ ....QZM.
0060  1e 02 00 00 0a 06 06 06  02 58 30 77 69 62 00 00   ........ .XOwib..
```

Figure 6.2 A packet capture showing a Router Advertisement with RIO carrying a IPv4-mapped IPv6 address and the IPv4 default gateway address (which Wireshark interprets as unknown).

6.1.4 Concluding Remarks

We discussed 3GPP-specified IP traffic offloading solutions and presented three IP-friendly offloading variations that are intentionally made to operate only at the IP level and make use of IETF protocols. A cellular operator can take advantage of the cellular network connection as a secure command channel to push offloading policies into the UE, while still only using standard IETF protocols. The implementation experience proved that IP level solutions using IETF-only technologies are feasible and lightweight to deploy both on the network side and specifically on the UE.

One of the major challenges we faced was the support for IPv4 traffic offloading in operating systems. While modern IP stacks offer a rich feature set for IPv6 to implement offloading in a multiple interface device, there is no clean solution available for IPv4. Another significant challenge is the resistance faced in the IETF community to using DHCPv6 or Router Advertisements for delivering offloading policies, and the persistent battle between DHCPv6 and IPv6 Neighbor Discovery Protocol for generic host configuration.

We believe that the final deployed IP traffic offloading solution will likely be a mixture of existing technologies standardized in 3GPP, in IETF, and what modern IP stacks can do. It is unlikely that mainstream operating systems' IP stack would implement 3GPP specific technologies – some third-party dialer software may then add those missing elements. For the future, however, we believe that it would be useful to do further research on how dedicated routing protocols could be adapted to provide routing information for the end nodes. Use of routing protocols might provide more scalable architecture, perhaps even providing improved multi-interfacing properties for the end nodes.

6.2   Evolving 3GPP Bearers to Multiple Prefixes and Next-hop Routers

6.2.1   Background and Motivation

The 3GPP Evolved Packet System (EPS) [1] Bearer and connectivity model from the IP point of view has remained unchanged practically since the birth of GPRS. The connectivity model boils down to a handful of technical and architectural assumptions, which have remained untouched for reasons such as thinking that what 3GPP created is something unique to the cellular access, and from a fear of threatening the established service models. For instance, to add a new IP address or IPv6 prefix, an additional connection between the UE and the network has to be created instead of just adding IP addresses to an existing interface. This essentially turns the UE into a multi-interfaced host. In 3GPP architecture a gateway node, a PGW, provides access to an external PDN. The gateway node is located in the mobile operator core network (home or visited) and anchors the UE and its connection both mobility and IP topology-wise.

The 3GPP's connectivity model worked fine as long as IPv4 was the only realistic IP version and UEs were *traditional phones* (see Section 2.6.1). The phones had a single radio access technology, a limited openness to the IP stack for applications and a top-down approach for IP connectivity. Applications controlled the activation of network resources and had a built-in knowledge of what kind of network access they needed (APNs, Quality of Service (QoS), IP versions, and such like). The noteworthy proliferation of *smart phones* with a bottom-up approach to IP connectivity and the emerging need to solve the practical challenges of multiple network access interfaces on consumer UEs, offers us an opportunity to revisit the 3GPP bearer and connectivity model. Furthermore, the use of APNs has turned out to be conceptually challenging for bottom-up IP connectivity. In modern UEs, applications are provided with less control on activating the network connectivity, and in general for managing the traffic routing decisions in the UE.

Enabling services and access isolation using multiple network connections has a cost. Each 3GPP Bearer and connection consumes both radio network and core network resources that are typically subject to licensing, while the equivalent functionality could be accomplished by just allowing multiple addresses (beyond dual-stack) on a single network interface. Yet there is an emerging desire to provide shortcuts to the Internet, for certain IP traffic, bypassing parts of the operator infrastructure. The solutions today build on the use of multiple interfaces, which in cellular-only access scenarios falls back to the activation of a dedicated connection, such as a LIPA enabled bearer (see Section 4.3.16 of [1]), or SIPTO enabled bearer (see Section 4.3.15 of [1]).

We have experimented with an idea for a new 3GPP bearer and connectivity model, which advocates multiple addressing on a single 3GPP network connection and abandons the use of multiple APNs. We propose a new abstraction of the 3GPP link, which essentially turns the existing point-to-point link into a Non-Broadcast Multiple Access (NBMA) resembling link that allows for multiple exit points for IP traffic in operators' IP network infrastructure. We call the bearer and connectivity model as the *evolved bearer*.

The new bearer and connectivity model is incremental to the existing 3GPP EPS with a fallback to bearer types available in the current EPS. Since multi-addressing is natural to IPv6, the enhancements we propose are only available for IPv6. The new bearer and connectivity model inherits the existing IPv4 address-on-a-connection assumption. The IPv4 address is always anchored to the PGW as it is today with EPS Bearers.

6.2.2 Multi-prefix Bearer Solution Proposal

Motivation and Design Goals

We started to develop the evolved bearer around nine fundamental design goals. These design goals are discussed next:

1) The number of APNs has to decrease. In a long run the need for any end user or a UE requested APN should become extinct. That would definitely have a positive impact on the mobile host design and ease the provisioning of the mobile hosts.
2) Configuring multiple IP addresses/prefixes on a UE should not require an activation of multiple PDN Connections and default bearers.
3) Change of the link model from a strict point-to-point link into an NBMA like link, where *multiple routers* can appear on the same conceptual link. The routers are logically on the same link but following the NBMA philosophy cannot talk to each other in any other way than using unicast traffic.
4) Not all addresses/prefixes need mobility, which would, with the help of multi-addressing and a new link model, introduce the concept of the *simple-IP* where certain IP addresses/prefixes are not topologically anchored to the PGW and do not necessarily survive handovers.
5) The UE IP host configuration should only follow IETF standard procedures. We acknowledge that there is a cost with this assertion, since lower layer mechanisms are usually significantly faster to configure an end host.
6) Do not put any effort into enhancing IPv4 functionality. In the evolved bearer concept IPv4 address is still anchored to a PGW and no other intermediate node on the new link can contribute with its own IPv4 addressing or IPv4 routers.
7) Do not change the existing gateway selection procedures and logic. It must still be possible to select a PGW from a visited or a home network, and whether a Serving Gateway (SGW) is collocated with the PGW must not affect the solution.
8) Realize local breakout functionality using a single bearer without a Network Address Translation (NAT).
9) Multiple address functionality should not increase the number of EPS Bearers and should be conservative on additional signaling.

Evolved Bearer Link Model

The evolved bearer 'single pipe' has several 'leaking points', next-hop routers, and topological areas. Each next-hop router can contribute with its own IPv6 resources and breakout traffic sent directly to it. Even if all the routers are on the same evolved bearer link, they do not really see or reach each other. Link scoped multicast traffic is a good example. There will always be a 'default' PDN Connection and bearer between the UE and the PGW, and this connection also provides network based mobility for the addresses anchored to it. The rest of the 'leaking points' attach to the activated evolved bearer if they so wish.

As stated earlier, the new evolved bearer mimics the NBMA link model for IPv6 [12]. Therefore, only unicast packet delivery at lower layers is supported, although all nodes consisting of the UE and the routers are on the same logical link. The UE learns the presence of all routers through normal IPv6 Neighbor Discovery Protocol means, and the routers on the link know the only end host on the link based on the bearer setup signaling. Obviously, there has to be means at the lower layers – namely at the Non-Access Stratum (NAS) protocol layer – to differentiate between nodes. Therefore, there is a need for link-layer addressing or an equivalent concept. Link-layer addressing is currently absent from 3GPP links. The link-layer address or its equivalent concept is needed for intermediate routers on the evolved bearer link to efficiently detect which traffic is destined to them without actively inspecting the traffic content at the IP layer.

The evolved bearer hooks into the Packet Data Convergence Protocol (PDCP) [13], which operates between the UE and the evolved Node B (eNodeB). There are still several unused PDCP Protocol Data Unit (PDU) type codes that can be utilized. We could reserve two or three more types for the evolved bearer purposes, that is one type for the eNodeB terminated/breakout traffic and one for the SGW terminated/breakout traffic. The PGW terminated traffic could use existing PDCP PDU type numbers. This approach would be conservative on the Radio Access Bearer (RAB) level, since no new bearer is needed for new addresses or eNodeB/SGW terminated traffic. We would still need one additional S1-U to represent the SGW terminated/breakout traffic. This could be a fair compromise to RAB consumption and also reduce the overall signaling. The eNodeB can do the traffic differentiation for local consumption or for forwarding to bearers heading to PGWs and SGWs without an IP lookup.

The issue of link-layer addressing, or rather the lack of it, still remains. Having a proper link-layer addressing makes the integration of IPv6 simpler from the UE IP stack point of view, since IPv6 was designed to work best over multicast enabled links with *link-layer* addresses. A straightforward solution would be mapping either PDCP PDU type codes (for data packets) to link-layer addresses or agree on a common link-layer addressing scheme for a UE, eNodeB, SGW, and PGW. The link-layer addressing does not mean that the address would be a real physical layer-2 address of each node on the evolved bearer link. The purpose is to allow easier mapping of IP flows and IPv6 prefixes to the proper PDCP PDU types and radio level constructs, thus the link-layer is merely a conceptual construct with known name space where the address implicitly contains the information of the network node where the traffic is to be terminated. Table 6.1 shows what the link-layer address mapping to PDCP PDU types could look like. We would map the link-layer address into an EUI-64 address. The Organizationally Unique Identifier (OUI) *'aa:bb:cc'* is fictional in our example. Requiring a PGW router to adhere the proposed EUI-64 can be loosened, as the PGW is not required to follow the new PDCP-based EUI-64 scheme. After all, the evolved bearer has no impact on the PGW.

The evolved bearer will still be modeled as an isolated per UE link; that is, there can be only one end host attached to the evolved link in addition to routers. Figure 6.3 illustrates

Table 6.1   PDCP PDU type mapping to a EUI-64 identifier

PDU Type	EUI-64 Identifier	Remarks
000b	none	Existing type for control PDUs.
001b	aa:bb:cc:01:00:xx:xx:xx	PGW terminated traffic using existing PDU type space.
010b	aa:bb:cc:02:00:xx:xx:xx	eNodeB terminated traffic, new PDU type value 2.
011b	aa:bb:cc:03:00:xx:xx:xx	SGW terminated traffic, new PDU type value 3.
100b	aa:bb:cc:33:xx:xx:xx:xx	Uplink multicast traffic, will reach eNodeB, SGW, and PGW, new PDU type 4.

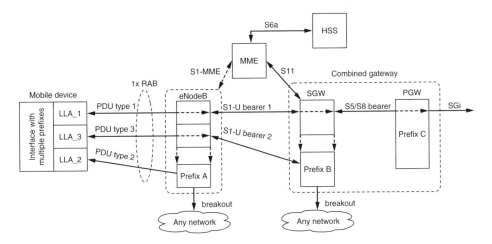

Figure 6.3   Mapping of PDCP PDU types and different 3GPP bearers within the *evolved bearer*.

how the evolved bearer looks like, and shows the mapping of PDU types, different bearers, and conceptual link-layer addresses.

The IPv6 prefix provided via the 'SGW router' or 'eNodeB router' does not provide mobility or any other advanced features such as Policy and Charging Control (PCC) [14] that are typically associated with 3GPP PDN Connections and EPS Bearers; the IPv6 prefix is just for plain *simple-IP* use. It should be noted that the handling of IPv6 prefixes that are terminated at the eNodeB do not need to be communicated with the Mobile Management Entity (MME) for their evolved bearer handling. The IPv6 prefixes are entirely local to the eNodeB. During handovers or SGW relocations, only the PGW anchored prefix is guaranteed to remain the same.

The evolved bearer still builds on top of the existing EPS Bearer but only for the part that concerns the default bearer. Therefore, following details need to be understood:

- The evolved bearer supports PDN Types IPv6 and IPv4v6, but not IPv4.
- The evolved bearer fate-shares the default EPS Bearer. When the default bearer is activated, the eNodeB and SGW 'routers' along with their respective setup procedures, depending on operator's configuration, may activate and start advertising their presence

and prefixes on the link. Also, when the default bearer gets deleted, anything related to the eNodeB and SGW routers and their addressing is deleted as well.

- There is no *dedicated bearer* concept for differentiated QoS handling for the evolved bearer, except for the bearer that concerns the PGW terminated/breakout traffic. That part remains unmodified from the EPS Bearer.
- IPv4 addressing is still possible but only for the PGW terminated traffic and when the PDN Type IPv4v6 is used. Neither eNodeB nor SGW terminated/breakout traffic have support for native IPv4.
- The evolved bearer inherits the 3GPP PDN Connection assumptions on how IPv6 addresses are configured [15, 16]. That means that the SLAAC [17] is the only way to configure an address and a single Prefix Information Option (PIO) [18] can be present in a Router Advertisement received from a router. The advertised /64 IPv6 prefix always has the 'L'-flag set to zero in a PIO.

The lifetimes in Router Advertisement Prefix Information Options (PIOs) originated by the eNodeB and SGW do not need to be set to infinity, because both of those termination/breakout points may be under different management from the PGW. The eNodeB part of the evolved bearer is the most independent, and changes in there can be implemented without impacting the rest of the 3GPP system.

Multicast Considerations

Although the evolved bearer is modeled as an NBMA resembling link, we need to define how multicast works over it at IP level, specifically from the UE point of view. This is important regarding the IPv6 NDP, which is piggybacked on top of Internet Control Message Protocol version 6 (ICMPv6) and makes extensive use of link scoped multicast.

From the evolved bearer routers' perspective, they do not necessarily ever see multicast traffic that other routers send. The router originated IP multicast is always unicast to the UE at the bearer level. Similarly, the UE originated IP multicast is always unicast at the bearer level to all routers on the evolved bearer link. There is no real multicast over the evolved bearer.

The IP level multicast over the evolved bearer link can be realized in two ways. First, the UE can always make multiple copies of the multicast packet and then unicast each copy to the respective router using the PDCP PDU type derived mechanism. Multiple copies implies unnecessary overhead over the radio interface, but from the forwarding point of view in the network there is no difference between IP multicast or unicast. Second, yet another PDCP PDU type could be reserved for the UE originated IP multicast – see Table 6.1. Using this approach only one copy of the packet would be sent over the radio interface and the eNodeB would take care of doing the required copies of the IP packet and delivering those further to other routers, namely to the SGW and PGW routers.

Having evolved bearer routers to respond to Duplicate Address Detection (DAD) and address resolution is little burden. Most of the Neighbor Discovery Protocol operations can be handled with precomputed answers anyway.

Supporting the NDP Redirect might now make sense in the evolved bearer. The assumption in that case is the evolved bearer routers should have some type of out-of-band knowledge of each others' presence, used link-local addresses, and other address management.

Multiple IPv6 Prefixes Per Bearer and Traffic Steering

The interesting question is how an evolved bearer appears to a UE at the IP layer. The UE sees the evolved bearer as one link with one or more routers on it. The PGW router is always present due to default bearer establishment to the PGW. The eNodeB and/or the SGW routers may be active based on the network side configuration. The UE learns about the evolved bearer routers by receiving a Router Advertisement from them, either unsolicited or after sending an Router Solicitation. Each of the Router Advertisements may contain at most one PIO including a prefix for the UE to configure its addresses. The Router Advertisement from the PGW router must contain a PIO with a prefix but the PIOs from other evolved routers may not. At minimum, the UE learns the link-local addresses of the evolved routers and their router preferences. Each evolved bearer router may have its own view of their preference.

In a case where the UE received multiple PIOs from different routers on an evolved bearer, the UE can obviously configure multiple addresses on its interface. When the UE has to select a source address for its outgoing traffic, the standard *default source address selection* as defined by IETF applies [10]. The UE may also have up to three default routers on the evolved bearer. The default router is entirely based on the existing IPv6 rules and practices [18, 9, 19]. In the selection of the Interface IDentifiers (IIDs) the UE can still follow the existing 3GPP specifications.

Using the evolved bearer solution, a network operator can easily utilize RFC 4191 [9] *router preferences* and *more specific routes* to choose which of the three routers visible to the UE is to be used as the most preferred default router. For instance, the operator can configure an eNodeB to be the high priority default, a SGW to be the mid-priority default, and a PGW to be the low-priority default gateway. This will easily, and using standard IPv6 machinery, tell the UE which route to use for default traffic.

There is a certain network deployment issue here. What happens if the UE decides to send traffic to an arbitrary destination via a next-hop router using a source address that was not advertised by that router? Ingress filtering is a common practice in leaf networks, which would cause the traffic to be dropped (with a possible ICMPv6 sent back to the UE). If ingress filtering is not in place, the return traffic would anyway be routed towards the network where the source address topologically belongs, causing asymmetric routing to take place. Asymmetric routing is known to be problematic when it comes to stateful middle-boxes such as firewalls.

We believe that the evolved bearer is a promising technology for localized content distribution and direct mobile-to-mobile communication. The evolved bearer is not only about breakout to the Internet. The breakout can definitely be into localized caches and information resources. Depending on the network architecture and the collocation properties of the SGW and the PGW, the content can be distributed at multiple levels in the access network.

Handovers

Only the PGW anchored prefix provides mobility. When either the eNodeB or the SGW changes as a result of a handover, there is no guarantee that the target eNodeB or SGW is in the same topological area. Therefore, the UE should always verify whether the old IP configuration on its multi-address interface is still valid after a handover [20]. It is expected that a UE that implements the evolved bearer also implements the simple procedure for Detecting Network Attachment DNA [21] in its IP stack and is able to provide the required link events to the IP stack from its network interface layer. Using DNA the UE can verify whether its evolved bearer IP configuration is still valid and renew the configuration when that is not the case. From the network point of view, both eNodeB and SGW routers must send Router Advertisements when a new mobile attaches to them.

Another interesting case is the inter-technology handover. Existing 3GPP specifications already have a job lot of text spent for this subject. The evolved bearer does not even try to solve or specify what happens to the eNodeB and the SGW prefixes after the inter-technology handover. The prefixes learned from the eNodeB and the SGW are simply lost in all cases known by the existing 3GPP specifications. The network needs to clean up the possible evolved bearer state from the MME and the SGW after a handover but that is again the expected behavior.

During the handovers the MME is responsible for the creation of a new S1-U tunnel for the SGW terminated/breakout traffic. The target eNodeB learns whether such a tunnel was already in place from the S1-MME signaling. We will discuss the required S1-MME enhancements for this purpose in Section 6.2.3.

Backward Compatibility with EPS

The evolved bearer proposal relies on a specific PDCP behavior and maintaining the existing 3GPP EPS Bearer as a base to guarantee backward compatibility. The evolved bearer does not define its own PDN Type. There are two cases to take a closer look at: what happens when a UE does not support the evolved bearer and what happens when the network does not support evolved bearer?

It could make sense to reserve a bit *evolved capability*, for example, in the UE network capability information element in the Attach Request NAS signaling message [22] where a mobile host could explicitly state that it supports the evolved bearer. Based on the new *evolved capability* bit the network may entirely skip the preparations for the evolved bearer and conserve signaling and resources. There is no need for an error cause code to the UE direction stating that evolved bearer is not supported. If the network has no support it has no means to respond anything meaningful.

If a UE does not support/implement the evolved bearer, it never expects to receive or send any PDCP PDU types other than those that are already specified in the existing PDCP specification [13]. The PDCP specification states that unknown PDCP PDU types are simply discarded.

Where a network does not support the evolved bearer the default EPS Bearer gets activated as specified in the existing 3GPP specifications. The UE uses the existing PDN Types IPv6 or IPv4v6 even when it wishes to initiate the evolved bearer. This comes from the design choice to build on top of the existing EPS Bearer. Obviously, if a network does not implement the evolved bearer neither does it initiate any traffic to the UE using other than the existing PDCP PDU types. If an eNodeB receives traffic from a mobile node that supports the evolved bearer, the unknown PDU types are simply discarded as per the existing specifications.

Since the evolved bearer router on the SGW breakout point does not survive a change of the SGW, there cannot be a case where the eNodeB would not support the evolved bearer but the SGW would have the evolved bearer activated. The MME is always the initiator of the activation of the SGW breakout/router on the evolved bearer. Note that the eNodeB can be in inactive mode on the evolved bearer while the SGW breakout/router is in active mode.

Regarding the newly evolved specific PDCP PDU types, PDCP data (PDUs) carry Robust Header Compression (RoHC) [23] encapsulated IP packets. The same assumption also applies to the evolved bearer.

6.2.3   Overall Impact Analysis

The evolved bearer has known implications for the 3GPP architecture and UEs, and additionally open issues. Here we will discuss about those briefly.

Impact on the 3GPP Architecture

Subscription profile currently contains service and even IP addressing provisioning information. There is probably a need to enhance the subscription profile and therefore also the S6a MME to Home Subscriber Server (HSS) interface [24] to carry an indication of whether a breakout/router is allowed in the eNodeB and/or the SGW. At the interface level the impact would be a single new enumerated Diameter Attribute Value Pair (AVP) stating the evolved subscription properties. The default behavior would be to not allow the evolved bearer.

Policy and charging control (PCC) is not impacted at all in our evolved bearer design. We have purposely left this part unchanged. The evolved bearer does not even support EPS dedicated bearers for QoS purposes on the eNodeB and the SGW parts, mainly because the Policy and Charging Control (PCC) system and the PGW has no knowledge of the addressing used on the eNodeB and the SGW. If QoS and PCC are desired, those apply only for the PGW terminated traffic.

Address management of the current 3GPP system is not impacted either. The evolved bearer design is such that an operator deploying those is responsible for configuring addressing at the eNodeB and the SGW level using non-3GPP means. Since the evolved bearer does not preserve prefixes during handovers, there is no need for address management and IP mobility enabling signaling either.

GTP-U [25] is entirely untouched. This applies to GTP User Plane (GTP-U) over S5/S8/S1-U interfaces. *GTP-C* [26] may need to be enhanced for S11-interface use between the MME and the SGW. We need to be able to tell the SGW that a specific S1-U tunnel does not have a corresponding S5/S8 GTP-U tunnel.

S1-MME interface [27] needs to be enhanced to carry the *evolved bearer capability* and provisioning information from the MME to the eNodeB. An enumerated information element would be included, for example, into the Initial Context Setup Request sent from the MME to the eNodeB. The enumerated information element indicates whether: 1) the eNodeB can be activated as a breakout/router, 2) the SGW can be activated as a breakout/router or 3) both can be activated. Additionally, there is a need for a new information element in the S1-MME signaling to indicate that the eNodeB actually supports the evolved bearer functionality. This information could be placed into the Initial UE message or into the Uplink NAS Transport message. Based on this indication and the subscription information, the MME knows whether further S1 and S11 procedures are needed for evolved bearer purposes.

NAS signaling would benefit from an explicit indication whether the UE supports the evolved bearer. As already discussed in Section 6.2.2 the NAS signaling could be enhanced, for example, with one *evolved capability* bit in the UE network capability information element in the Attach Request NAS message. The eNodeB and the MME would use this information to determine whether anything needs to be prepared in the network side for the evolved bearer.

An eNodeB and a SGW will have a larger impact. First, both nodes need to be enhanced with an IP router functionality and relevant IPv6 stack features at the user plane. Furthermore, the eNodeB takes an additional impact as it needs to have modified PDCP implementation as well as enhancement on the S1-MME interface. It should be noted that the SGW does not need to support the evolved bearer if the eNodeB does, which allows for reducing the impact into a single network node out of two. Regarding the SGW impact there would now be a 'new type' of bearer. More specifically the S1-U bearer for SGW terminated/breakout traffic has no matching S5/S8 bearer towards the PGW.

A PGW is not impacted at all unless the SGW and the PGW are the same combined node and the SGW implements the evolved bearer.

PDCP is impacted because the whole evolved bearer concept relies on its extensions.

An MME is impacted since we need to enhance the S1-MME interface for two purposes: carrying the evolved capability/provisioning indication and activating the S1-U GTP tunnels for the SGW terminated/breakout traffic. The default behavior is that no evolved bearers get established and the additional evolved bearer related information elements are just ignored.

X2 interface [28] between evolved Node Bs (eNodeBs) during the handover has no impact since our evolved bearer does not need anything from it.

The amount of signaling does increase in a number of messages when the SGW breakout/router gets activated. However, that still only concerns S1 signaling and only the part that is sufficient to set up an S1-U tunnel. There is no additional PCC, S5/S8, DNS name resolving, or subscription handling related signaling, which is rather lightweight compared to conventional EPS Bearer setup.

Impact on the UE Implementation

The UE modifications for the evolved bearer type will concentrate around the interface between the IP stack and the cellular modem, be the IP stack and cellular modem inside the same or separate devices. Essentially, after the initial setup of the new type of bearer, the modem has to be able to map uplink packets arriving from the IP stack to correct PDCP PDU types, and differentiate the downlink packets arriving with different PDCP PDU types to the IP stack. The IP stack itself will see multiple routers and will select among them using standard procedures already used on multi-router links, and hence the IP stack itself does not need protocol changes. Of course the IP stack should have the standard features required for multi-router links, such as support for RFC 4191. However, UE mobility may cause changes on eNodeBs and/or SGWs, and hence the information that the IP stack has learned through those routers, with ICMPv6 or DHCPv6, can become obsolete. Hence changes in eNodeB and SGW need to cause 'link-up' events from the cellular modem to the IP stack so that the IP stack can perform information refresh procedures.

There could be several approaches for presenting different PDCP PDU types to the IP stack. One viable approach, which follows the NBMA link design, is to present the IP stack with one network interface with multiple routers, each having different link-layer addresses (see Table 6.1). In this approach the cellular modem would use the specified link-layer addresses as source addresses when encapsulating downlink IPv6 packets to Ethernet frames of the UE's internal interface. The above implementation approach has the benefit of being considerably transparent to the applications, especially if the UE architecture style binds applications to use a single network interface at a time. Furthermore, if the IP stack is on a separate device, this approach is practically the only choice (e.g., if the UE is tethering, sharing, the cellular connection to other devices).

6.2.4 Open Issues and Future Work

We have left several areas out of scope in this study or for partial attention. The first is the *Idle Mode* mobility. The second is the possible introduction of the evolved bearer to the GPRS and 3G.

We stated that each Router Advertisement can contain at most one PIO. This follows the existing 3GPP specifications, but there could be justified use cases to relax the restriction. This topic is also left for further study, mainly because we currently have no robust way to differentiate between 'roles' or 'service' associated to prefixes; that is, there is no *prefix coloring* readily available [29].

The current design of the evolved bearer still only supports stateless address autoconfiguration [17], which follows the existing 3GPP standards method of address configuration [15, 16]. However, investigating how to support DHCPv6 [30] and stateful address auto-configuration on the evolved bearer would admittedly be useful regarding the future developments of wireless broadband access. DHCPv6 has a number of administrative aspects that operators are keen on.

Finally, we have entirely overlooked the evolved bearer when Proxy Mobile IPv6 (PMIPv6) [31] is used instead of GTP in the EPC [32]. However, we believe it is doable but definitely has additional challenges compared to GTP due a different link realization between the UE and the PGW.

6.3 LTE as the Uplink Access for Home Networks

6.3.1 Homenet at IETF

IETF has worked on an interesting topic called *Homenet* (Home Networking), please see *http://tools.ietf.org/wg/homenet/* for the latest status. The work is about introducing IPv6 in small residential home networks. The protocol work results may also find applications in other small networks outside residential domains. IPv6-only home networks are possible but the coexistence of IPv4 is not ruled out. However, IPv4 specific work and enhancements are out of the scope of Homenet. Borrowing the IETF working group charter, topics that will be addressed include:

- (Automated) prefix configuration for routers; the configuration concerns the home network internal routers as well as the customer edge router [33]. There are multiple approaches based on prefix delegation and DHCPv6 [34, 35], possibly extending IPv6 NDP, or routing protocols [36].
- Managing routing; when there are multiple subnets, the home network internal routing has to be accomplished somehow [37]. It is not possible to rely on just default routes all the time.
- Name resolution; how can we discover name servers and get that provisioned to every host within the home work, naming internal nodes, and how can we manage possible naming delegation and reverse zones [38, 39]?
- Service discovery; there are ways to discover network services on a link scope but how can we extend that to multiple subnets and non-trivial topologies [40]?
- Network security; there needs to be at least minimum level of built-in security properties in a home network [41]. Also it may not be trivial to dynamically discover 'security boundaries' within the home network.

It has been envisioned that future IPv6 enabled residential home networks are not topologically trivial, and the services that people would like to deploy intentionally or unintentionally on those actually lean towards properly segmented networks with all the complexity that it brings. Current IPv4 and often NAT-based home networks tend to work out of box, even when there are nested NAT devices in the network (but obviously these networks suffer from the connectivity limitations caused by translation, such as limited visibility between network segments). IPv6 is still a relatively young technology, especially when it comes to 'clueless' consumers and their home networks with all their variety of (badly behaving) networking devices. And all has to work just fine with zero configuration from the consumer point of view. There is still some work to do.

The first task that the IETF working group is tasked to do is to lay out an architecture document that outlines how to construct home networks involving multiple routers and subnets/segments, and without having NAT devices in between subnets. The architecture document [42] is expected to implement the IPv6 addressing architecture [43], prefix delegation [44], global and Unique Local Address (ULA) addresses so that the home network could internally be still functional when access to the global Internet is temporarily lost, source address selection rules [10], internal routing, autoconfiguration (such as one proposal based on Open Shortest Path First version 3 (OSPFv3) [45]), and other existing components of the IPv6 architecture.

Changes and improvement to existing standards are possible, when seen necessary. The router connecting the residential home network to the Internet is supposed to follow existing guidelines and requirements [41, 46].

6.3.2 Homenet and 3GPP Architecture

What does (wireline) residential home networks have in common with 3GPP wireless network access? It is likely, and has already happened on several occasions, that cellular access technology will replace the wireline cable in Consumer Premises Equipments (CPEs). LTE specifically has enough bandwidth and reasonable roundtrip times to satisfy large groups of consumers, especially in emerging markets where deploying wireline Internet access might turn out to be far more troublesome than wireless.

The addressing in home networks is very likely to be built on top of DHCPv6. The 3GPP added DHCPv6 Prefix Delegation support to its architecture starting from Release-10. As we saw in Section 4.4.6, 3GPP DHCPv6 Prefix Delegation has some peculiarities [47] that might not be obvious for people with previous experience from deploying IPv6 on wireline networks. Furthermore, the lifetime of the delegated prefixes fate-share the wireless connection lifetime. As we know, wireless connections, even when stationary, tend to have hiccups. If the wireless part (the PDN Connection) is provisioned dynamically then the delegated prefixes are also dynamically assigned, which most likely leads to a frequent disturbing renumbering in the home network. This would be unacceptable. Therefore, static PDN Connection address assignment linked to static prefix delegation is more likely to be a requirement when using, for example, LTE connectivity as a cable replacement. According to Section 4.4.5 the static addressing in 3GPP specifications is not entirely a 'done deal' as of now. There is still some work to do, at least in 3GPP.

3GPP Release-10 is still quite ahead of normal mobile operator release cycles, which means that the DHCPv6 Prefix Delegation may take some time before it actually materializes. Of course it is always possible that a specific feature gets *cherry picked* to earlier software releases. In a meanwhile, it is very possible that the LTE or 3G connections as cable replacements for CPEs are stuck with a single /64 IPv6 prefix and no mobile network side support for DHCPv6 for any parameter configuration. This would make an 'interesting' (read challenging) residential home network deployment scenario, for example, for the following reasons:

- The 3GPP link model requires the /64 IPv6 prefix to be also configured in the Wide Area Network (WAN) link side of the CPE. There has been related discussion in the context of tethering with 3GPP UEs, where a single /64 IPv6 prefix is used on both sides of a UE [48].
- The Neighbor Discovery Proxy [49] is not really usable in the home network side, since loop free topologies cannot be guaranteed.
- DNS server information comes dynamically from the mobile operator using non-IP technologies where it then needs to be propagated further into the home network.
- The /64 IPv6 prefix may change every time the cellular WAN link goes down and comes up again.

The inadequate numbering resources/possibilities might lead to fallback solutions and promote a deployment of IPv6 NAT, which would be unfortunate at best.

6.3.3 Additional 3GPP Deployment Options

Section 4.2.4 discussed external PDN access that was realized using a L2TP tunnel between the PGW in the role of L2TP Access Concentrator (LAC) and a LNS in an external PDN. The same approach could be used for Homenet purposes today. Figure 6.4 shows a deployment option using L2TP for the PGW and the external network connectivity.

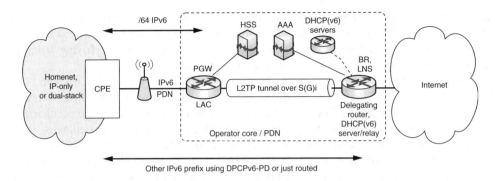

Figure 6.4 Homenet deployment option using L2TP.

There would not necessarily be a need for a DHCPv6 Prefix Delegation for delivering the subscriber prefix to the CPE. The IPv6 prefix assigned to the subscriber would be 'statically' routed to the residential home network, when the PDN Connection from the CPE to the PGW is up. Alternatively, the LNS can have a role of a DHCPv6 server or a relay, and let the CPE in the residential network request delegated prefixes using DHCPv6 Prefix Delegation. There is no similar aggregation requirement for the IPv6 prefix assigned to the PDN Connection and the assigned residential home network IPv6 prefix than there is for 3GPP Release-10 defined DHCPv6 Prefix Delegation. Such 'mis-alignment' of the requirements originates from the fact that the L2TP-based solution is not a 3GPP standard but just a common vendor supported feature. On the other hand, the described L2TP-based deployment would then lose possible PCC integration benefits, since PCC cannot handle non-aggregated prefixes.

Actually, it would also be possible for the residential home network CPE to speak one of the common interior routing protocols towards the LNS, and in that way get the routing in place between the home network and the mobile operator PDN.

6.4 Port Control Protocol

Hosts in the Internet very commonly run applications that utilize long-lived connections or need to be able to receive incoming connections. Typical applications include instant messaging, VoIP, file sharing, games, Internet of Things (IoT) or Machine-Type Commu-nications (MTC) or Machine-to-Machine (M2M) nodes, and operating systems' generic notification channels.

In the ideal Internet the applications would not have significant problems on having longlasting connections or on ability to receive incoming transport sessions. However, in the real operational Internet these features are often unavailable. The networks deploy IPv6/IPv4 Network Address Translations (NAT64s) and Network Address Translation from IPv4 to IPv4s (NAT44s) (both referred in this section as NAT), and firewalls or other middle-boxes, which both place restrictions on how long transport layer sessions can stay idle, and whether incoming connections are allowed or not. Without any control on the network, the host implementations have to send frequent keep-alive signaling to keep NAT/firewall mappings open, and this can be a significant power consuming factor in 3GPP UEs. Furthermore, without control for port forwarding rules of Network Address Translations (NATs) and firewalls, it is often impossible to receive unsolicited incoming connections.

Automated solutions for the connectivity problems caused by NATs and firewalls in local area networks, such as home networks, have existed for a long time. The two most common solutions are Universal Plug and Play (UPnP) Internet Gateway Device (IGD) Device Control Protocol (version 1.0 [50], and 2.0 [51] that added IPv6 support) and NAT Port Mapping Protocol (NAT-PMP) [52]. Both of these protocols allow hosts to request a local NAT or firewall, a device usually in the host owner's control, to create port forwarding rules that will help applications on hosts. It is worth noting that manual configuration of port mapping rules to local NAT or firewall is often possible, even though it may be a cumbersome process and unscalable if mapping rules change often.

While the UPnP and NAT-PMP have served local use cases, the emergence of carrier grade NAT44s and NAT64s has caused challenges. People are expecting to be able to use services like they could when they had public IP addresses that were not address translated, although introduction of NAT into the network makes that a troublesome goal. As technologies, the UPnP and NAT-PMP are not suitable for controlling a network entity in the operator's core networks. To bridge this gap, the IETF is actively developing a protocol called Port Control Protocol (PCP), which could provide end users some control of NAT or firewall in the operators' core network [53]. At the time of writing, the PCP has not yet been published as an RFC. Therefore, it is recommended to check out the latest status of IETF's PCP working group.

6.4.1 Deployment Scenarios

The most important deployment scenarios from the 3GPP networks point of view are illustrated in cases A and B of Figure 6.5. In scenario A, the UE implements the PCP client, and the NAT/firewall implements the PCP server in itself. This is the simplest of deployments.

The difference of scenario B from A is that the PCP server is not located at the NAT/firewall, but instead on a separate network entity. From the UE point of view scenario B does not really differ from A. Obviously, from the network point of view, the difference is significant. The split deployment model is interestingly allowed by the PCP proposal, but the interface between the PCP server and the NAT/firewall is left undefined and to be implemented by proprietary solutions.

Scenario C presents the deployment model that is the main use case for PCP. In this case, a CPE is implementing an interworking function between Local Area Network (LAN) and uplink Internet connection – which could be LTE as discussed in Section 6.3. Essentially, if the CPE would implement a NAT/firewall and had a public IP address on its uplink, the existing UPnP and NAT-PMP (NAT-PMP is not shown in the figure) protocols would be enough. However, if the CPE has a private IP address from the network provider, some means for extending the reach of UPnP and NAT-PMP are required. Hence, the CPE will implement a piece of software that converts UPnP and NAT-PMP messages on the LAN to PCP messages on the CPE's uplink connection [54].

In all of these scenarios the NAT can be any kind of NAT, for example, the usual IPv4 to IPv4 kind, or also the NAT64 IPv6 to IPv4 protocol translator.

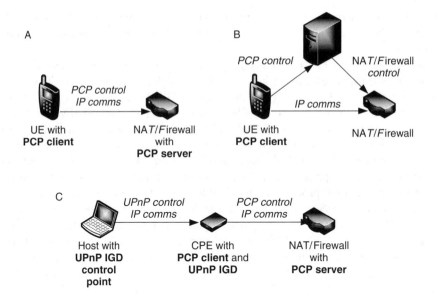

Figure 6.5   Three deployment scenarios for PCP.

6.4.2   Protocol Features

The core features of the PCP are the following:

Reduction of keep-alive traffic: By allowing control of the lifetime of existing or new NAT mapping, PCP allows clients to influence the frequency at which they need to send keep-alive packets.

Port forwarding control: The PCP allows applications to create NAT port mapping rules from an internal IP address, protocol, and port to an external IP address, protocol, and port. Mixes between address families are possible.

Support for transport layer protocols: The transport layer protocols that utilize 16-bit port numbers are supported. These include TCP, User Datagram Protocol (UDP), Stream Control Transmission Protocol (SCTP) and Datagram Congestion Control Protocol (DCCP). The transport layer protocols not using ports are only supported with firewall and prefix translation scenarios, but not with NATs. These include, for example, Internet Control Message Protocol (ICMP), ICMPv6, and Encapsulating Security Payload (ESP).

While the PCP is not able to control all routers and boxes on the path from the client to the destination, it is usually enough to be able to control the access network where the client is. If the destination is a service in the public Internet, the NATs and firewalls of significance are on the client's access network. In the case of peer-to-peer communications, it is enough for both peers to control NATs and firewalls in their respective access networks, and then signal the public-facing IP addresses and ports to the peer using a rendezvous protocol of a specific application, such as Session Initiation Protocol (SIP). The PCP does not care which rendezvous protocol is used.

6.4.3   PCP Server Discovery

The PCP client can learn the IP address or the fully qualified domain name of a PCP server by manual configuration, provisioning, or with help of Dynamic Host Configuration Protocol (DHCP) [55]. Additionally, the client can simply attempt to send PCP requests to the default IPv4 and IPv6 routers that the client has configured for itself, or perhaps utilize anycast addresses as recently proposed [96].

6.4.4   Protocol Messages

The PCP protocol is based on the idea that PCP clients indicate which kind of services they would like, and PCP servers inform clients what the clients get. The control of what is allowed is fully on the server side [53]. In the basic operation the PCP client is sending requests for itself, but the PCP also supports requests sent by third parties when the PCP is used in fully trusted networks.

PCP messages are transported using UDP with a maximum message size of 1100 bytes, which is small enough to allow some tunneling headers and still fit into the IPv6's minimum Maximum Transmission Unit (MTU) of 1280 bytes.

The IP address in PCP messages is always transported in 128-bit fields for the sake of simplicity. When the fields include an IPv4 address, the IPv4-mapped IPv6 address format is used – see Section 3.1.8.

In the following sections we look at the overview of the PCP. Full protocol specification features and details are available in the base specification [53].

PCP Requests

All PCP requests include the request type, 32-bit requested lifetime, PCP client's own IP address, and then possible optional information elements.

The following two requests are defined in the base specification [53]:

MAP: The MAP request can be used for creating explicit *endpoint independent* mappings between internal and external addresses, protocols, and ports, before a transport layer connection is established. In the MAP request the client provides information about a random nonce, an upper layer protocol, and an internal port, and suggest an external IP address and port.

PEER: The PEER request can be used to create a new *endpoint dependent* mapping, or control the lifetime of an existing mapping created by already established transport level connection, to a remote peer's IP address and port.

Both request messages include a suggested external IP address, which is useful, for example, when a client is restoring an old, perhaps lost, state to the NAT or firewall. This is also a useful property with applications such as active mode File Transfer Protocol (FTP) that require different transport layer flows, within a single application layer session, to use the same IP addresses.

If a request is for a protocol that does not have 16-bit port numbers, the request is considered to apply to all messages of that protocol.

The protocol itself allows mapping for all protocols and all ports, in which case the NAT/firewall would essentially have 1:1 mapping between external and internal IP addresses.

PCP Responses

All PCP responses include the response type a result indicating success or failure reasons, the server's epoch time value, optional information elements, and then a 32-bit lifetime indicating the lifetime of a mapping or in case of error, how long the response is likely to remain the same.

The responses are defined for the two specified requests:

MAP: The PCP server returns to the client, in the case when mapping was allowed, the same nonce, protocol, the internal port, and assigned external IP address and port, which may or may not be the same as the client requested.

PEER: The PCP server returns to the client, in the case of success, information about the external address and port assigned for the client, and also the remote peer's port and IP address.

The epoch lifetime value provides hints for PCP clients about the state of the PCP server; if client notices surprisingly low epoch value, it is an indication of server reboot and hence could trigger the client to refresh its mappings.

6.4.5   Cascaded NATs

It can happen that the network has cascaded NATs. The PCP client can detect the presence of multiple NATs by sending a PCP request to the PCP server that the client is configured with, and observing if the PCP response shows a different IP address for the client than the client used in sending the request. In this case, the client has to talk also to the inner, newly detected, NAT, in order to have the rules set correctly all the way from the client to the Internet. A specific Recursive PCP solution proposal for this scenario has been published [97].

This kind of scenario could appear, for example, if a UE is participating in a double translation scheme: a PCP client on a computer using tethering with a UE to access the Internet could first detect the outer NAT on the network, a NAT64, and then the inner NAT, a NAT46, on the UE. The client on the computer may need to create mappings to both NAT46 and NAT64 to have the full path covered.

6.4.6   Relation to IPv6 Transition

Even though the main need for PCP is in the IPv4-only domain, where IPv4 address shortage is complicating deployments and perhaps some more lifetime for IPv4 can be bought with PCP, there is also clear use in IPv6:

Firewall control: The need for firewalls continues to exist when the Internet moves from IPv4 to IPv6. While some 3GPP deployments may manage without firewalls, others will use them. Hence there continues to be the need for controlling session timeouts and to allow selected incoming connections to get through.

NAT64 control: The IPv6-only deployment with NAT64 is happening – see Section 5.4.3. UEs will have a need to manage NAT timeouts for IPv4-only destinations, and also for allowing incoming connections from the IPv4-only peers.

IPv6 NAT control: Although IPv6-to-IPv6 NAT is a heresy, even without port translation, the world has a bad habit of going in unwanted directions. If these entities were to become a reality, port control needs would be there for IPv6 just as they are today for IPv4.

6.5 Internet of Things

The Internet of Things (IoT) is a topic that has been discussed for decades, often with different names but with more or less in the same context, for example: Machine-to-Machine (M2M) type of communications, smart homes, and ubiquitous computing. In essence, IoT is about connecting billions of inexpensive small computing and communicating devices into the Internet – we refer to these individual objects as *smart objects*. The typical examples include smart light bulbs, automated fridges, temperature sensors, and such like that are not part of the traditional set of computers such as desktop, laptops, or smart phones. In this section we will refer to this domain generally as the *Internet of Things*.

Despite the slow start to introduce Internet connectivity to everyday objects, the IoT is now an area of strong growth. A recent Organisation for Economic Co-operation and Development (OECD) report says that according to some estimates, as many as 50 billion mobile wireless smart objects may be connected to the Internet by 2020, and the total number of devices could even reach 500 billion [56]. Even if such figures were to end up being overly optimistic, it would be safe to say that there will be a lot of smart objects connecting to the Internet during the next decade.

6.5.1 Typical Use Cases

While use cases for IoT are limited only by human imagination, a set of typical use cases exists that are helpful for understanding the problems, designing the architectures and protocols, and just to grasp what this area is about. More information about some use cases are available, for example, from European Telecommunications Standards Institute (ETSI) TS 102 689 [57, 58], IETF RFC 6568 [59], and IPSO Alliance 'IP for Smart Objects' white paper [60].

Below we list a set of typical use cases for 'Smart Something' with short descriptions and a few examples:

Electricity metering: One area where smart objects are already being deployed is electricity meters. Both consumers and electricity companies benefit from real time information regarding electricity consumption. It helps people to control the electricity usage and allows, for example, much more fine-grained electricity pricing. In the cases where households even contribute electricity into the grid, from solar cells, windmills, or electric cars, or the households wish to consume electricity at optimal pricepoints, the smart objects can be used to control when and how much electricity is transferred.

Healthcare and fitness: In the healthcare domain smart objects can be used to measure and report the status of patients in hospitals, elderly people in homes or care centers, and babies in their sleep. The smart objects can be measuring vital signs, or for example, generate an alert if elderly persons fall and cannot get up. The smart objects can also monitor the state of medicine in storage and organ transplants in transit. In the area of fitness, typical use cases are heart rate monitors used during exercising.

Home automation: In the home, sensors and actuators can monitor and control the building maintenance functions and hence things such as lighting levels, heating, cooling, moisture, energy consumption, and water consumption (to detect possible leaks). Security use cases are also typical, such as movement detection, intruder detection, lock control, and so forth.

Industrial automation: In the industrial automation arena smart objects can be used, for example, to monitor the state and location of cargo containers or postal packages, the movement of objects inside factories, lighting, process monitoring and controlling, monitoring of the condition of machinery to get early warnings for failures, and so on.

Civic automation: Cities and civil infrastructure are full of places where smart objects can find a home. For example, smart lighting with capability to turn on and off depending on need and reporting of near-future, or already occurred, device malfunctions to allow more efficient maintenance. Safety and security applications are also there, for cities to monitor places more efficiently and to monitor the condition of infrastructure, such as bridges, pipes, or buildings.

Intelligent Transportation Systems: In the traffic, smart objects can help to control traffic signals and speed limits based on need, in order to control congestion and, for example, city air condition. Smart objects can measure local driving conditions based on weather or direct measurements from the road, and control speed limits or issue warnings to drivers. Cars can communicate with others and with traffic signs to alert drivers, and train and bus traffic may be optimized based on people present on stations and at stops.

Agriculture: Smart objects deployed outdoors in the fields or indoors in greenhouses can measure variables affecting plant growth. These include soil and air moisture levels, lighting conditions, and temperatures. The sensors can drive actuators that help plant growth, and alert farmers to problems. The measurement data may also help in estimating crop yields.

6.5.2 Standardization Organizations Working with IoT

As the IoT is a global phenomenon, it is understandable that many standardization organizations and industry alliances are interested in the topic. In Table 6.2 we list some of these organizations, including the most relevant from the cellular networks point of view: IETF and 3GPP. However, other organizations also influence this large domain, and hence in this section we will give an overview of what different organizations are working on. In some of the technical areas these organizations complement each other, but in some aspects overlapping and competing technologies are being standardized. These overlaps

Table 6.2 Some standardization organizations working with IoT

Organization	Focus Area
3GPP	Cellular access changes to support MTC
IETF	Network, transport, and application layer protocols
OMA	M2M device management standards
ETSI	Standards for M2M communications
OneM2M	Creation of common M2M service layer
IPSO Alliance	Promotes use of IP for IoT
ZigBee Alliance	Protocols on top of IEEE 802.15.4
Bluetooth SIG	Standards for Bluetooth

include domains such as the device management and configuration domain, but also on other domains such as information reporting.

The work areas of some of these organizations are described in more depth in the following sections.

3GPP Standardization Work

The 3GPP *http://www.3gpp.org* is working on significant systems improvements for IoT, referred to as Machine-Type Communications (MTC) in 3GPP vocabulary. The continuously evolving work was ongoing, at the time of writing, in multiple 3GPP areas (see the 3GPP organization in Section 2.1.1). Hence, this section should be read with caution and just to see what kinds of issues IoT is causing for many aspect of 3GPP standards.

On the System Aspects area, the SA1 working group is working on *requirements*, such as: service requirements for MTC [61], studies for enhancements for MTC [62], studies for alternatives to E.164 [63], and more. The SA2 working group is working on *architectural* issues such as: architecture enhancements [64], system improvements for MTC [65], and also analyzing implications caused to other areas such as to packet radio service (from 2nd Generation (2G) to LTE), architectural requirements, and so on. As usually is the case with major features, *security* aspects of MTC requires special attention from the SA3 working group [66]. The MTC has implications also for charging, to the SA5 working group's *telecom management*, for example: Charging Data Record (CDR) parameters [67], packet switched domain charging [68], and diameter charging applications [69]. The charging aspects are handled by the SA5 working group.

At least four of the Core Network and Terminals area working groups are involved in the MTC work as well. The CT1, responsible of defining UE – the core network layer-3 radio protocols, is working on updates to *NAS configuration management object* [70], although the set of CT1's other documents will also be affected. Quite obviously the smart objects on 3GPP networks need *interworking with external networks* and hence CT3 is working on MTC interworking function and service capability server [71] alongside lesser updates to their other specifications. The CT4 is working for system improvements for *network internal protocols* and also on the location management procedures [72]. The CT4 has a particularly long list of affected documents from Diameter to GTP, Proxy Mobile IPv6 (PMIPv6), numbering, addressing, packet radio service, organization of subscriber's data, and WLAN interworking to name a few. Finally, the CT6 is working on implications to *smart card application aspects*.

IETF Standardization Work

The main standardization organization for network and transport layer protocols, and the important organization for some application layer protocols, is the IETF *http://www.ietf.org*. The IoT devices that are not really constrained can easily utilize existing protocols, such as IPv4, IPv6, TCP, and HyperText Transfer Protocol (HTTP), defined by the IETF a long time ago. However, for a class of IoT devices, the existing protocols are too resource consuming. The constrained categories of devices can be, for example, those that perform industrial, structural, or agricultural monitoring or are used for home automation, healthcare, or vehicle telematics [59]. For those kinds of *constrained nodes* IETF is defining new lightweight protocols. We list some of the most interesting standards or standard drafts below.

Constrained Application Protocol (CoAP): The CoAP is a binary encoded protocol that allows HTTP style mechanisms and REpresentational State Transfer (REST)-mechanisms to work over UDP. The protocol is designed to be as lightweight to implement and transport as possible. The CoAP is actually a set of documents built over the core CoAP specification [73]. These include, for example, 'Observing Resources in CoAP' that allows a CoAP client to request a CoAP server to send data without continuous polling [74] and 'Constrained RESTful Environments (CoRE) Link Format' that defines how servers can describe their resources and attributes [75]. There are also several other CoAP related enhancements in the pipeline. At the time of writing, it seems that the CoAP is becoming the protocol for constrained IoT node communications, while in general in the M2M domain HTTP seems to be prevalent (e.g., Open Mobile Alliance (OMA) Converged Personal Network Service (CPNS) uses HTTP).

IPv6 over IEEE 802.15.4: The RFC 4944 specifies how IPv6 packets can be transported over IEEE 802.15.4 networks [76].

IPv6 over Bluetooth Low Energy: An Internet-Draft (work in progress) specifies how IPv6 packets can be transported over Bluetooth Low-Energy networks [77].

Compression of IPv6 on IPv6 over Low power Wireless Personal Area Networks (6LoWPANs): RFC 6282 defines how IPv6 packets can be compressed over 6LoW-PANs, with a focus on IEEE 802.15.4 networks, but with a solution that also works with Bluetooth Low-Energy networks [78].

IPv6 Neighbor Discovery Optimizations: RFC 6775 specifies optimizations for Neighbor Discovery Protocol on 6LoWPANs [79].

Routing over low power and lossy networks: A family of RFCs describes a Routing Protocol for Low-power and Lossy Networks (RPL) protocol, the core of which is RFC 6550 [80]. The RPL describes how IPv6 packets can be routed in 6LoWPANs, and the metrics and algorithms needed for the work. See also Section 3.10.4.

Mobile Ad hoc NETworking (MANET): The MANET has been a long-term activity in the IETF. The first RFC 2501 from this space was published back in 1999 [81]. RFC

2501 recognized the need for low-power operations, but has focus in particular on high mobility of both nodes in the network and the network itself. The MANET includes a set of protocols, currently 12 RFCs, for setting up static and dynamic mesh networks and related routing systems using different routing techniques.

Implementation guidance: A recent work has been ongoing to describe implementation guidelines in an attempt to help software developers make as light-weight implementations of the IP suite as possible [82].

OMA Device Management Work

The OMA *http://www.openmobilealliance.org* specifications are already very widely used for device management (see Section 4.7.4) and therefore it is natural for OMA to be also interested in managing IoT nodes. OMA has been defining CPNS, lightweight M2M systems, and device classification technologies.

The CPNS is described in depth in the OMA's requirements [83], architecture [84], and technical specifications [85]. The CPNS defines personal networks to include, for example, those in homes, cars, or body area networks. The supported radio technologies are, among others, WLAN and Bluetooth. The nodes in the personal networks are enabled to produce and consume data to and from other nodes in the same, or different, personal networks. Furthermore, connectivity to service providers outside of the personal networks is supported as well. The CPNS framework includes a server, a gateway, and end nodes. The end nodes talk to the server and others via the gateway that is acting as an intermediary. These specifications specifically note that a 3GPP UE could play the role of the CPNS gateway, and hence if OMA's approach gains popularity it may incur changes to UEs.

OMA is currently designing device management and service enablement for lightweight M2M of resource constrained nodes. The requirements document [86], architecture diagram [87], and currently very draft technical specification [88] describe the system in depth. In short, this framework includes definitions for clients, servers, and lightweight application layer M2M communication protocols to perform device discovery, registration, bootstrapping, device management, service enablement, and information reporting.

The third work item ongoing in OMA is device classification, for which a candidate white paper is available [89]. The paper introduces means and attributes to classify devices based on horizontal attributes, such as local communications interface type (e.g., wide area, local area), protocol stack properties, human interface devices, persistent configuration storage, and such like. The classification could then be used for categorizing different devices together for easier management and provisioning purposes and for analyzing the suitability of different device provisioning and management tools for each class.

ETSI Standardization Work

ETSI *http://www.etsi.org* has established an M2M technical committee to develop standards for M2M communications. The committee has developed a set of specifications for M2M listed below, and is working on additional documents such as for M2M interfaces.

Service requirements: Defines requirements for M2M communications, management (e.g., configuration), functionality (e.g., data collection), security (e.g., authentication, integrity), and for naming, numbering and addressing – allowing for both IPv4 and IPv6 [57].

Functional architecture: Defines high level architecture, functional architecture, key reference points, identification and addressing, security, bootstrapping, provisioning, and resource management, among many other smaller topics. This is the main document for ETSI's M2M documentation [90].

mIa, dIa, and mId interfaces: Specifies in detail the mIa, dIa, and mId reference points introduced in the 'Functional architecture' document [91].

Smart metering use cases: Describes the M2M application for smart metering in a very detailed manner [58].

In this section we will not describe in more depth how ETSI sees the M2M, as that could be a subject of a dedicated book. The standards by ETSI are available for free download, and hence we recommend interested readers to follow the references for more details.

OneM2M

OneM2M *http://www.onem2m.org* is a newly founded organization, launched on 24 July 2012, whose purpose is to gather M2M related organizations together to develop technical specifications for a common M2M service layer that would enable M2M devices to interoperate at a global scale. The OneM2M was launched by seven standards defining organizations: Association of Radio Industries and Businesses (ARIB), Alliance for Telecommunications Industry Solutions (ATIS), China Communications Standards Association (CCSA), ETSI, Telecommunications Industry Association (TIA), Telecommunications Technology Association (TTA), and Telecommunication Technology Committee (TTC). This brand new organization did not have documents to share at the time of writing. The new organization's impacts on the IoT domain will be seen in the future.

IPSO Alliance

The IPSO Alliance *http://www.ipso-alliance.org* is a non-profit organization focused on promoting the use of IP in smart objects and on supporting interoperability aspects. IPSO Alliance has arranged several interoperability events, often collocated with IETF meetings. In these events vendors have been able to verify interoperability of their implementations. On the documentation side IPSO Alliance, for example, has produced white papers describing insights into lightweight operating system implementations, security, low-power network use, RPL protocol, and more. IPSO Alliance has also published 'The IPSO Application Framework' document, which is not standard as such, but that describes a REST-based design for IP-based smart object systems [92]. The document specifies interfaces for presenting available resources that can be used for interaction between various nodes and backend systems.

ZigBee Alliance

The ZigBee Alliance *http://www.zigbee.org* defines standards on top of the IEEE's 802.15.4 radio. These standards define communication protocols, for example, for building management, energy management, healthcare, telecommunications, consumer electronics, and energy management. The ZigBee Alliance also has a 'ZigBee Certified' system in place, which provides test suites and logos for products that are compliant with ZigBee's certifications. The focus on ZigBee-using low-power smart objects is mostly on those requiring low-traffic through secure mesh networks. The ZigBee networks are usually reasonably static, such as those present in homes, industries, smart grid, and buildings.

Bluetooth Special Interest Group

Bluetooth Special Interest Group (SIG) *http://www.bluetooth.com* is a non-profit group founded at 1998 for developing standards, profiles, and certification around Bluetooth and then licensing the technologies and trademarks for manufacturers. This group is not focusing on IoT in particular, but some of the old standards such as transporting IPv4 and IPv6 over Bluetooth with 'Bluetooth Personal Area Networking'-profile [93] and newer ones such as 'Bluetooth Low-Energy' [94], are possible technologies to be used on smart objects. This is true especially for smart objects that people carry with them, as in the 3GPP handsets Bluetooth support is almost ubiquitous, while, for example, ZigBee is non-existent. Hence the handsets can be providing Internet connectivity to the Bluetooth enabled smart objects.

6.5.3 IoT Domain from the 3GPP Point of View

In Figure 6.6 we illustrate one view of the IoT domain from the 3GPP point of view. The IoT domain is centered around the Internet, which is furthermore connected to different types of networks: cellular network and fixed home, civil, industrial, and the generic 'other' types of networks as illustrated in this figure. All of these networks can contain sensors and actuators, and of course any infrastructure required for providing connectivity for those (not shown in the figure). We have expanded the 3GPP network part a bit and we show there the sensors and actuators as well – these can be directly connected to 3GPP access and hence be typical machine-to-machine style of entities. It is worth realizing that this figure is one presentation – in another presentation the 3GPP access could be presented as wireless connectivity technology used by wireless home, civil, industrial, and other networks to connect to the Internet.

In the 3GPP network in Figure 6.6 we are also showing a connected UE, which is performing tethering for a LAN that contains sensors and actuators. This illustrates the way the networks can be connected to other networks, and how a UE is not only accessing sensors and actuators all around the Internet, but can also be providing Internet connectivity to IoT nodes 'below' it. Effectively, the UE is performing tethering of the cellular connection to any type of LAN technology, for example to WLAN or to some low-power radio technology.

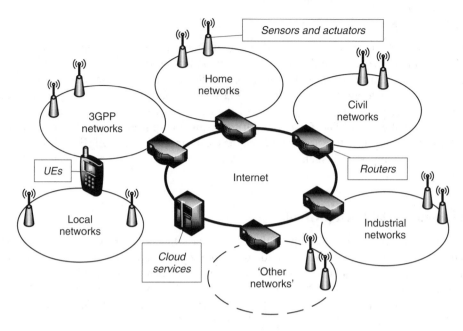

Figure 6.6   Illustration of the world of IoT.

The IoT domain very often comes with cloud-based services, also illustrated in Figure 6.6. These cloud servers can act as rendezvous points, data collection servers, social networking sites, and command and control centers to name a few.

From the 3GPP point of view the most interesting things are the IoT nodes directly accessing through 3GPP networks, the nodes using 3GPP access through UEs, services that the 3GPP network may provide to IoT nodes, and of course accessing local and remote IoT nodes, and cloud services, by the UEs.

6.5.4   Implications to UEs

The implications of the world of IoT to UEs depends fully on the level that a UE wants to participate, or contribute, to IoT. If the UE is a smart object itself, then obviously it needs to support IoT/M2M/MTC standards from the organizations it is compliant with. The UE may need to implement CoAP and be compliant with REST APIs defined by IPSO Alliance, the ones defined by 3GPP, or, for example, the gateway role of OMA CPNS. If this sounds unclear – it currently is. As the competing standardization landscape is fast evolving in this area, the requirements and expectations that real markets will place for UEs are not yet clear.

UEs of mobile handset type, however, do not necessarily need to be aware of IoT at all. The handsets can perfectly well continue to exist and ignore IoT. However, in some scenarios a mobile handset may want to provide Internet connectivity to nearby, often inexpensive, but usually connectivity-limited IoT nodes. Such smart objects could be, for example, IPv6-only capable wearable heart rate monitors, temperature sensors, cameras, and so forth. This can place requirements for supported radios and tethering software.

Even though connectivity to smart objects may be provided by handset type of UEs, there will also be UEs that are more of the gateway-like entities (e.g., fixed to a building, to a vehicle, or such like). In any case, the smart objects may be using many kinds of (low-power) radios, like WLAN, IEEE 802.15.4, Bluetooth Low Energy, or Near Field Communications (NFC), in addition to wired connectivity methods, that may bring additional hardware requirements for the UEs. The IEEE 802.15.4, though, is not intended for mobile use and hence such radio would be supported effectively only by the UEs of fixed type. To provide Internet connectivity IP-wise, a UE has to provide some sort of IPv6 tethering services, as described in Section 4.7.5, and in some cases also IPv4 tethering using the common means such as IPv4 NAT. Providing connectivity for IPv6-only smart objects, however, brings a possibly significant problem: what if the mobile UE does not have IPv6 connectivity due to home, or visited, network not supporting IPv6? The set of solutions for those cases includes at least host-based IPv6 to IPv4 protocol translation, host-based double protocol translation, and IPv6-over-IPv4 tunneling. Active research is ongoing to tackle this issue and there are currently no definite solutions.

The UEs interested in accessing services provided by IoT may need to be able to talk new protocols, such as CoAP. This can bring additional requirements for the IP suite UEs implement, and for the applications as well. For example, a web browser may need to be enhanced with CoAP in order to to be able to access smart objects.

6.5.5   Implications to 3GPP Networks

The millions, if not billions, of new smart objects with 3GPP modems will definitely have implications for 3GPP networks. Even more, if the smart objects themselves do not happen to have 3GPP modems, those can still access the 3GPP networks via local gateways such as mobile handsets or CPEs. The 3GPP is working on many 3GPP standards updates to accommodate smart objects, or support MTC, as we described in Section 6.5.2. In this section we will talk about some of the issues in more detail.

The 3GPP access is a lucrative access for smart objects due to very widespread outdoors and indoors coverage, especially on 2G networks, and also quite reasonable power consumption figures for idle nodes (especially when the 3GPP modem is turned off). This means that while smart phones are perhaps not such keen data users in 2G accesses (including IPv6 use), the smart objects may very well be. Hence it can happen that IPv6 sees widespread adoption on 2G due to IoT!

If 2G becomes a very popular access for IoT, it can have significant implications for cellular networks. In particular, it could ossify the upgrade strategies in a way that 2G may not be that easy to turn off – in an attempt to replace it with 3G or LTE – if there are huge numbers of (inexpensive) smart objects using 2G and without 3G capability [56].

Address Space Consumption

If the estimates stated by the OECD materialize, and the 50 billion devices connect to wireless networks and the 500 billion in total to the Internet, it is quite easy to see that numbering that device population with a mere 4 billion IPv4 addresses would be challenging to say the least. For the IPv6 address space the numbering of IoT is not a major challenge, as even the 500 billion addresses is just a minuscule part of the whole address space.

Even for a network operator in practice, remembering the 64-bit prefix per PDN Connection approach used in 3GPP and described in Section 4.1.2, we can calculate that one /37 prefix can number one hundred million smart objects having in total one hundred million unique PDN Connections. Considering a typical /32 prefix allocation for a cellular operator, the one hundred million smart objects would only consume 1/32 of the operator's IPv6 address block (and operator would get more if the /32 proved to be too little).

While IPv4 is seriously problematic for numbering IoT, IPv6 will have no issues. The need to deploy IPv6 for IoT is obvious.

Signaling Will Increase

It has been observed that already existing UEs, especially smartphones, can cause 3GPP networks to overload due to too much signaling traffic. The reasons can be, for example, the ones identified by 3GPP specification 23.843 [95]:

1. Special mobility events causing registration floods: these can be caused by *network attaches* or *location updating*, for example when an airplane lands and people turn on their phones.
2. Radio Access Technology (RAT)-reselections caused by scattered 2G, 3G and 4th Generation (4G) network coverage causing (very) frequent intersystem handovers.
3. Restarting network nodes, such as Base Station Controller (BSC), Radio Network Controller (RNC), Serving Gateway Support Node (SGSN), or MME can trigger large numbers of registration attempts.
4. UEs causing a flood of events, such as bearer activations, Short Message Service (SMS) sending/receiving, or content polling/pushing by/to large numbers of nodes in the same area.

Now, with the plans to add numerous smart objects into the 3GPP networks, the risks of signaling overload situations on the network grows even larger. The smart objects might be configured to transmit measurements at the same time, or there could just be a very large number of smart objects in the same network segments and hence problems caused by network element restarts or intersystem handovers due to bad coverage can exacerbate the signaling overload scenarios. Furthermore, signaling overload risks increase if IoT command and control node decides to poll information from large numbers of smart objects at the same time.

To avoid signaling overloading, the IoT architecture itself, design of individual smart objects, deployment of IoT, and controllers of large smart object networks have to take mitigating actions. For example, in many of the listed problematic cases signaling overload risk can be reduced by adding long enough random delay for smart objects' actions, and thus spreading massive events across time more widely (and definitely avoid instantaneous communications of many smart objects). Signaling can also be decreased by idle smart objects turning off their cellular modem – if they are not waiting for inbound communications. The 3GPP 23.843 also proposes some solutions, such as optimizations to periodic area update signaling, new status codes to indicate network overload to UEs and communicable back-off timers for retries, signaling load reductions by smarter network configuration, and other tweaks and improvements. The work to find solutions to signaling problems is ongoing, and the implications of the IoT are being taken into account [95].

Subscriptions Will Increase

With the introduction of the IoT and the many types of smart objects that will make use of the 3GPP access, the network operators will consequently have a significantly larger installed base of connected UEs, and there will be an increased number of subscriptions and UEs per subscriber. This can have implications for subscriber management, but probably this is one of the positive problems.

Privacy and Security Issues May Arise

Until now, the emergency services have been perhaps the most critical service for the commercial 3GPP networks, and it has been of utmost importance to have those services working in all conditions. The ordinary use of handsets by people: voice calls, messaging, and Internet use, have been less critical, although of course very high availability is expected there as well. With the emergence of IoT and the resulting use of sensors to monitor, and actuators to modify, the real world, the importance of service availability, security, and privacy are on the rise. The smart objects are producing data that has high privacy requirements, and the actuators may be driving systems where strict access control is of upmost importance. Issues of security can lead to misuse of camera or presence sensors, for example, and unwanted access to actuators such as heaters or water valves may result in loss of money or property damage.

New Opportunities Can Arise

In the world where mobile handset subscriptions have reached or surpassed one hundred percent of the population in many countries, and competition has focused on providing faster data or cheaper calls, the emergence of IoT will bring additional business opportunities. It is already clear that cellular operators are, for example, selling simple property surveillance systems to ordinary people – and more services will surely come to market in the coming years.

For equipment manufacturers the opportunities are also huge. 3GPP packet data access is becoming present in most places and for many business ideas is already virtually always available, and this combined with low cost 3GPP modems and data packages, is enabling the creation of all imaginable (and still unimaginable) Internet connected objects and, of course, also gateways for the smart objects that are too simple or power constrained to directly use cellular access.

6.6   Chapter Summary

In this chapter we touched upon two enhancements to 3GPP architecture. The first dealt with network steered (bulk) Internet traffic offloading from the 3GPP access to alternative access technologies. The offloading solutions were built entirely on top of IPv6 functionality. We described both DHCPv6- and NDP-based solution approaches. For the NDP-based solution we also described and analyzed how to extend offload to also handle IPv4 traffic. The solutions make use of the 3GPP access as the trusted channel for offloading policies pushed by the operator from the network to the UE. All these solutions have a certain set of challenges, issues caused by possible multiple provisioning domains being one.

The second enhancement is about evolving the 3GPP bearer model. Our humble attempt was to define a backward compatible approach to implementing a new style 3GPP evolved bearer that supports both multiple IPv6 prefixes and next-hop routers. Multiple next-hop routers imply the possibility of achieving local breakout for the traffic, and letting the address selection in the UE make the decision where to send packets based on the router preferences and more specific routes received from the network. The impact of the evolved bearer on the existing 3GPP architecture was also presented.

The next topic concerned residential home networks that make extensive use of IPv6. Our motivation on this topic was the envisioned development where the cellular connectivity eventually replaces wired connectivity to home networks and their CPEs. We looked into issues originating from the 3GPP connectivity model and the current IPv6 address management. Possible solution approaches that make use of both standardized and non-standard features of the 3GPP packet core were discussed and presented.

In this chapter we also introduced PCP, which is currently in the final stages of standardization in IETF, and already being proposed for 3GPP adoption. The PCP may find its way into future 3GPP deployments, as the developments in 3GPP networks and UEs are heading exactly towards the main use cases of PCP. NATs, both NAT44 and NAT64, are increasingly making their way into 3GPP networks and at the same time a class of UEs is being used as home CPEs and handset types of UEs are going to great lengths to conserve power. The PCP can help in both cases: for CPE type of UEs PCP provides facilities to convert UPnP and NAT-PMP protocols of the home network to protocol understood by the NAT at the operator network, and for handsets PCP provides a means to reduce keep-alive signaling and hence decrease energy consumption. However, the history of protocols designed for hosts to control core network elements has not been successful. It remains to be seen whether PCP will be adopted by the market on a large scale.

We discussed the strong growth expected to occur in the domain of IoT, and how the smart objects may become major drivers of IPv6 in 3GPP networks. The 3GPP standards and networks need updates to be able to handle the billions of new UEs connecting to networks, and the mobile handsets will in some cases need updates to be able to provide Internet connectivity to smart objects and also to be able to consume services provided by the smart objects. Some UEs will also become, more or less fixed, gateways for smart objects that are using low-power radios for Internet access. We also saw that the standardization in relation to IoT, M2M, and MTC is very active and likely to continue evolving the domain and producing standards that also influence 3GPP access. The IoT has a big promise for new business, and the authors are anxious to see how the field develops and what the future in this regard will bring.

References

1. 3GPP. *General Packet Radio Service (GPRS) enhancements for Evolved Universal Terrestrial Radio Access Network (E-UTRAN) access*. TS 23.401, 3rd Generation Partnership Project (3GPP), March 2012.

2. 3GPP. *IP flow mobility and seamless Wireless Local Area Network (WLAN) offload; Stage 2*. TS 23.261, 3rd Generation Partnership Project (3GPP), March 2012.

3. 3GPP. *Study on S2a Mobility based On GTP & WLAN access to EPC (SaMOG)*. TR 23.852, 3rd Generation Partnership Project (3GPP), July 2012.

4. 3GPP. *Access to the 3GPP Evolved Packet Core (EPC) via non-3GPP access networks; Stage 3*. TS 24.302, 3rd Generation Partnership Project (3GPP), September 2011.

5. 3GPP. *Operator Policies for IP Interface Selection (OPIIS)*. TR 23.853, 3rd Generation Partnership Project (3GPP), August 2011.

6. Blanchet, M., Seite, P. *Multiple Interfaces and Provisioning Domains Problem Statement*. RFC 6418, Internet Engineering Task Force, November 2011.

7. Savolainen, T., Kato, J., and Lemon, T. *Improved Recursive DNS Server Selection for Multi-Interfaced Nodes*. RFC 6731, Internet Engineering Task Force, December 2012.

8. Dec, W., Mrugalski, T., Sun, T., Sarikaya, B., and Matsumoto, A. *DHCPv6 Route Options*. Internet-Draft draft-ietf-mif-dhcpv6-route-option-05, Internet Engineering Task Force, August 2012. Work in progress.

9. Draves, R. and Thaler, D. *Default Router Preferences and More-Specific Routes*. RFC 4191, Internet Engineering Task Force, November 2005.

10. Thaler, D., Draves, R., Matsumoto, A., and Chown, T. *Default Address Selection for Internet Protocol Version 6 (IPv6)*. RFC 6724, Internet Engineering Task Force, September 2012.

11. Korhonen, J., Savolainen, T., and Ding, A. *Controlling Traffic Offloading Using Neighbor Discovery Protocol*. Internet-Draft draft-korhonen-mif-ra-offload-05, Internet Engineering Task Force, August 2012. Work in progress.

12. Atkinson, R., Haskin, D., and Luciani, J. *IPv6 over NBMA Networks*. Internet-Draft draft-ietf-ion-ipv6-nbma-00, Internet Engineering Task Force, June 1996. Work in progress.

13. 3GPP. *Evolved Universal Terrestrial Radio Access (E-UTRA); Packet Data Convergence Protocol (PDCP) specification*. TS 36.323, 3rd Generation Partnership Project (3GPP), January 2010.

14. 3GPP. *Policy and charging control architecture*. TS 23.203, 3rd Generation Partnership Project (3GPP), March 2012.

15. Korhonen, J., Soininen, J., Patil, B., Savolainen, T., Bajko, G., and Iisakkila, K. *IPv6 in 3rd Generation Partnership Project (3GPP) Evolved Packet System (EPS)*. RFC 6459, Internet Engineering Task Force, January 2012.

16. 3GPP. *Interworking between the Public Land Mobile Network (PLMN) supporting packet based services and Packet Data Networks (PDN)*. TS 29.061, 3rd Generation Partnership Project (3GPP), December 2011.

17. Thomson, S., Narten, T., and Jinmei, T. *IPv6 Stateless Address Autoconfiguration*. RFC 4862, Internet Engineering Task Force, September 2007.

18. Narten, T., Nordmark, E., Simpson, W., and Soliman, H. *Neighbor Discovery for IP version 6 (IPv6)*. RFC 4861, Internet Engineering Task Force, September 2007.

19. Hinden, R. and Thaler, D. *IPv6 Host-to-Router Load Sharing*. RFC 4311, Internet Engineering Task Force, November 2005.

20. Choi, J. and Daley, G. *Goals of Detecting Network Attachment in IPv6*. RFC 4135, Internet Engineering Task Force, August 2005.

21. Krishnan, S. and Daley, G. *Simple Procedures for Detecting Network Attachment in IPv6*. RFC 6059, Internet Engineering Task Force, November 2010.

22. 3GPP. *Non-Access-Stratum (NAS) protocol for Evolved Packet System (EPS); Stage 3*. TS 24.301, 3rd Generation Partnership Project (3GPP), March 2012.

23. Bormann, C., Burmeister, C., Degermark, M., Fukushima, H., Hannu, H., Jonsson, L-E., Hakenberg, R., Koren, T., Le, K., Liu, Z., Martensson, A., Miyazaki, A., Svanbro, K., Wiebke, T., Yoshimura, T., and Zheng, H. *RObust Header Compression (ROHC): Framework and four profiles: RTP, UDP, ESP, and uncompressed*. RFC 3095, Internet Engineering Task Force, July 2001.

24. 3GPP. *Evolved Packet System (EPS); Mobility Management Entity (MME) and Serving GPRS Support Node (SGSN) related interfaces based on Diameter protocol*. TS 29.272, 3rd Generation Partnership Project (3GPP), March 2012.

25. 3GPP. *General Packet Radio System (GPRS) Tunnelling Protocol User Plane (GTPv1-U)*. TS 29.281, 3rd Generation Partnership Project (3GPP), June 2010.

26. 3GPP. *3GPP Evolved Packet System (EPS); Evolved General Packet Radio Service (GPRS) Tunnelling Protocol for Control plane (GTPv2-C); Stage 3*. TS 29.274, 3rd Generation Partnership Project (3GPP), March 2012.

27. 3GPP. *Evolved Universal Terrestrial Radio Access Network (E-UTRAN); S1 Application Protocol (S1AP)*. TS 36.413, 3rd Generation Partnership Project (3GPP), December 2011.

28. 3GPP. *Evolved Universal Terrestrial Radio Access Network (E-UTRAN); X2 general aspects and principles*. TS 36.420, 3rd Generation Partnership Project (3GPP), April 2011.

29. Korhonen, J., Patil, B., Gundavelli, S., Seite, P., and Liu, D. *IPv6 Prefix Mobility Management Properties*. Internet-Draft draft-korhonen-dmm-prefix-properties-02, Internet Engineering Task Force, July 2012. Work in progress.

30. Droms, R., Bound, J., Volz, B., Lemon, T., Perkins, C., and Carney, M. *Dynamic Host Configuration Protocol for IPv6 (DHCPv6)*. RFC 3315, Internet Engineering Task Force, July 2003.

31. 3GPP. *Proxy Mobile IPv6 (PMIPv6) based Mobility and Tunnelling protocols; Stage 3*. TS 29.275, 3rd Generation Partnership Project (3GPP), March 2012.

32. 3GPP. *Architecture enhancements for non-3GPP accesses*. TS 23.402, 3rd Generation Partnership Project (3GPP), March 2012.

33. Baker, F. and Droms, R. *IPv6 Prefix Assignment in Small Networks*. Internet-Draft draft-baker-homenet-prefix-assignment-01, Internet Engineering Task Force, 2012. Work in progress.

34. Grundemann, C. and Donley, C. *Home Network Autoconfiguration via DHCPv6 Relay*. Internet-Draft draft-gmann-homenet-relay-autoconf-01, Internet Engineering Task Force, March 2012. Work in progress.

35. Nordmark, E., Chakrabarti, S., Krishnan, S., and Haddad, W. *Evaluation of Proposed Homenet Routing Solutions*. Internet-Draft draft-chakrabarti-homenet-prefix-alloc-01, Internet Engineering Task Force, 2011. Work in progress.

36. Arkko, J., Lindem, A., and Paterson, B. *Prefix Assignment in a Home Network*. Internet-Draft draft-arkko-homenet-prefix-assignment-02, Internet Engineering Task Force, July 2012. Work in progress.

37. Howard, L. *Evaluation of Proposed Homenet Routing Solutions*. Internet-Draft draft-howard-homenet-routing-comparison-00, Internet Engineering Task Force, December 2011. Work in progress.

38. Cloetens, W., Lemordant, P., and Migault, D. *IPv6 Home Network Naming Delegation Architecture*. Internet-Draft draft-mglt-homenet-naming-delegation-00.txt,.ps,, Internet Engineering Task Force, July 2012. Work in progress.

39. Cloetens, W., Lemordant, P., and Migault, D. *IPv6 Home Network Front End Naming Delegation*. Internet-Draft draft-mglt-homenet-front-end-naming-delegation-00.txt,.ps,, Internet Engineering Task Force, July 2012. Work in progress.

40. Lynn, K. and Sturek, D. *Extended Multicast DNS*. Internet-Draft draft-lynn-homenet-site-mdns-01, Internet Engineering Task Force, September 2012. Work in progress.

41. Woodyatt, J. *Recommended Simple Security Capabilities in Customer Premises Equipment (CPE) for Providing Residential IPv6 Internet Service*. RFC 6092, Internet Engineering Task Force, January 2011.

42. Chown, T., Arkko, J., Brandt, A., Troan, O., and Weil, J. *Home Networking Architecture for IPv6*. Internet-Draft draft-ietf-homenet-arch-04, Internet Engineering Task Force, July 2012. Work in progress.

43. Hinden, R. and Deering, S. *IP Version 6 Addressing Architecture*. RFC 4291, Internet Engineering Task Force, February 2006.

44. Troan, O. and Droms, R. *IPv6 Prefix Options for Dynamic Host Configuration Protocol (DHCP) version 6*. RFC 3633, Internet Engineering Task Force, December 2003.

45. Lindem, A. and Arkko, J. *OSPFv3 Auto-Configuration*. Internet-Draft draft-ietf-ospf-ospfv3-autoconfig-00, Internet Engineering Task Force, October 2012. Work in progress.

46. Singh, H., Beebee, W., Donley, C., and Stark, B. *Basic Requirements for IPv6 Customer Edge Routers*. Internet-Draft draft-ietf-v6ops-6204bis-11, Internet Engineering Task Force, September 2012. Work in progress.

47. Korhonen, J., Savolainen, T., Krishnan, S., and Troan, O. *Prefix Exclude Option for DHCPv6-based Prefix Delegation*. RFC 6603, Internet Engineering Task Force, May 2012.

48. Byrne, C. and Drown, D. *Sharing/64 3GPP Mobile Interface Subnet to a LAN*. Internet-Draft draft-byrne-v6ops-64share-03, Internet Engineering Task Force, October 2012. Work in progress.

49. Thaler, D., Talwar, M., and Patel, C. *Neighbor Discovery Proxies (ND Proxy)*. RFC 4389, Internet Engineering Task Force, April 2006.

50. UPnP Forum. *InternetGatewayDevice:1 Device Template Version 1.01*. Standardized DCP 1.0, UPnP Forum, November 2001.

51. UPnP Forum. *InternetGatewayDevice:2 Device Template Version 1.01*. Standardized DCP (SDCP) 1.0 and 1.1, UPnP Forum, December 2010.

52. Cheshire, S. and Krochmal, M. *NAT Port Mapping Protocol (NAT-PMP)*. Internet-Draft draft-cheshire-nat-pmp-05, Internet Engineering Task Force, September 2012. Work in progress.

53. Wing, D., Cheshire, S., Boucadair, M., Penno, R., and Selkirk, P. *Port Control Protocol (PCP)*. Internet-Draft draft-ietf-pcp-base-28, Internet Engineering Task Force, October 2012. Work in progress.

54. Boucadair, M., Dupont, F., Penno, R., and Wing, D. *Universal Plug and Play (UPnP) Internet Gateway Device (IGD)-PortControl Protocol (PCP) Interworking Function*. Internet-Draft draft-ietf-pcp-upnp-igd-interworking-04, Internet Engineering Task Force, September 2012. Work in progress.

55. Boucadair, M., Penno, R., and Wing, D. *DHCP Options for the Port Control Protocol (PCP)*. Internet-Draft draft-ietf-pcp-dhcp-05, Internet Engineering Task Force, September 2012. Work in progress.

56. OECD. *Machine-to-Machine Communications: Connecting Billions of Devices*. OECD Digital Economy Papers 192, OECD Publishing, January 2012.

57. ETSI. *Machine-to-Machine communications (M2M); M2M service requirements*. TS 102 689, European Telecommunications Standards Institute (ETSI), August 2010.

58. ETSI. *Machine-to-Machine communications (M2M); Smart Metering Use Cases*. TR 102 691, European Telecommunications Standards Institute (ETSI), May 2010.

59. Kim, E., Kaspar, D., and Vasseur, JP. *Design and Application Spaces for IPv6 over Low-Power Wireless Personal Area Networks (6LoWPANs)*. RFC 6568, Internet Engineering Task Force, April 2012.

60. IPSO Alliance. *IP for Smart Objects*. White Paper 1.1, IPSO Alliance, July 2010.

61. 3GPP. *Service requirements for Machine-Type Communications (MTC); Stage 1*. TS 22.368, 3rd Generation Partnership Project (3GPP), March 2012.

62. 3GPP. *Study on Enhancements for MTC*. TR 22.888, 3rd Generation Partnership Project (3GPP), March 2012.

63. 3GPP. *Study on alternatives to E.164 for Machine-Type Communications (MTC)*. TR 22.988, 3rd Generation Partnership Project (3GPP), March 2012.

64. 3GPP. *Architecture enhancements to facilitate communications with packet data networks and applications*. TS 23.682, 3rd Generation Partnership Project (3GPP), March 2012.

65. 3GPP. *System improvements for Machine-Type Communications (MTC)*. TR 23.888, 3rd Generation Partnership Project (3GPP), March 2012.

66. 3GPP. *Security aspects of Machine-Type Communications*. TR 33.868, 3rd Generation Partnership Project (3GPP), March 2012.

67. 3GPP. *Telecommunication management; Charging management; Charging Data Record (CDR) parameter description*. TS 32.298, 3rd Generation Partnership Project (3GPP), March 2012.

68. 3GPP. *Telecommunication management; Charging management; Packet Switched (PS) domain charging*. TS 32.251, 3rd Generation Partnership Project (3GPP), December 2011.

69. 3GPP. *Telecommunication management; Charging management; Diameter charging applications*. TS 32.299, 3rd Generation Partnership Project (3GPP), December 2011.

70. 3GPP. *Non-Access Stratum (NAS) configuration Management Object (MO)*. TS 24.368, 3rd Generation Partnership Project (3GPP), September 2011.

71. 3GPP. *Tsp interface protocol between the MTC Interworking Function (MTC-IWF) and Service Capability Server (SCS)*. TS 29.368, 3rd Generation Partnership Project (3GPP), September 2012.

72. 3GPP. *Location management procedures*. TS 23.012, 3rd Generation Partnership Project (3GPP), October 2010.

73. Shelby, Z., Hartke, K., Bormann, C., and Frank, B. *Constrained Application Protocol (CoAP)*. Internet-Draft draft-ietf-core-coap-11, Internet Engineering Task Force, July 2012. Work in progress.

74. Hartke, K. *Observing Resources in CoAP*. Internet-Draft draft-ietf-core-observe-06, Internet Engineering Task Force, September 2012. Work in progress.

75. Shelby, Z. *Constrained RESTful Environments (CoRE) Link Format*. RFC 6690, Internet Engineering Task Force, August 2012.

76. Montenegro, G., Kushalnagar, N., Hui, J., and Culler, D. *Transmission of IPv6 Packets over IEEE 802.15.4 Networks*. RFC 4944, Internet Engineering Task Force, September 2007.

77. Nieminen, J., Savolainen, T., Isomaki, M., Patil, B., Shelby, Z., and Gomez, C. *Transmission of IPv6 Packets over Bluetooth Low Energy*. Internet-Draft draft-ietf-6lowpan-btle-11, Internet Engineering Task Force, October 2012. Work in progress.

78. Hui, J. and Thubert, P. *Compression Format for IPv6 Datagrams over IEEE 802.15.4-Based Networks*. RFC 6282, Internet Engineering Task Force, September 2011.

79. Shelby, Z., Chakrabarti, S., Nordmark, E., and Bormann, C. *Neighbor Discovery Optimization for IPv6 over Low-Power Wireless Personal Area Networks (6LoWPANs)*. RFC 6775, Internet Engineering Task Force, November 2012.

80. Winter, T., Thubert, P., Brandt, A., Hui, J., Kelsey, R., Levis, P., Pister, K., Struik, R., Vasseur, J., and Alexander, R. *RPL: IPv6 Routing Protocol for Low-Power and Lossy Networks*. RFC 6550, Internet Engineering Task Force, March 2012.

81. Corson, S. and Macker, J. *Mobile Ad hoc Networking (MANET): Routing Protocol Performance Issues and Evaluation Considerations*. RFC 2501, Internet Engineering Task Force, January 1999.

82. Bormann, C. *Guidance for Light-Weight Implementations of the Internet Protocol Suite*. Internet-Draft draft-ietf-lwig-guidance-02, Internet Engineering Task Force, August 2012. Work in progress.

83. Open Mobile Alliance. *Converged Personal Network Service Requirements*. Requirements Document 1.1, Open Mobile Alliance (OMA), March 2012.

84. Open Mobile Alliance. *Converged Personal Network Service Architecture*. Architecture Document 1.1, Open Mobile Alliance (OMA), May 2012.

85. Open Mobile Alliance. *Converged Personal Network Service Core Technical Specification*. Technical Specification 1.1, Open Mobile Alliance (OMA), September 2012.

86. Open Mobile Alliance. *Lightweight Machine to Machine Requirements*. Requirements Document 1.0, Open Mobile Alliance (OMA), September 2012.

87. Open Mobile Alliance. *Lightweight Machine to Machine Architecture*. Architecture Diagram 1.0, Open Mobile Alliance (OMA), September 2012.

88. Open Mobile Alliance. *Lightweight Machine to Machine Technical Specification*. Technical Specification 1.0, Open Mobile Alliance (OMA), September 2012.

89. Open Mobile Alliance. *White Paper on M2M Device Classification*. White Paper candidate, Open Mobile Alliance (OMA), June 2012.

90. ETSI. *Machine-to-Machine communications (M2M); Functional architecture*. TS 102 690, European Telecommunications Standards Institute (ETSI), October 2011.

91. ETSI. *Machine-to-Machine communications (M2M); mIa, dIa and mId interfaces*. TS 102 921, European Telecommunications Standards Institute (ETSI), February 2012.

92. Z. Shelby and C. Chauvenet. *The IPSO Application Framework*. draft draft-ipso-app-framework-04, IPSO Alliance, August 2012.

93. Bluetooth SIG. *Personal Area Networking Profile*. Bluetooth Profile 1.0, Bluetooth SIG, December 2002.

94. Bluetooth SIG. *Specification of the Bluetooth System*. Bluetooth Core 4.0, Bluetooth SIG, June 2010.

95. 3GPP. *Core Network Overload Study*. TR 23.843, 3rd Generation Partnership Project (3GPP), July 2012.

96. Cheshire, S. *PCP Anycast Address*. Internet-Draft draft-cheshire-pcp-anycast-00.txt, Internet Engineering Task Force, February 2013. Work in progress.

97. Cheshire, S. *Recursive PCP*. Internet-Draft draft-cheshire-recursive-pcp-00.txt, Internet Engineering Task Force, February 2013. Work in progress.

Index

Deploying IPv6 in 3GPP Networks: Evolving Mobile Broadband from 2G to LTE and Beyond, First Edition.
Jouni Korhonen, Teemu Savolainen and Jonne Soininen.
© 2013 John Wiley & Sons, Ltd. Published 2013 by John Wiley & Sons, Ltd.